TASEK BERA
THE ECOLOGY OF A FRESHWATER SWAMP

MONOGRAPHIAE BIOLOGICAE

VOLUME 47

Editor

J. ILLIES †

Schlitz, F.R.G.

Dr W. Junk Publishers The Hague-Boston-London 1982

TASEK BERA

The Ecology of a Freshwater Swamp

Edited by

J. I. FURTADO and S. MORI

Dr W. Junk Publishers The Hague-Boston-London 1982

Distributors:

for the United States and Canada

Kluwer Boston, Inc.
190 Old Derby Street
Hingham, MA 02043
USA

for all other countries

Kluwer Academic Publishers Group
Distribution Center
P.O. Box 322
3300 AH Dordrecht
The Netherlands

Library of Congress Cataloging in Publication Data
Main entry under title:

The Ecology of a tropical freshwater swamp, the
Tasek Bera, Malaysia.

(Monographiae biologicae ; v. 47)
Includes bibliographical references.
1. Swamp ecology--Bera Lake (Negri Sembilan and Pahang) 2. Lake ecology--Bera Lake (Negri Sembilan and Pahang) 3. Bera Lake (Negri Sembilan and Pahang)
I. Furtado, J. I. II. Mori, Shuichi. III. Series.
QH185.E27 1982 574.5'26325'095951 82-14182

ISBN-13: 978-94-009-7982-6 e-ISBN-13: 978-94-009-7980-2
DOI: 10.1007/978-94-009-7980-2

Softcover reprint of the hardcover 1st edition 1982

Contents

1. Introduction

J. I. Furtado and S. Mori

This volume, 'Tasek Bera, the ecology of a tropical freshwater swamp', is a monograph based on research at Tasek Bera, under the auspices of the International Biological Programme (IBP) involving Japanese and Malaysian scientists under the sponsorship of the Malaysian National Committee for the IBP (MNC-IBP) which was initiated by the Malaysian Scientific Association (MSA). This collaboration was initiated by one of us (S. Mori) in 1966 at the Scientific Committee for Freshwater Productivity (PF) of the International Biological Programme (SCIBP) of the International Council of Scientific Unions (ICSU). It involved discussions between the Japanese National Committee for the IBP (JIBP) and the Malaysian National Committee for the IBP (MNC-IBP) and their respective sub-committees for freshwater productivity; a brief visit to Tasek Bera on 8 May 1968; a reappraisal of the proposal at the South East Asian Regional Meeting of Inland Water Biologists held on 5-11 May 1969 at the University of Malaya, Kuala Lumpur and the Tropical Fish Culture Research Institute, Batu Berendam, Malacca, organised during the Pacific Science Association Inter-Congress at the University of Malaya; a further prolonged field visit by two Japanese scientists (I. Ikusima and T. Mizuno) from August 1970 to March 1971; funding by the Japan Society for the Promotion of Science (JSPS); and, finally, an agreement signed between the Japan Society for the Promotion of Science and the Government of Malaysia through the Economic Planning Unit, Prime Minister's Department on 26 September 1970, for a joint project of scientific investigations into freshwater lakes of Malaysia. This preparatory phase had a gestation period of four years due, largely, to the inequalities of scientific resources (viz. personnel, expertise, organizations, networks, institutions, equipment, technology, interest, commitment and funds) between a developing country like Malaysia and an advanced country like Japan to launch a collaborative scientific investigation.

Furtado, J. I. and Mori, S., eds., Tasek Bera. ISBN 90 6193 100 2.

The joint scientific investigation of Tasek Bera lasted four years. The Tasek Bera was selected as the research site because it represents a typical and yet endangered natural ecosystem in Malaysia, and probably in the southeast Asian humid tropics. It is a lowland alluvial riparian swamp system, and not a true lake, which is rare in this region. The swamp system occupies about 10% of the catchment area, and is essentially forested (i.e. about 99% of the swamp) while the flowing water channels only occupy about 1% of the swamp. The study site was thus unusual in comparison with other IBP (PF) sites in being neither a lake nor a river, posing several challenging technical problems in methodology and logistics for this investigation. The study focussed on the metabolic processes of production and decomposition in relation to the environment. These are itemised in the Contents. Due to logistical problems, especially accessibility, this study was limited essentially to the open water channels and the *Lepironia-Pandanus* littoral region which is a successional phase of the swamp-forest and is derived from it, and to the vicinity of Pos (or Fort) Iskandar. The study involved four Malaysian scientists (J. I. Furtado, R. P. Lim, R. J. Morley and M. Ratnasabapathy) and 15 Japanese scientists (S. Mori, T. Mizuno, I. Ikusima, O. Sato, T. Koyama, Y. Tezuka, T. Okino, S. Kumano, T. Watanabe, K. Gose, N. Mizuno, Y. Shiraishi, M. Nagai, K. Nishiyama and M. Yoshimi) (Table 1) and was thus unequal in effort. Besides the inequalities mentioned earlier, differences in language and cultural perspectives made collaboration difficult but not insurmountable.

This monograph on the tropical freshwater swamp, the Tasek Bera, Malaysia, attempts to synthesize the essential results emanating from this study as they relate to the ecological metabolic processes studied. Because it has been necessary to economize data and interpretations in this synthesis, only a limited selection of data is presented and further data may be found in Data Book 1 (Mizuno 1972) and Data Book 2 (Furtado 1974) (Appendix I). Some of the results have been synthesized partially and preliminarily in papers presented at the East Asian Regional Seminar for the IBP, Kyoto (30 January–1 February 1973) and the Malaysian IBP Synthesis Meeting, Kuala Lumpur (12-18 August 1974); others have formed the postgraduate theses of several participants (R. P. Lim–M.Sc., R. Morley–Ph.D., O. Sato–D.Sc., S. Kumano–D.Sc., M. Nagai–M.Sc., K. Nishiyama–M.Sc. and M. Yoshimi–M.Sc.), and still others have been published in a different form as primary papers in scientific journals. This synthesis was initiated in 1974 through a visit of one of us (J. I. Furtado) to the Otsu Hydrobiological Station, University of Kyoto for discussions with Japanese scientists involved in the Tasek Bera project. This visit was made possible through the award of a Senior

Table 1 Directory of participants

Furtado, Professor Dr. J. I., Department of Zoology, University of Malaya, Kuala Lumpur, Malaysia

Gose, Dr. Kyuemon, Oyodo Senior High School, Oyodo-cho, Yoshino-gun, Nara Prefecture, Japan

Ikusima, Professor Dr. Isao, Department of Marine Ecology, Chiba University, Yayoi-cho, Chiba 280, Japan

Koyama, Professor Dr. Tadashiro, Water Research Institute, Nagoya University, Furo-cho, Chikusa-ku, Nagoya 464, Japan. Present address: Kameiri 22-66, Tashiro-cho, Chikusa-ku, Nagoya 464, Japan

Kumano, Dr. Shigeru, Faculty of Science, Kobe University, Rokkodai, Nada-ku, Kobe 657, Japan

Lim, Dr. Richard P., Department of Zoology, University of Malaya, Kuala Lumpur, Malaysia

Mizuno, Dr. Nobuhiko, Department of Biology, Ehime University, Bunkyo-cho, Matsuyama-shi, Ehime 790, Japan

Mizuno, Professor Dr. Toshihiko, Department of Biology, Osaka Kyaiku University, Minami-Kawahori-cho, Tennoji-ku, Osaka 543, Japan

Mori, Emeritus Professor Dr. Syuiti, President, Shizuoka Women's University, Yata 409, Shizuoka 422, Japan. Present address: Higashikubotacho 76, Kitashirakawa, Sakyo-ku, Kyoto 606, Japan

Morley, Dr. Robert J., Robertson Research International, Tyn-y-coed, Llanrhos, Llandudno, Gwynedd LL30 ISA, Wales

Nagai, Mr. Motoichiro, c/o Department of Biology, Osaka Kyoiku University, Minami Kawahori-cho, Tennoji-ku, Osaka 543, Japan

Nishiyama, Mr. Koichi, c/o Department of Biology, Osaka Kyoiku University, Minami Kawahori-cho, Tennoji-ku, Osaka 543, Japan

Okino, Dr. Tokio, Suwa Hydrobiological Station, Shiushu University, 5-2-4 Kogan-Dori, Suwa-shi 392, Nagano-ken, Japan

Ratnasabapathy, Mr. M., Department of Botany, University of Malaya, Kuala Lumpur, Malaysia

Sato, Dr. Osamu, Water Research Institute, Nagoya University, Furo-cho, Chikusa-ku, Nagoya 464, Japan. Present address: Research Institute for Hazards in Snowy Areas, Niigata University, Igarashi-2 no-machi, Niigata 950-21, Japan

Shiraishi, The late Dr. Yoshikazu, Freshwater Fisheries Research Laboratory, Chuguji, Nikkon City, Tochigi-ken, Japan

Tezuka, Professor Dr. Yasuhiko, Otsu Hydrobiological Station, Kyoto University, Otsu, Shiga-ken, Japan

Watanabe, Dr. Toshiharu, Department of Biology, Nara Women's University, Kitanoyanishi-machi, Nara 630, Japan

Yoshimi, Mr. Masatoshi, Department of Biology, Osaka Kyoiku University, Minami Koborecho, Tennoji-ku, Osaka 543, Japan

Fellowship by the Japan Society for the Promotion of Science. The gestation period of about six years for finalisation of this monograph is due largely to differences in comprehension and communication in the English language, to differences in the choice and synthesis of data obtained using different techniques, and to the lack of financial support for this aspect of the synthesis. This monograph has been accomplished by the sheer perseverence and persistence of all scientists involved in the spirit of the international collaboration. Most of this monograph embodies original data and interpretation, and those scientists responsible for each section are clearly listed.

This monograph would not have been possible without the support and effort of all participants especially Dr. R. P. Lim, Professor Isao Ikusima and Professor Toshihiko Mizuno for their untiring editorial assistance. This international research collaboration was made possible through the efforts of Mr. K. D. Menon (former Director-General of Forestry) as Chairman of MNC-IBP, Mr. Soong Min-Kong (former Director-General of Fisheries) and Professor A. J. Berry as former Chairman of the MNC-IBP (PF) Sub-Committee, various members of the MNC-IBP (PF) subcommittee (Professor A. Johnson, Mr. K. Daljeet Singh, Mr. Baharin Kassim, Dr. J. A. Bullock, Professor D. S. MacLusky, Mr. A. A. Jothy, the late Dr. G. A. Prowse, Mr. Anwar Mahmud, Tengku Ubaidillah bin Tengku Abdul Kadir, Mr. Tan Yee Thong, the late Mr. Abu Bakar Othman Merican, Dr. D. S. Winoto, Professor W. R. Stanton, Mr. M. Ratnasabapathy, Dr. A. B. Viner, Dr. Ang Kok-Jee and Mr. Lee Chan-Lui), Dr. H. Tamiya as Chairman of JIBP, Professor Tatuo Kira representing JIBP and JSPS and leader of Malaysian projects, the Economic Planning Unit (Prime Minister's Department), especially Mr. Amiruddin Hussain, the Immigration Department, the Royal Customs and Excise Department, the Royal Malaysian Police, the Department of Aboriginal Affairs, the District Officer (Temerloh), the Manager of the Dunlop Bahau Estate, the Director of the Tropical Fish Culture Research Institute, Batu Berendam (Malacca), and the Chairman and staff of the School of Biological Sciences and the Vice Chancellor of the University of Malaya; and without the special assistance of Mr. Eric R. Alfred (formerly Curator, Raffles Museum, Singapore), Mr. Edwin Arthur, Mr. Lee-Thian-Heng, Mr. George Liew Kim-Seng, Mr. Mohd. Nasir Alang, Professor E. Soepadmo, Dr. B. C. Stone, and Nonek and Jassny and their families.

The spirit of international collaboration that made this research and monograph possible is best expressed by these few verses from Japanese haiku.

References

T. Mizuno (ed.). 1972. Data on Malaysian-Japanese IBP (PF) Research at Tasek Bera, Malaysia. No. 1. 147 pp.
Department of Biology, Osaka Kyoiku University, Osaka.

J. I. Furtado (ed.). 1974. Malaysian-Japanese IBP (PF) Research Data on Tasek Bera, Malaysia. No. 2. 304 pp.
School of Biological Sciences, University of Malaya, Kuala Lumpur.

The vision (Issa, 1762–1826)
'In its eye
are mirrored far-off mountains –
dragonfly!'
'To-yama no me-dama ni utsuru tombo kana'.

Marvels (Kito, 1740–1788)
'"Marvelous!" I say,
as I watch, now this, now that –
and springtime goes away'.
'Mezurashi to miru-goto-ni haru ya yuku'.

In the field (Issa, 1762–1826)
'Tilling the field? – Just so
does he act as he walks there,
the crow!'
'Hata-uchi-no mane wo shite aruku karasu kana'.

Persistence (Basho, 1644–1694)
'Did it yell
till it became *all* voice?
Cicada-shell!'
'Koe ni mina naki-shimate ya semino-kara!'

In the paddy field (Raizan, 1653–1716)
'Women rice-planting:
all muddy, save for one thing –
that's their chanting'.
'Sa-otome ya yogorenu mono wa uta bakari'.

2. General description of Tasek Bera

R. P. Lim, J. I. Furtado and R. J. Morley

2.1 Locality and Physiography
(R. P. Lim and J. I. Furtado)

Tasek Bera is an alluvial peat swamp ecosystem located 3°5′ N, 102°38′ E in the southwest of Pahang State, and about 25.6 km from the nearest town of Bahau, in adjacent Negeri Sembilan State (Fig. 1). It measures 34.62 km at its longest and 25.28 km at its broadest point, and has an area of 61.50 km^2 in a watershed measuring 613.83 km^2. The watershed of this swamp lies on the plains between the main and eastern ranges of Peninsular Malaysia, within the 250 ft a.s.l. contour (Fig. 2). The plains forming this watershed are complex undulations and gently rolling hills, traversed by several small streams and rivers. The Sungai Pahang ('Sungai' is Malay for 'River') – Tasek Bera complex is less than 100 ft a.s.l. The watershed is composed predominantly of argillaceous strata which weather to pale-coloured, firm, heavy textured soils representing 'various stages of dissection of an ancient erosion surface' (Leamy and Panton 1966).

The swamp is situated in a north-south longitudinal axis, and drains northwards through a single confluence, the Sungai Bera, a tributary of the Sungai Pahang. The Sungai Bera originates east of the swamp watershed and passes very close to the main channel of the swamp near its outlet near Tanjong Kuin before draining northwards into the Sungai Pahang (Fig. 1). Tasek Bera is a dendritic complex occupying extensive sinuous tracts or arms of water, and wide areas of reed and forest-swamp in between patches of raised ground. A number of first, second, third and fourth order streams drain from the high ground into these sinuous tracts or arms which merge approximately midway in a north-south axis forming the main channel system, known as Tasek Bera (*sensu stricto*). Tasek Bera (*sensu lato*), as considered herein, includes Tasek Bera (*sensu stricto*), Tasek Dampar, Paya (the Malay term for 'swamp')

Furtado, J. I. and Mori, S., eds., Tasek Bera. ISBN 90 6193 100 2.

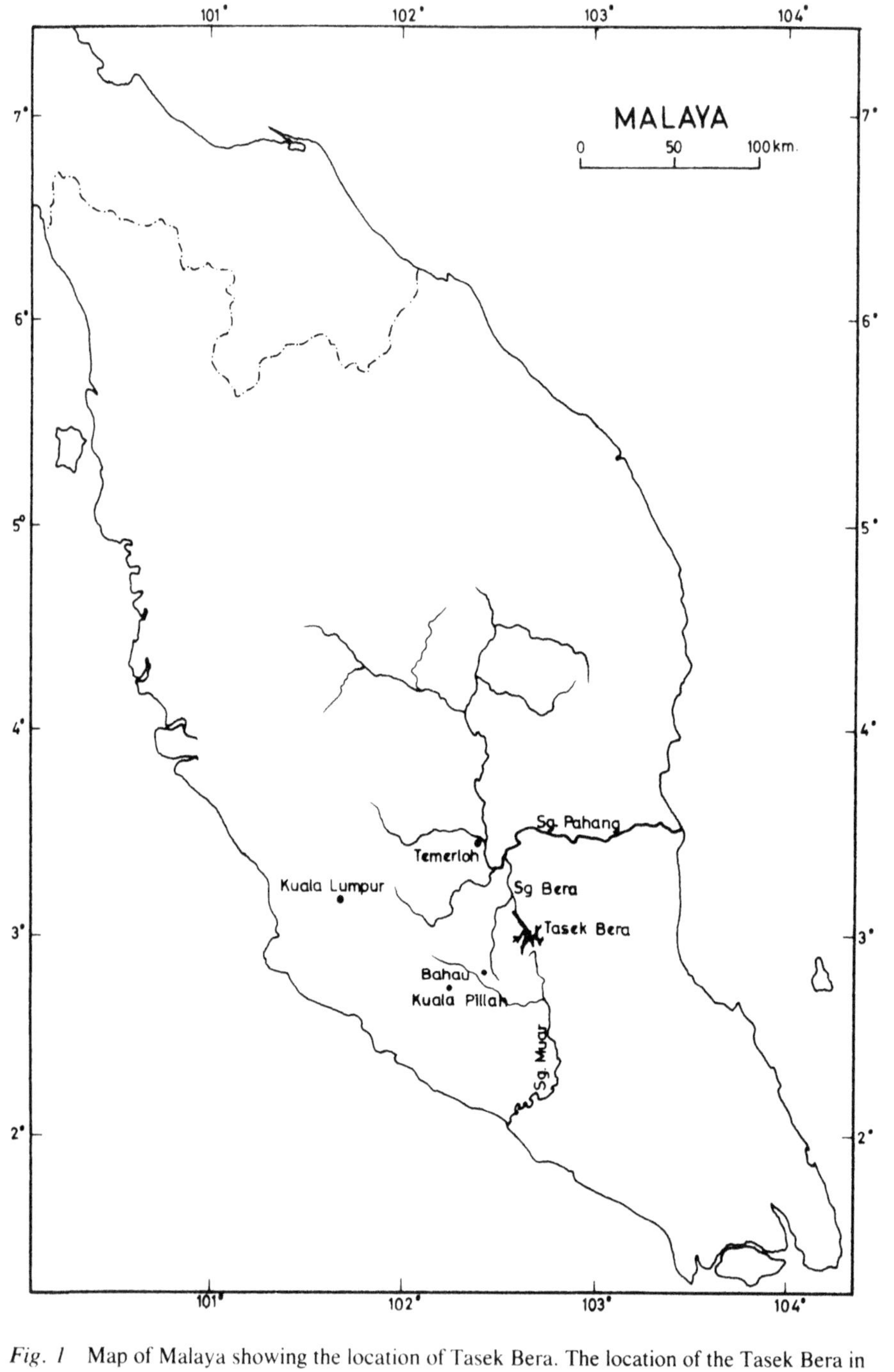

Fig. 1 Map of Malaya showing the location of Tasek Bera. The location of the Tasek Bera in relation to the Sungai Bera-Pahang River complex, and to the Sungai Palong-Muar River complex in Peninsular Malaysia (or Malaya). (R. P. Lim and J. I. Furtado)

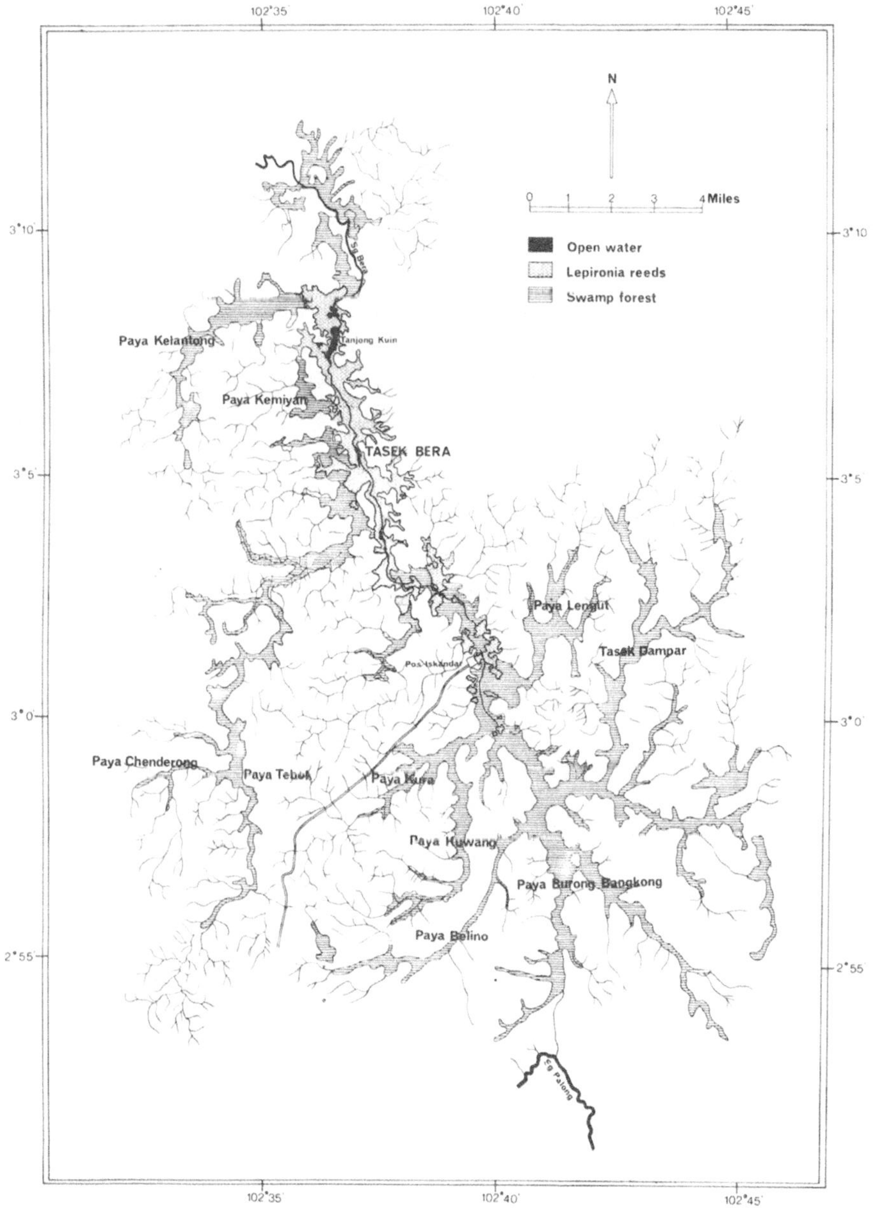

Fig. 2 The Tasek Bera and its watershed showing the distribution of the swamp (littoral region) and water (limnetic region), the 'belukar' or secondary vegetation, the streams and various swamp components, the outlet into the Sungai Bera, the proximity of Sungai Palong, and the location of Pos Iskandar. (R. P. Lim and J. I. Furtado)

Burong Bangkong, Paya Kuwang, Paya Belino, Paya Kura, Paya Chendrong, Paya Langut and the Paya Kelantong, all of which contribute to the dendritic form of the swamp and drain into the main channel system.

Tasek Bera is composed of three major habitat types: (1) the limnetic or open water region which is fringed by stands of *Utricularia* in the surface waters; (2) the *Lepironia* reed and *Pandanus* clump stands forming part of the littoral region; and (3) the *Eugenia* swamp-forest stands which form the major part of the littoral region (Plate 1). The swamp-forest stands are the most extensive forming about 67% of Tasek Bera, the *Lepironia* and *Pandanus* stands are less extensive forming about 32%, while the limnetic region is least extensive forming only 1% of the swamp (Fig. 2; Table 2). The littoral region is thus the most dominant (99%) feature of this swamp ecosystem, while the limnetic region and the main water channel are mainly in the northern half of the swamp.

The benthic substrate of the swamp is of peat comprising undecayed plant material and pieces of wood and silt. In the littoral region this peat is up to 7 m deep. In 1972, W. S. Null (pers. comm.) obtained several benthic cores, two of which were from Tanjung Kuin near the swamp outlet. Here, in the

Plate 1 Aerial view of the study area of Tasek Bera showing the main vegetational components. (R. P. Lim)

Table 2 Major habitat types forming Tasek Bera, as evaluated from aerial photographs of the Survey Department, Malaysia, 1966. (R. P. Lim and J. I. Furtado)

Regions	Area (km^2)	%
1. Limnetic or open water (including *Utricularia* zone)	0.75	1.2
2. *Lepironia* reed and *Pandanus* tree stands in the littoral	19.56	31.8
3. *Eugenia* swamp-forest stands (including 20.56 km^2 at Tasek Bera *sensu stricto*) in the littoral	41.19	67.0
Total	61.50	100.0

limnetic region a 2.5 m core was obtained before reaching sand, with its profile showing a detritus-mud layer underlain by clay and finally by sand. In the littoral region, however, a 7 m core was obtained in a *Pandanus* clump, with its profile showing a detritus-clay layer underlain by a mud layer, a fine sand layer and finally coarse sand. The shallower peat layer in the limnetic in comparison to the littoral is due to the flowing water conditions of the limnetic channels which often contain large pieces of fallen trees. In the *Lepironia* and *Pandanus* littoral, tree stumps occur suggesting that this was once swamp-forest (see later).

The water in the swamp moves at varying speeds with maximum flow in the limnetic channels (see later). The benthic substrate is thus continually shifting and unstable because of this flowing water condition, and this is reflected in a relatively poor development of benthos. Where the benthic substrate is relatively stable in some stream channels, the macrophyte *Cryptocoryne griffithii* occurs in rich stands.

Although Tasek Bera drains into the Sungai Bera by a main channel, this outlet is narrow and unable to cope with surface flow in the monsoonsal seasons, especially the northeast monsoon (September-January), resulting in marked fluctuations in the water level. During these monsoons, especially the northeast monsoon, the Sungai Bera and the Sungai Pahang are flooded and cause a reverse flow of water into Tasek Bera through several channels near the main outlet near Tanjong Kuin. As a consequence, the water level in the swamp rises between 1 and 5 m during the northeast monsoon. However, the littoral has an average depth of 0.8 m, while the limnetic region of 2.0 m with an average of 2.5 m in channels in the swamp-forest and a maximum of 7.0 m in the main channel near the outlet.

2.2 Origin and history
(R. J. Morley)

2.2.1 *Introduction*

Inland lakes are temporary landscape features, being the result of interruptions or complications in the normal fluvial cycle of erosion. Given time, barriers holding back lakes will be breached through back-cutting and, by the slow process of peneplanation, any remnants of lake sediments will be scoured out by erosion and washed downstream. Lakes are therefore normally associated only with areas affected by geologically recent tectonic movements, volcanism, glacial activity or other processes which result in irregularities in river profiles. The occurrence, therefore, of Tasek Bera, an extensive inland freshwater system within the supposedly geologically stable Malayan massif, an area generally characterised by eroded contours, and a landscape approaching peneplanation, appears at first sight somewhat incongruous.

In this chapter, an attempt has been made to explain this incongruity. Firstly, the geological framework of the area is briefly considered. This is followed by a discussion of the part Tasek Bera has played in the history of the drainage system of the central Malay Peninsula. The more recent history has been studied by examination of the sedimentary record of Tasek Bera. The complex nature of the recent history of the site has been elucidated by an attempt to interpret the history of vegetation in the area through the method of pollen analysis. A time scale for the vegetational and stratigraphic history has been provided by radiocarbon dates.

2.2.2 *Geological framework*

The nature of bedrock in the area is poorly known, owing to the thick soil cover and the absence of adequate exposures. The 7th edition of the Geological Map of West Malaysia (scale 1:500,000, ed. Cung Sooi Keong 1973) shows the majority of sediments in the area to consist of arenaceous and argillaceous rocks of Jurassic and Triassic age. These grade from a predominantly marine facies in the west to continental in the east. The hills in the Bukit Chermingat area, to the northeast, consist predominantly of Permian deposits. Sedimentary rocks in the area are thought to be highly folded (Gobbett and Hutchison 1973).

The argillaceous rocks which cover most of the western watershed consist mainly of shales mixed with minor bands of sandstones and conglomerates, the entire formation being intensely folded and impregnated with quartz veins. The weathered products form a heavy, coarse textured and mainly pale-coloured soil of firm consistency. From west to east of the region the proportion of arenaceous sandstones increases very gradually so that, in the vicinity of the Sungai Bera east of the Tasek Bera watershed, the lithology is one predominantly of arenaceous rocks interbedded with subordinate bands of shale. Correspondingly, the soils gradually change to more highly coloured sandy, friable types.

The remaining upland areas in the vicinity are formed of igneous rocks. Bukit Palong, to the west, consists of granite, whereas Karisek, to the east of the upper S. Palong catchment, is formed of intermediate to basic volcanics. The lowland area to the south of Bukit Charmingat is also formed of granitic rocks.

The local alluvial deposits were sedimented subrecent to the formation of the swamp, and form gently undulating terraces along many of the drainage channels (Dumanski and Ooi 1966).

Three main series of soil associations occur in the region, namely the Batu Anam, Malacca and the Tavy series; all of which are of the order Ultisol (Soil Survey Staff, Malaysia 1960).

Bedrock in the Tasek Bera area plays only a small part in determining the form of the present landscape, owing to the very long period of erosion and weathering which the area has undergone. The course of flow of the main rivers, and the effect of geologically recent fault movements in the area are much more significant in determining the form of the present landscape.

2.2.3 *Tasek Bera as part of the Sungai Pahang-Sungai Muar drainage system*

The evolution of Tasek Bera cannot be properly understood unless viewed in a wide context as part of the main drainage system of Peninsular Malaysia. Despite its low altitude (c. 25 m a.s.l.), it lies on the primary watershed of the peninsula, which divides eastward from westward flowing streams (Fig. 3). Further examination of stream patterns in southern Pahang and Johore indicates that a second watershed, dividing the lower Sungai Pahang catchment from the S. Rompin catchment, joins the main divide at Tasek Bera. Tasek Bera, therefore, occurs on a col at the radiating centre of three major drainage systems.

A glance at river patterns within the vicinity of this col shows numerous anomalies, especially barbed tributaries, which suggest that river capture has

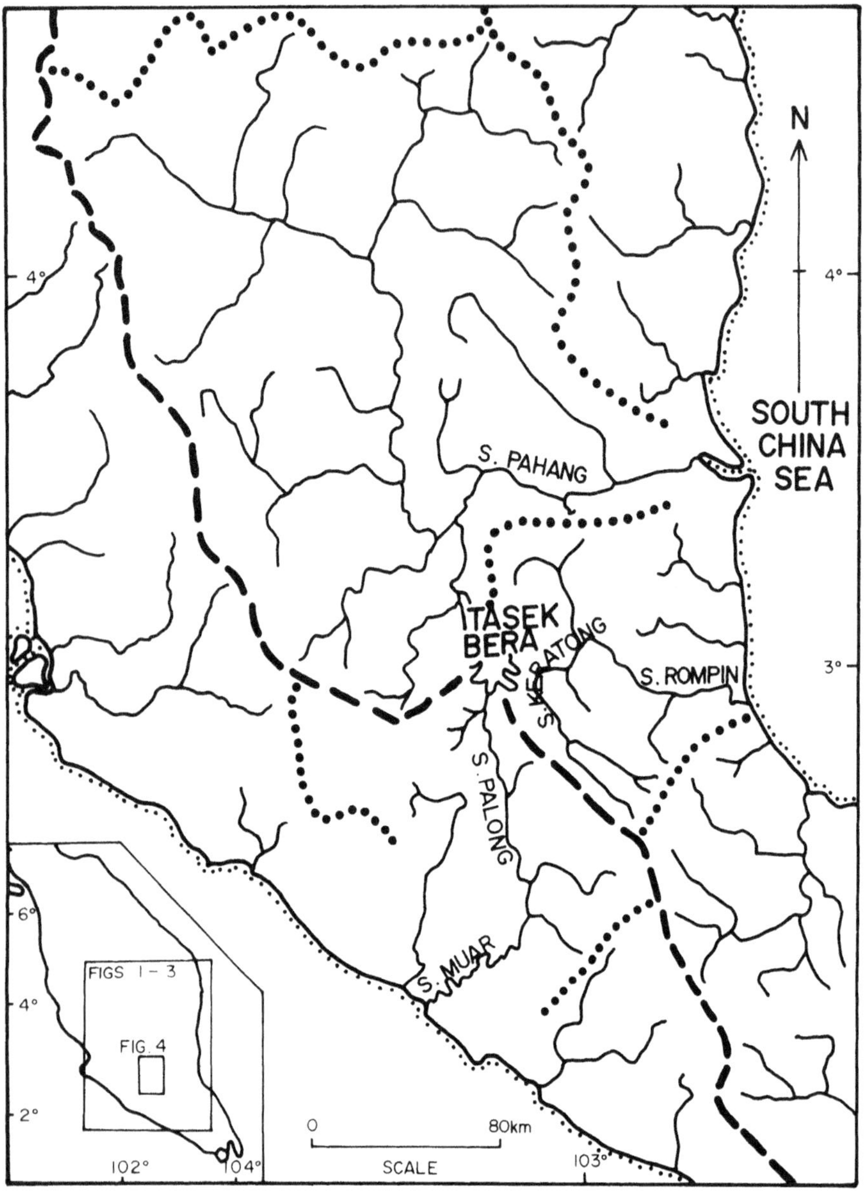

Fig. 3 The main drainage systems in the area surrounding Tasek Bera. Primary watershed shown by dashed line (– – –), secondary watersheds by dotted lines(. . .).

played an important part in determining the present configuration of the drainage channels in the area, and that a closer study of the history of the drainage channels themselves is necessary to form a basis for the discussion of Tasek Bera as we see it today.

Richardson (1947) considered that the present drainage system has envolved at least during the course of the entire Tertiary period. The original drainage pattern would have consisted of north-south trending streams, following the grain of the main structural trends of the geological basement. It is thus probable, as first pointed out by Scrivenor (1931), that the original outlet of the S. Pahang headwaters was not in the South China Sea but the Malacca Straits, by way of the Tasek Bera area and the present S. Palong-S. Muar valley. Backcutting by the lower S. Pahang eventually cut through the Gunung Berantai-Bukit Chermingat range to capture the southward flowing stream (Fig. 4), thus diverting the main flow of the ancestral S. Pahang headwaters into the South China Sea. Owing to the present magnitude of the S. Pahang delta, the degree of establishment of the present course, and the number of events, particularly to the south of the present river, which have since occurred (see below), it is probable that this capture occurred moderately early in the history of the river, probably in the earliest Pleistocene or Late Tertiary.

The gradient advantage of the new Lower Pahang must then have been considerable, as part of the southward flowing stream below the point of capture was also diverted; a considerable catchment area thus developed to the south of the elbow of capture (Fig. 5). This consisted of Tasek Bera, flowing into the S. Pahang by the present S. Bera. Also incorporated into the catchment were the headwaters of the present S. Palong, which drained into the Tasek Bera through the Ayer Kuning channel, and, with most probability, the headwaters of the S. Keratong-S. Rompin. The headwaters of this former tributary probably included the whole of the S. Karatong headwaters above Bukit Serok, consisting of the present S. Jekatih and S. Pukin.

The final stage in the evolution of the Tasek Bera channel, leading to its present configuration, consisted of the gradual capture of the southernmost headwaters of this catchment. At a fairly early stage, backcutting by the S. Rompin - S. Karatong diverted the southeastern headwaters, the elbow of capture being at Bukit Serok. Secondly, the southwestern headwaters were recaptured by the S. Palong - S. Muar, the major elbow of capture being at the southern end of the Ayer Kuning channel, the 'wind gap' between the two river systems consisting of the Ayer Kuning channel itself. This recapturing most probably took place during the late Quaternary or Holocene. Strong evidence for the very recent capture at this point is provided by the fact that water still flows through this channel, downcutting within the southern channel not having progressed below the level at capture (Fig. 7). A further

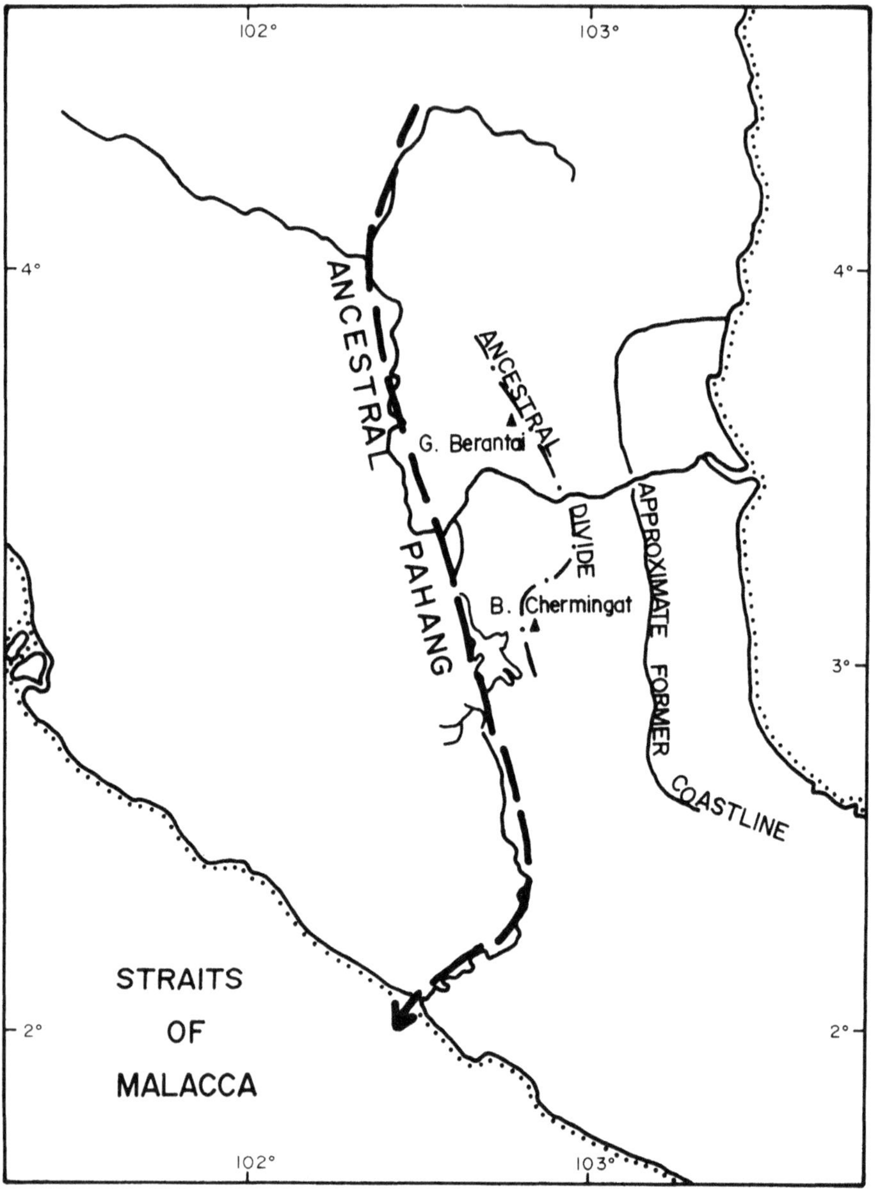

Fig. 4 Probable original course of ancestral Sungai Pahang.

imminent capture is probable of a small stream among the headwaters of the S. Serting (which flows into the S. Bera below Tasek Bera) by the S. Muar at Bahau (Fig. 6).

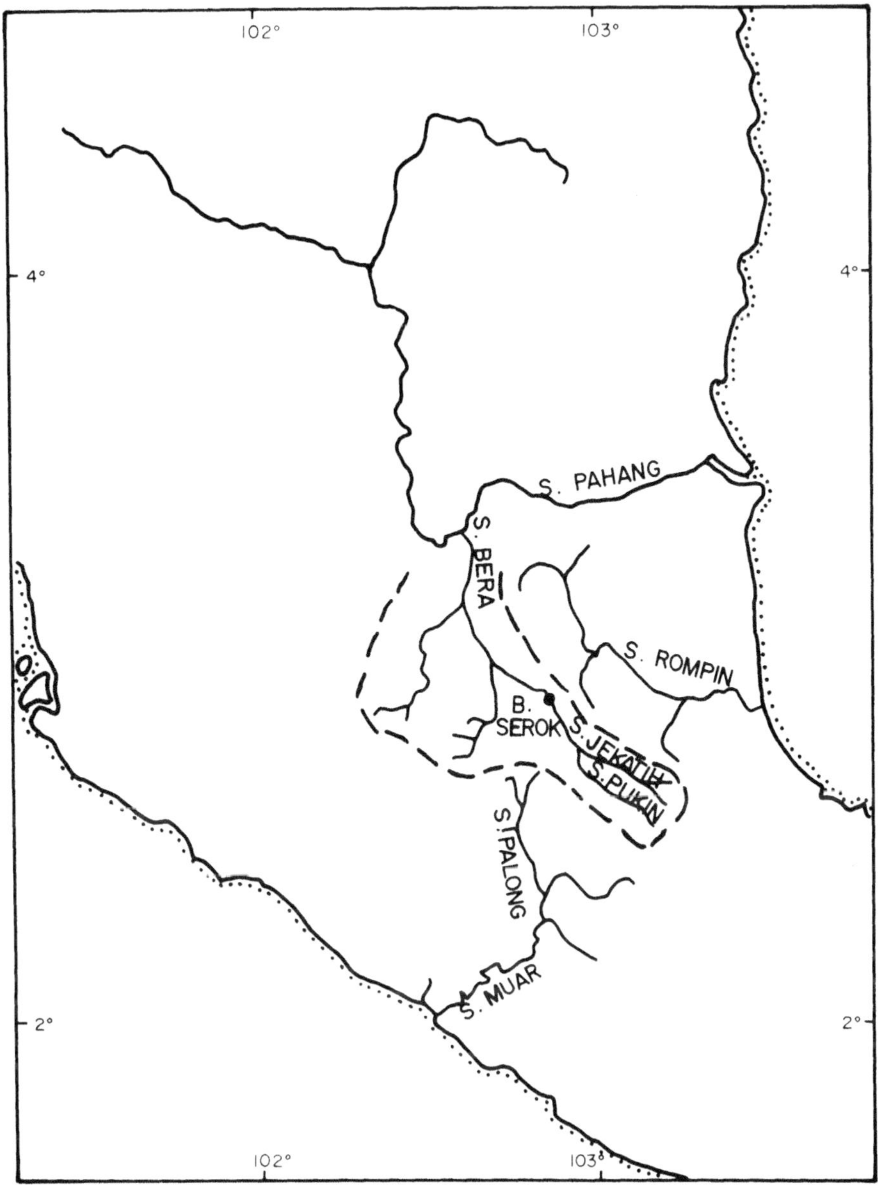

Fig. 5 Former southern catchment of Sungai Bera.

It is difficult to arrive at an explanation of the cyclical capture and recapture of streams in the Tasek Bera area. Barton (unpublished) considered that the course of the Lower Pahang, during backcutting, would have been much shorter than it is today, the former coastline probably being some 50 km

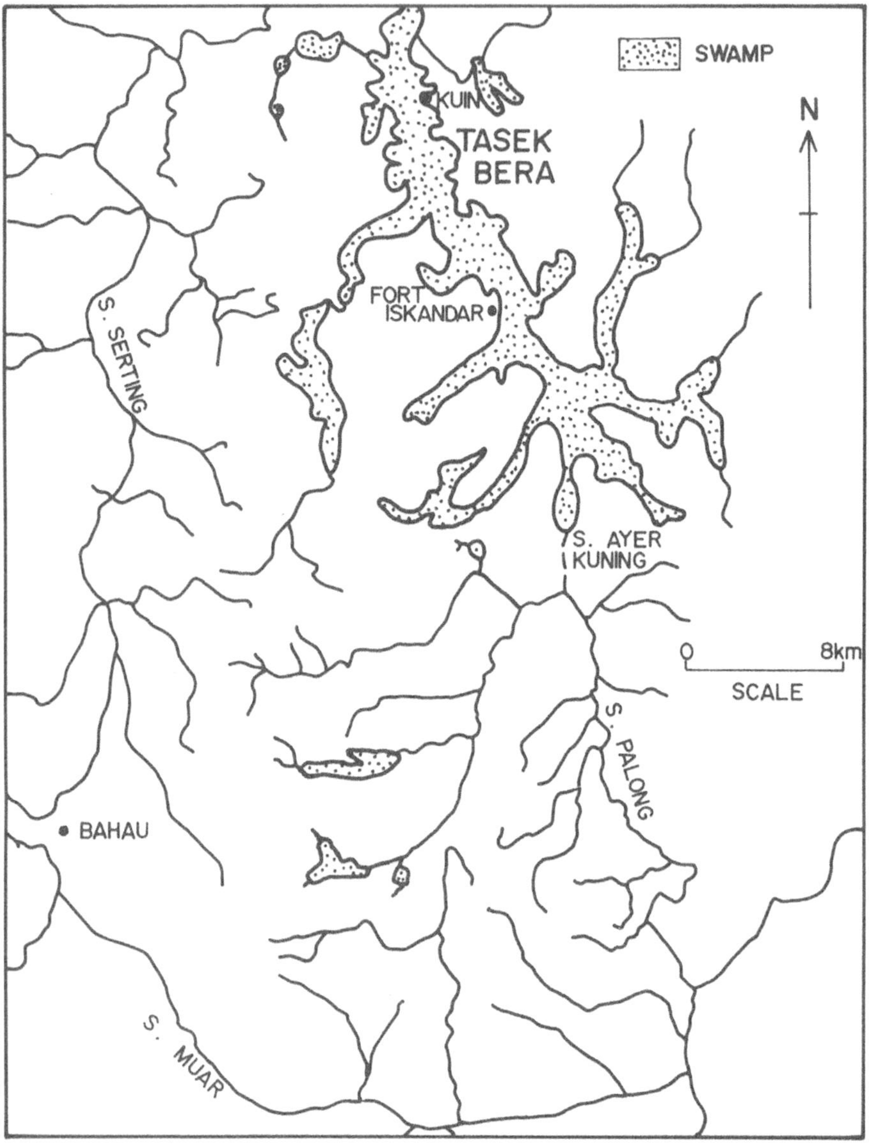

Fig. 6 Tasek Bera and the nature of river channels to the south.

further inland. This shorter course may have given the S. Pahang a sufficient gradient advantage to capture the ancestral S. Pahang headwaters. The reason for the recapture of the Tasek Bera headwaters by the S. Palong-S. Muar is obscure. The gradient profiles of these streams show that they have a

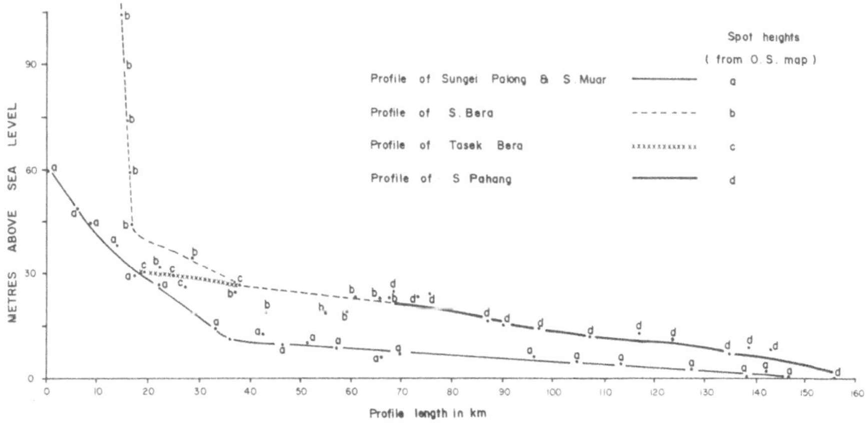

Fig. 7 Profile diagrams of rivers associated with Tasek Bera. Profiles have been determined from spot heights shown along river courses on ordnance survey maps.

considerable base level advantage over the S. Pahang up to 25 km from the point of capture (Fig. 7). This advantage may be the result of the fact that, with the presently restricted headwaters of the S. Palong - S. Muar, insufficient sediment has been carried to the lower reaches to build up the flood-plain to any significant degree, whereas the flood-plain build up of the S. Pahang, with its extensive headwaters, has been considerable, giving rise to an extensive delta and considerable coastal advance.

A slight gradient advantage of the southern streams may well have been of great significance in recapturing these headwaters. An additional factor of major importance in this area may have been that of slight crustal movement.

A prominent line of dislocation extends from Kuala Lumpur in the west, to Ulu Endau in southeast Pahang (Shu 1969). This fault zone must traverse the southern part of the Tasek Bera area. Evidence suggests that this zone has been active in the recent geological past (Stauffer 1968). Tertiary (? Late Miocene-Pliocene) beds in the Batu Arang Basin in Selangor exhibit steep folding near the northern edge of the basin (Law 1961), where the western extension of this zone if believed to be located (Gobbett and Tjia 1973). Movement along this fault zone during Pleistocene times is therefore highly probable. This movement may have been crucial in determining the direction of flow, at various times, through the Tasek Bera area.

The cause of damming of Tasek Bera is also open to question. It is possible that damming may have been initiated by levée development along the S. Pahang. This is thought unlikely, however, owing to the distance (25 km) of the S. Pahang from the confluence of Tasek Bera with the S. Bera. Merton (1962) considered that the second phase of river capture may have been implemental. He argued that the reduced flow of water through the river channel could lead to a reduction in speed of the throughflow, which in turn could cause settling out of seston. Neither of these hypotheses, however, explains the occurrence of riverine swamps in the upper tributaries of the S. Palong, upstream from the confluence with the S. Ayer Kuning (Fig. 6). The presence of riverine swamps at both Tasek Bera and in the S. Palong head-waters suggests an origin common to both. This may have occurred independently of river capture. The only mechanism which is thought to be able to account for a drainage impediment in both river systems is that of a general tilting caused through movement along the Kuala Lumpur fault zone.

2.2.4 *Sedimentary history*

Stratigraphy was studied in two areas by means of borehole transects (Fig. 8), one being located near Pos Iskandar, positioned transversely across the swamp; the other was aligned longitudinally with the channel in the vicinity of Kuin (Fig. 9). Sampling was largely undertaken with the Livingstone sampler (Livingstone 1955). Additional material was obtained from the Kuin area using a 'Russian' design sampler (West 1968).

Field descriptions of the appearance, properties and physical components of sediment types encountered follow the system of Troels-Smith (1955). Simple laboratory tests were also undertaken on one of the cores, which was studied in detail for its pollen content (see below). These tests included determination of wet weight, dry weight and weight following ignition at 550°C (all per cc wet sediment). Percentages of material remaining after drying and after ignition were calculated from these measurements. During palynological investigations, absolute minimum estimates of numbers of pollen grains and spores per unit weight and unit volume of sediment were determined using a modification of the method of Jorgensen (1967). (For details of this method, see Morley 1976.) In the discussion of sedimentary history, changes of pollen influx into sediments may yield information in changes in the rate of sediment deposition, especially if it is thought that pollen rain, during the time of deposition of the sediments, has remained uniform.

The southern transect revealed the site to be of regular cross-section. The deepest borehole in this area penetrated 5.30 m of organic and inorganic sediments. The lowest horizon recorded was a brown clay (physical compo-

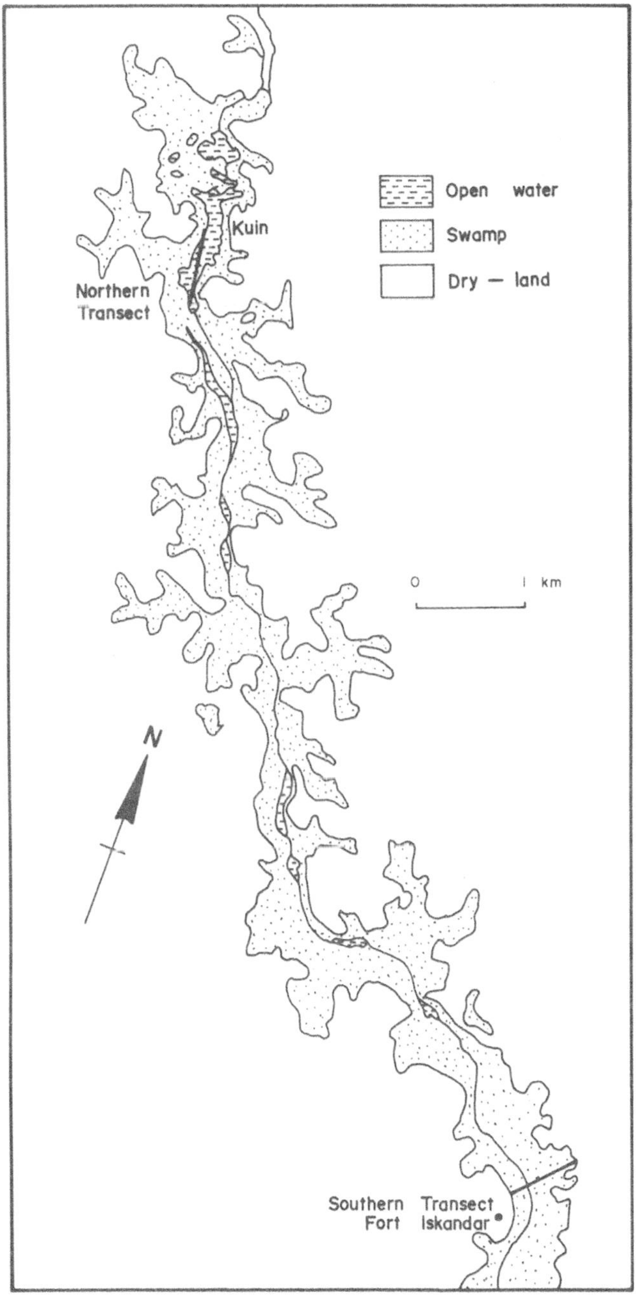

Fig. 8 Location of borehole transects at Tasek Bera. Distribution of swamp-forest plotted from aerial photos.

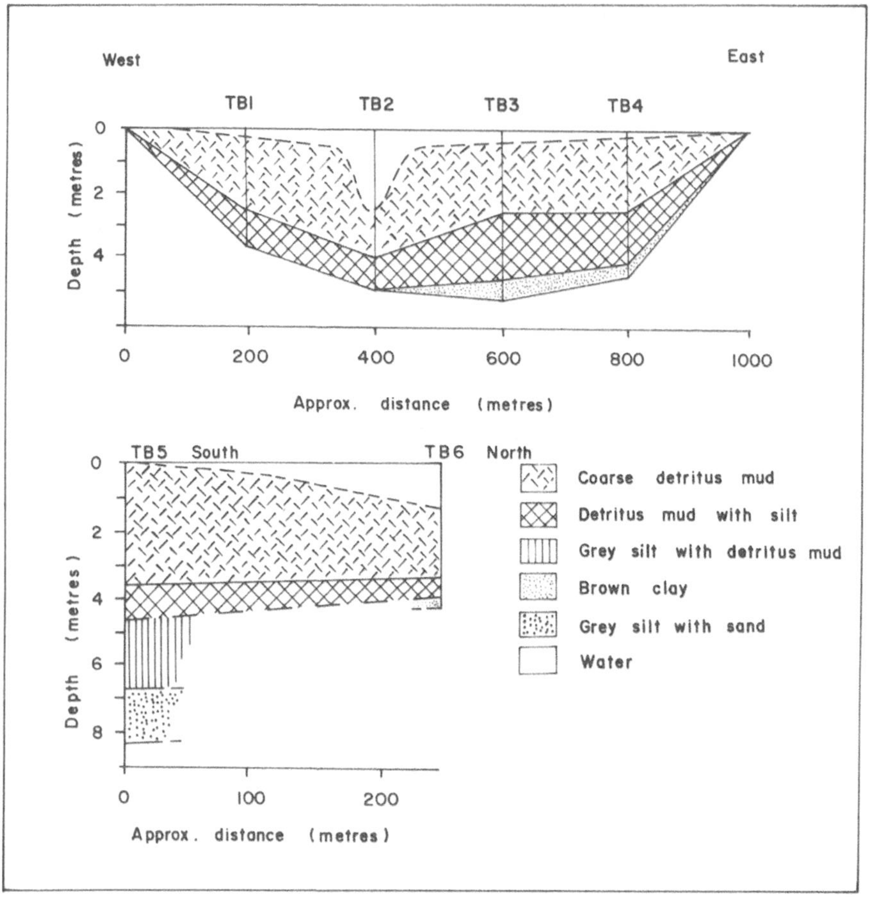

Fig. 9 Stratigraphic sections at Tasek Bera. Southern transect above, northern transect below. Different vertical scales are used for the two profiles.

nents Ag_4, using Troels-Smith's notation). This was recorded only on the eastern side of the transect and was of maximum thickness (0.80 m) at borehole TB.3. Above this layer were sediments which consisted predominantly of fine grey silt (Ag_3 Dh_1 Dl_+), which contained varying amounts of organic matter. These sediments were thicker on the eastern side of the transect, reaching a maximum thickness of 2.00 m at borehole TB.3.

Overlying the grey silts were up to 3 m of soft, coarse detritus muds (Dh_2 Dl_2). These muds contained many woody fragments, but preservation of plant material was rather poor. The muds were of a dark brown colour (Nigror

3) and were very moist (Siccitas 1). They were neither stratified nor elastic (Stratificato 0, Elasticitas 0) but were to some degree humified (Humocitas 2). They were overlain by up to 3 m of water.

The longitudinal section near Kuin revealed a succession similar to that observed near Pos Iskandar; however, at core TB.5, three additional sediment types were present below the grey silt layer. The detailed field characteristics of sediment in core Tb.5 are shown in Table 3. The lowermost layer present consisted of sand and silt, together with a considerable proportion of woody fragments and some fine organic debris, reaching a thickness of 1.00 m. Overlying this were 1.15 m of uniform, fine grey silts with occasional wood fragments. Microscopic examination of these silts revealed the presence of predominantly fine, well sorted silica fragments ranging in size from 5-20 μm (very fine silt-fine silt). Above these silts were 1.55 m of organic detrital deposits which contained numerous silty lenses. The nature of the silt in this layer was similar to that of the layer below. In the longitudinal transect, the brown clay observed near Pos Iskander was present only at core TB.6.

Core TB.5 was selected for detailed palynological study as this core contained the most complete sedimentary record observed at Tasek Bera. Sediment characteristics revealed during laboratory tests, and minimum estimates of the pollen and spore content per g and per cc sediment, are given in Fig. 10.

Radiocarbon assays were obtained for the basal sample (8.55 m-8.65 m) and also at 3.85 m-3.96 m. Dates obtained were, respectively, 4500 ± 80 years (Birm 205) and 660 ± 75 years (SRR 467). These dates are in radiocarbon years BP.

The presence of abundant sand particles and coarse debris in the lowermost layer suggests that the basal sediment may have been deposited at a time when the current through the tasek was considerably greater than at present, perhaps prior to the capture of the western headwaters by the S. Palong, or prior to the drainage impediment which resulted in the development of swamp in the area. Pollen deposition rates within this section are quite low, suggesting rapid deposition. The radiocarbon assay suggests that these sediments were deposited only during the mid-Holocene, c. 4500 radiocarbon years BP. This date may indicate the time of origin of the tasek. However, the stratigraphic studies at Tasek Bera were by no means sufficient to rule out the possibility of there being older sediments. It is possible that the date may be a slight underestimate, as *Pandanus helicopus* roots may have penetrated the sediment a considerable time after deposition.

The presence of fine silt layers at various intervals in each of the cores examined suggests that sedimentary conditions at the tasek were inter-

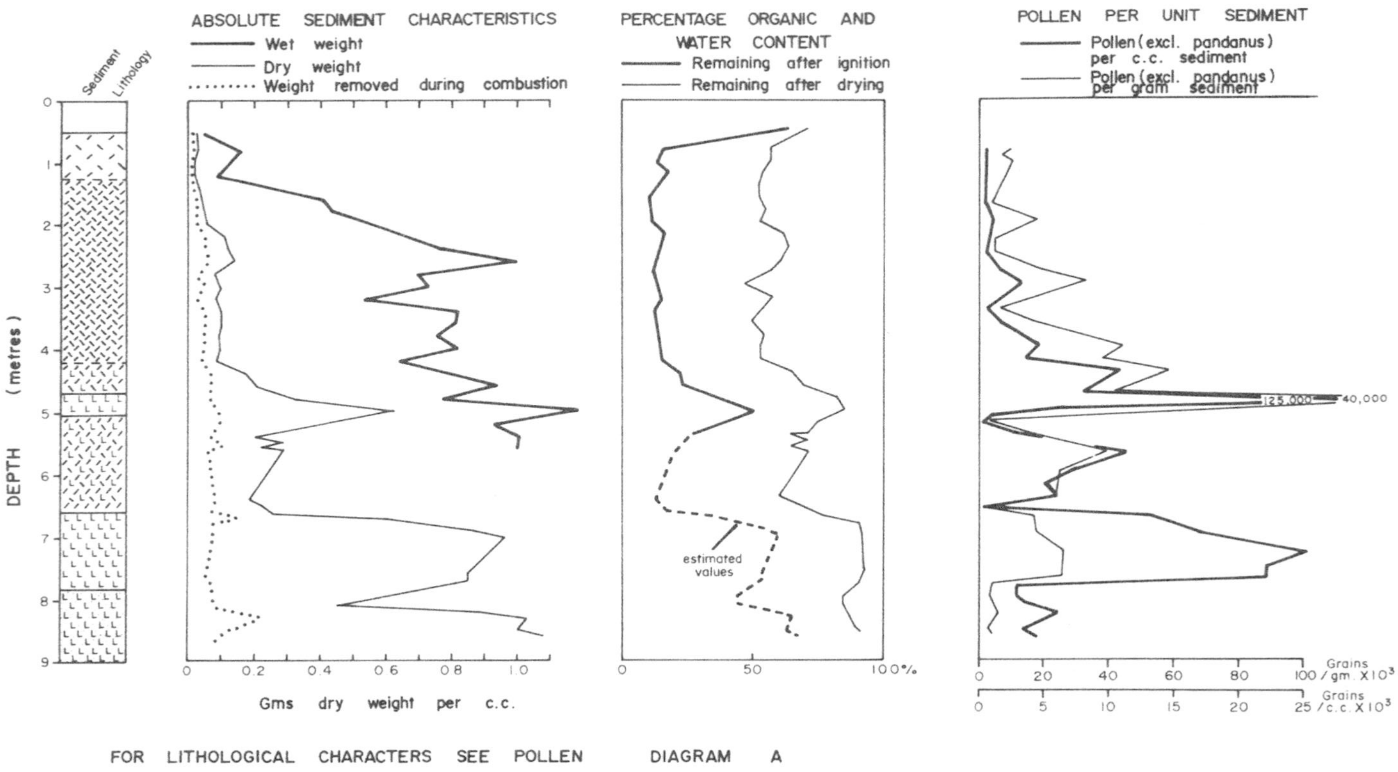

Fig. 10 Physical characteristics of core TB 5.

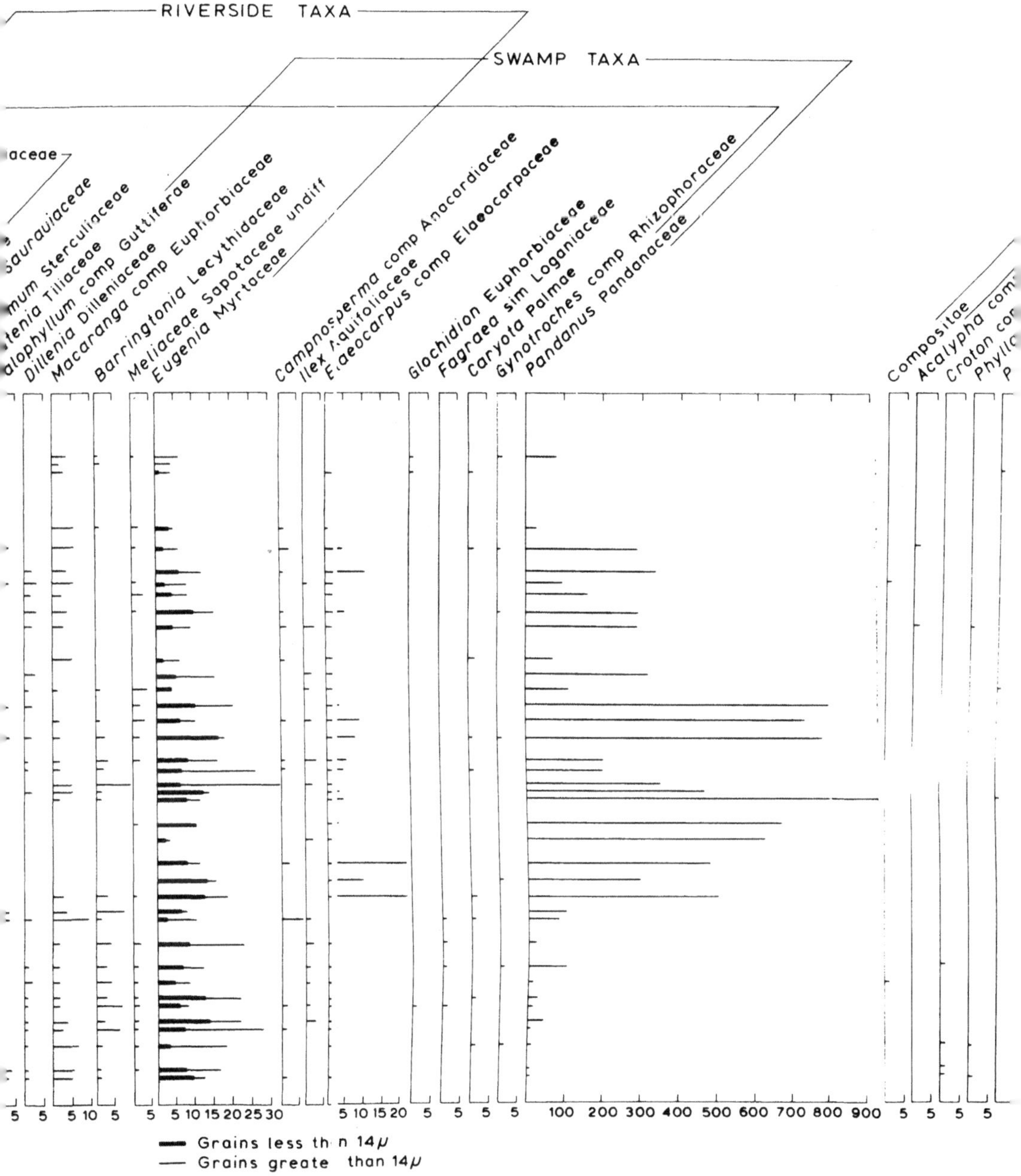
RIVERSIDE TAXA
SWAMP TAXA
Sauraulaceae
Sterculiaceae
Tiliaceae
comp Guttiferae
Dillenia Dilleniaceae
Macaranga comp Euphorbiaceae
Barringtonia Lecythidaceae
Meliaceae Sapotaceae undiff
Eugenia Myrtaceae
Campnosperma comp Anacardiaceae
Ilex Aquifoliaceae
Elaeocarpus comp Elaeocarpaceae
Glochidion Euphorbiaceae
Fagraea sim Loganiaceae
Caryota Palmae
Gynotroches comp Rhizophoraceae
Pandanus Pandanaceae
Compositae
Acalypha
Croton
5 5 5 10 5 5 5 10 15 20 25 30 5 10 15 20 5 5 5 5 100 200 300 400 500 600 700 800 900 5 5 5 5
Grains less than 14μ
Grains greater than 14μ

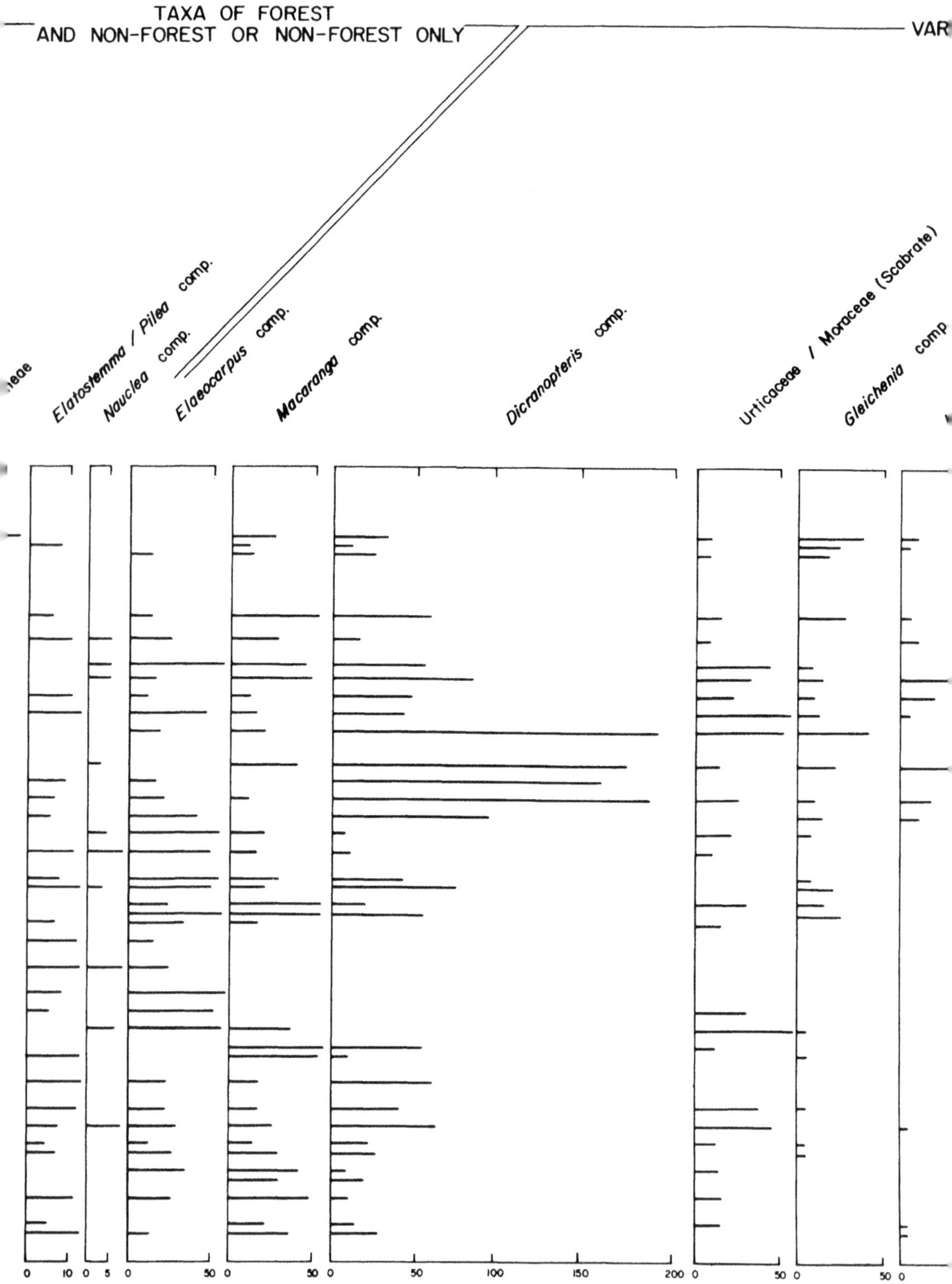

TAXA OF FOREST
AND NON-FOREST OR NON-FOREST ONLY
VAR
Elatostemma / Pilea comp.
Nauclea comp.
Elaeocarpus comp.
Macaranga comp.
Dicranopteris comp.
Urticaceae / Moraceae (Scabrate)
Gleichenia comp
0 10 0 5 0 50 0 50 0 50 100 150 200 0 50 0 50 0

Table 3 Details of stratigraphy of pollen analysis core.

Depth (in m below raft)	Description	Nig[a]	Str[b]	Elas[c]	Sicc[d]	Hum[e]	Components
0 –0.30	Air						
0.30–0.80	Open water						
0.80–1.75	Coarse brown detritus and with many root remains	3	0	1	1	1	$Dh_3Dl_1Sh_+$
1.75–1.85	Very fibrous unhumified detritus mud with abundant root fragments	3	0	1	1	0	Dh_3Dl_1
1.85–2.00	Dark brown detritus mud, leaf and root fragments abundant. More humified	3–4	0	1	2	1–2	$Dh_2Dl_1Sh_1$
2.00–3.40	Brown detritus mud with common leaf and root fragments. Humified	3	0	1	2	1–2	$Dh_2Dl_1Sh_1$
3.40–4.18	Dark brown coarse detritus mud with leaf and root fragments. Humified	3–4	0	1	2	1–2	$Dh_2Dl_1Sh_1$
4.18–4.76	Brown detritus mud with many leaf and root fragments. Humified. Increasing proportion of grey silt toward base	2–4	1	1	2	1–2	$Dh_2Sh_1Dl_+Ag_1$
4.76–5.05	Fine grey silt with occasional wood fragments	1	0	2	2	0	Ag_4Dl_+
5.05–5.15	Coarse detritus mud, many plant fragments present. Humified. Silt fraction present	3–4	0	1	1	2	$Dh_2Dl_1Sh_1Ag_+$
5.15–5.50	Black-brown silty detritus mud with abundant plant fragments	3	0	2	2	1–2	$Dh_2Dl_+Sh_1Ag_1$
5.50–6.50	Black-brown silty detritus mud with abundant plant fragments. Silty lenses present	3	1–2	2	2	1–2	Dh Dl Sh Ag
6.50–6.60	Black coarse detritus mud with many large plant fragments	4	0	1	1	1–2	$Dh_2Sh_1Dl_1$
6.60–7.85	Fine grey silt with occasional wood fragments	1	0	2	2	0	Ag_4Dl_+
7.85–8.65	Silty sand with many woody and leaf remains, becoming more sandy toward base	2	1	1	2	1	$Ga_2Ag_2Dl_+Dh_+$

[a] Nigror [b] Stratificatio [c] Elasticitas [d] Siccitas [e] Humocitas

mittently different from those at present. The abundance of mineral sediment might be explained by higher currents. Also, the possibility cannot be ruled out that mineral material may have been derived from areas outside the present catchment. The high density of pollen and spores within the mineral layers might be explained by relatively low deposition rates. Extensive river transport of pollen might also be responsible for these high values.

The radiocarbon date from 3.85 m - 3.95 m (660 ± 70 radiocarbon years BP) indicates that the uppermost sedimentation here was very rapid. Again, *Pandanus helicopus* roots may cause this date to be rather young; pollen deposition rates within this section, however, are very low, substantiating very rapid deposition of the upper three metres of sediment.

2.2.5 *Vegetational history*

A glimpse of the history of vegetation at Tasek Bera is provided by the record of pollen and spores which have become incorporated into the sedimentary sequence during deposition. The palynological record in no way provides a mirror-like image of vegetational history, but, taking into account the various processes of pollen production, pollen transfer, incorporations into the sediments and pollen preservation, a tentative picture can be reconstructed of some of the major vegetational changes which have occurred within the vicinity of the site.

The process of palynological analysis from an extremely floristically rich area such as lowland Malaysia is necessarily slow. Consequently, the pollen record of only a single core, located near Tanjong Kuin, has so far been studied.

2.2.5.1 *Palynological methodology.* Samples for pollen analysis were obtained from core TB.5, collected by means of the Livingstone sampler. Owing to the possibility that the outer part of the core may be contaminated by extraneous material, only sediment from the innermost part was used. In all cases, samples were restricted to a 1 cm slice of sediment.

Palynological preparations were made with the aim of obtaining absolute estimates of pollen deposition per unit weight or volume of sediment, as information on pollen influx may be of considerable help in interpreting the nature of past vegetation, and also, in assessing sedimentation rates (see Section 2.2.4). Of the available methods (Peck 1974), the method of Jorgensen (1967) is thought to be most applicable to the estimation of pollen influx into sediments of the type present at Tasek Bera, and so was used here exclusively.

Sample maceration followed the standard method of deflocculation in 10% potassium hydroxide solution, followed by acetolysis (Erdtman 1943), which removes cellulose. Silicates were removed by boiling in hydrofluoric acid. Samples were mounted in silicone oil. This mounting medium is preferred as it permits turning of grains under the coverslip thus aiding observation of the nature of pollen grains, and hence improving the likelihood of accurate determinations. Silicone oil has also been shown to cause a minimum of size changes of mounted pollen with time (Reitsma 1969), thus making preparations more permanent compared to other mounting media.

Pollen and spores were counted using a Zeiss (Oberkochen) Photomicroscope at a magnification of ×400, using a ×25 plan-apochromatic objective, ×8 Kpl. eye-pieces, and ×2 Optovar magnification. Initial determinations were verified using a × 100 plan-apochromatic oil immersion objective, giving a magnification of × 1600.

Pollen and spores were determined by comparison with modern reference specimens, mostly collected from named herbarium material. The degree of certainty of the determinations should be defined clearly so that, in interpretation, equal weight is not given to both certain and doubtful determinations. A modification of the system outlined by Benninghoff and Kapp (1962) has been used here. When the pollen or spore type is similar to, but differs slightly from the named taxon, it is followed by the suffix 'sim'. When the pollen type is morphologically identical to the taxon indicated but the type originates from more than one taxon, or some doubt exists about the conclusiveness of the determination, the name is followed by the suffix 'comp'. When the determination is certain, no suffix is given. For binomial names, the suffix refers only to the section of the name which immediately precedes it.

A list of the pollen and spore types determined during this study, and an indication of the plant taxa from which they were probably derived, is included in Appendix II.

It is pointed out by Faegri and Iversen (1964) that the statistical significance of changes in pollen diagrams depends chiefly on the size of the pollen counts. They postulate, from experience, that, at least in temperate areas, a minimum pollen count of 150 grains is necessary, and point out that considerable errors may result if fewer grains are counted. In tropical areas, where many more pollen types may be encountered, each represented by only low frequencies, the size of count required to give reliable data needs to be considerably higher (e.g. Hamilton 1972; Martin 1963). Analysis of the reliability of representation of pollen in assemblages from Malaysia (Morley 1976) suggests that, in order to yield data of sufficient reliability, counts of between 300 and 500 grains should be carried out in each sample.

During this study, this objective was reached for the majority of samples; some, however, contained too little pollen to permit the requisitive number to be counted. These samples do not, however, affect the general interpretation of the pollen assemblages in terms of vegetation.

The interpretation of vegetational history from a pollen analysis requires the construction of pollen diagrams, for which a pollen sum must be selected. Previous experience has shown that in order to obtain as full a picture as possible of vegetational change from pollen spectra, a number of pollen diagrams using different pollen sums must be examined (e.g. Flenley 1976). In this study, two pollen diagrams were constructed using different sums.

For the interpretation of the history of hydrophytic vegetation, a pollen diagram was constructed using total pollen and spores as sum. Changes in the total abundance of hydrophytic* vegetation would then be illustrated by changes in the ratio of pollen of hydrophytic origin to that from dry-land vegetation, which, it is thought, varied little during the period of time covered by the pollen profile. It became apparent that pollen of *Pandanus* completely overwhelmed the assemblages and distorted considerably pollen spectra of other hydrophytic taxa. Much more reconcilable changes were observed when *Pandanus* pollen was excluded from the pollen sum. The interpretation of hydrophytic vegetation succession was therefore made using the pollen sum 'total pollen and spores excluding *Pandanus*'. This constitutes Pollen Diagram A (Fig. 11).

The succession of hydrophytic vegetation has been clarified on the diagram by the construction of a summary diagram, using the same pollen sum, showing relative changes of pollen and spores derived from major ecological groups (the attribution of taxa to ecological groups is discussed and presented in Appendix I). Four such groups have been illustrated, i.e. pollen derived mainly from swamp trees (groups NF + S, F + S, excluding *Pandanus*), swamp herbs (groups S, F + S, NF + S), dry-land taxa (group F, NF + F, NF) and taxa of various habitats (all other group combinations). The ecological group and life form of the source taxa of all pollen types found in the samples analysed are shown to the right of the summary diagram. Pollen types are arranged according to the life form of source taxa. Three life form groups are illustrated: a) trees, b) shrubs, herbs and epiphytes, and c) taxa of various life forms. Pteridophyte spores are grouped separately. Within each life form grouping, pollen types are arranged according to the ecology of the source taxa. All are categorised into three ecological groups: taxa restricted to

* In the context of this paper, the term 'hydrophytic' refers to all aquatic and swamp vegetation.

dry-land vegetation, those restricted to hydrophytic vegetation and thirdly, those common to both. An additional category is included for arboreal taxa, i.e. trees common to river sides.

The approach of Wright and Patten (1963) is used for the illustration of dry-land vegetation changes (Diagram B, Fig. 12). The pollen sum is restricted to pollen and spores derived from taxa considered to be restricted to dry-land areas, and includes only taxa occurring within ecological groups F, NF + F and NF. This selection imposes some limitation on the degree of interpretation of this diagram, as the pollen sum, for some levels, is rather low. However, when interpreted with caution, useful conclusions can be drawn. The diagram has been constructed in three sections: a summary diagram, selected pollen taxa included in the pollen sum, and selected taxa excluded from the sum. The summary diagram shows the relative proportions of pollen and spores attributable to arboreal forest taxa (group F), to herbaceous and shrub taxa (groups F, NF and F + NF) and to arboreal non-forest taxa (groups NF and F + NF). Only common or ecologically significant taxa, which are included in the pollen sum, have been incorporated into the diagram. A number of taxa excluded from the pollen sum have been incorporated on the diagram. Although derived from taxa characteristic of a number of habitats, it is thought that these pollen types, at various times, may have been derived from taxa which formed a significant part of the dry-land vegetation of the area.

2.2.5.2 *Modern pollen rain.* The interpretation of fossil pollen spectra in terms of extant vegetation becomes a matter of speculation unless the pollen production of existing plant taxa and vegetation types is examined for comparison. This can be accomplished by analysis of the pollen content of soils, sediments or pollen preserving traps from areas where the detailed floristic content of vegetation is known. In order to determine the nature of the present pollen rain at Tasek Bera, the pollen content of sediment samples from beneath each of the major hydrophytic vegetation types present, and also from beneath several transitional plant communities was studied. Space does not permit the detailed presentation and discussion of these analyses here; a detailed account can be found elsewhere (Morley 1976). The following conclusions were drawn, however, concerning the nature of pollen rain at Tasek Bera, which are particularly relevant to this discussion.

1) *Pandanus helicopus* is a most prolific pollen producer. In all samples examined, the representation of *Pandanus* pollen was far greater than the representation of *Pandanus* spp. at the sampling localities. *Elaeocarpus* spp. are also large producers of pollen. Pollen of *Eugenia* was much less

common than expected. Little pollen of the other numerous arboreal swamp taxa present there was recorded. Among herbaceous taxa, *Lepironia articulata* produces moderately low frequencies of pollen, and *Utricularia* spp. are very low producers. No pollen of *Cryptocoryne* was found.
2) Water transport of pollen was shown to be considerable.
3) The majority of pollen from dry-land taxa was thought to be derived from within 2 or 3 km of the site of deposition.
4) The main hydrophytic vegetation types could be tentatively identified by their pollen spectra, recognition being possible, by examination of the proportions of the dominant pollen types present in the assemblages, e.g. *Pandanus*, *Elaeocarpus* comp, *Eugenia* comp and Cyperaceae.

Pure *Pandanus* swamp could be recognised by the presence of very high frequencies of *Pandanus* pollen (over 200%, pollen sum all pollen excluding *Pandanus*). Closed swamp-forest upstream from *Pandanus* swamp is characterised by the combined presence of high frequencies of *Elaeocarpus* comp and *Eugenia* comp pollen, together with very low frequencies of *Pandanus* pollen (10-20%). Swamp-forests also yielded much higher percentages of *Lycopodium cernuum* spores than other hydrophytic vegetation. Mixed swamp-forest with *Pandanus* swamp differed little from pure *Pandanus* swamp. Recognition may be possible through the combined presence of medium percentages of *Pandanus* pollen, abundant *Lycopodium cernuum* spores and the presence of high *Eugenia* comp values.

Lepironia articulata swamp may be recognised by the presence of c. 20% Cyperaceae pollen, and moderate frequencies of *Pandanus* (c. 100%).

No data were obtained concerning the recognition of open water communities by their pollen spectra.

2.2.5.3 *Zonation and description of pollen assemblages derived from hydrophytic and riparian vegetation* (Diagram A. Fig. 11). For the convenience of description and discussion, the pollen diagrams have been zoned. The zones established here are purely local and must not be considered to be of any regional significance unless proven to apply elsewhere.

Five zones have been established on the basis of major changes in the representation of pollen derived from hydrophytic or riparian sources. One zone has been subdivided into two subzones on the basis of minor assemblage changes.

These zones and subzones are described below in ascending order of depth.

Zone TB-V 8.66 m - 6.45 m. This zone is characterized by low frequencies of pollen of *Pandanus* (values less than 100%, calculated outside pollen sum), and high values of pollen derived from taxa characteristic of riparian communities, e.g. *Aglaia* comp (up to 5%), *Canthium dicoccum* comp (up to 15%). Pollen of other riparian taxa are persistently represented throughout, e.g. *Engelhardia, Gardenia* and *Pometia*. Occasional grains of *Pterospermum* and *Schoutenia* were also most probably derived from this source. Pollen of taxa characteristic of riverside communities, but by no means restricted to that habitat, were also well represented, e.g. *Eugenia* comp, *Calophyllum, Dillenia, Barringtonia* and *Macaranga* comp. Pollen of arboreal swamp taxa was infrequent: that of the following taxa, characteristic of swamp-forest, was present in restricted frequencies throughout, *Campnosperma* comp, *Ilex, Elaeocarpus* comp, *Fagraea* sim and *Caryota*. The major part of *Eugenia* comp pollen is also thought to be derived from swamp sources.

Pollen derived from herbaceous taxa within this zone is rare. Cyperaceae pollen occurs in low frequencies of up to 7%. It is most common towards the upper contact, being absent from the base of the zone. Pollen of *Utricularia* sp. occurs rarely towards the upper contact. Grains of *Flagellaria indica* occur scattered throughout.

None of the pteridophyte spores which were determined were thought to be derived from definite hydrophytic sources. The following types showed distinct concentrations within this zone and may therefore by derived from taxa of wide ranging ecological tolerance, which have occurred within the hydrosere, e.g. Filices psilate and areolate, *Dicranopteris* comp, *Hymenophyllum* comp, *Platycerium* comp and *Lycopodium cernuum*. Of undetermined pollen types, Indet. Nos. 62, 63, 58, 60 and 197 are abundant. Few significant changes or trends are evident in pollen assemblages within this zone. The numerous lithological changes present are not reflected in the pollen spectra.

Zone TB-W 6.45 m - 5.00 m. Very high values of *Pandanus* pollen characterise this zone. Values rise rapidly from over 400% at the base of the zone to over 900% at the upper contact (percentages of *Pandanus* pollen calculated outside the pollen sum). Also striking is a marked absence or rarity of pollen of many taxa characteristic of river side communities, e.g. *Aglaia, Barringtonia, Calophyllum, Canthium dicoccum* comp, *Dillenia, Engelhardia, Lophopetalum* and *Macaranga* comp. Similarly, pollen of Cyperaceae is poorly represented. This pattern is followed also by numerous pteridophytes, e.g. *Gleichenia* comp and *Lycopodium cernuum*, and by pollen spectra of the

Urticaceae/Moraceae group and several undetermined pollen types, e.g. Indet. Nos. 58, 60, 62, 63 and 179. *Eugenia* comp pollen was abundant throughout, values ranging from c. 10% to 20%. Most abundant, however, were small tricolporate grains, comparable mostly with pollen of *Elaeocarpus* and *Uncaria*. Pollen comparable to *Elaeocarpus* occurred in frequencies ranging to 25%, whereas *Uncaria* comp reached 60% in some spectra.

Numerous changes occur within certain pollen spectra, which it is thought are derived from hydrophytic sources. On the basis of these changes, two subzones are described.

Subzone TB-W one 6.45 m - 5.65 m. High frequencies (up to 25%) of pollen comparable to *Elaeocarpus* characterise this section. *Pandanus* pollen occurs in frequencies of up to 500%. *Uncaria* comp shows a distinct increase towards the upper contact. Undifferentiated pteridophyte spores exhibit much lower values than in the overlying subzone.

Subzone TB-W 2. 5.65 m - 5.00 m. *Elaeocarpus* comp pollen is present in low frequencies only (less than 5%), *Pandanus* pollen being very abundant (percentages between 600% and 900%). *Uncaria* comp shows a steady decline. Undifferentiated pteridophyte spores contribute up to 30% of the assemblage.

Zone TB-X 5.00 - 3.85 m. Values of *Pandanus* pollen range from 200% to 800% within this zone. *Eugenia* comp pollen is abundant (up to 35%). Pollen of riverside taxa returns, together with other pollen or spore types, with which they are associated in zone TB-V. The relative frequencies of these types differ considerably, however, from the values present in that zone. Taxa characteristic of riverside communities, but uncommon in swamp vegetation, include *Saurauia* comp, *Canthium dicoccum* comp and *Pometia* (the latter two types occurring only rarely). Taxa common to both riverside and swamp communities e.g. *Barringtonia, Dillenia* and *Macaranga* comp, are abundantly represented. Pollen of taxa characteristic of swamp vegetation occur in low frequencies throughout, e.g. *Campnosperma* comp, *Elaeocarpus* comp and *Ilex*. *Uncaria* comp pollen witnesses a steady decline through this zone.

Pteridophyte spores are well represented; undifferentiated pteridophyte spores comprise c. 15% and *Dicranopteris* comp between 5% and 10% of the assemblage. Other spores commonly present include *Gleichenia* comp, Filices areolate, *Lindsaya* comp, *Stenochlaena palustris* and *Lycopodium cernuum* comp.

Zone TB-Y 3.85 m- 1.80 m. *Pandanus* pollen percentages of c. 300% and up to 10% Cyperaceae pollen characterise this zone. Many taxa typical of riverside communities are less well represented than previously, only *Engelhardia* occurring throughout. *Calophyllum*, *Macaranga* comp and *Dillenia* are common. *Eugenia* comp witnesses a steady decline through the zone. Toward the upper contact, pollen of the aquatic *Utricularia flexuosa* is present in low frequencies. Urticaceae/Moraceae (scabrate) show frequencies of up to 15% in the upper part.

Spores of pteridophytes are abundant. *Dicranopteris* comp reaches values of 30% near the base of the zone. Undifferentiated pteridophyte spores occur in frequencies of 15% - 25% throughout. *Gleichenia* comp, *Lindsaya* comp, *Stenochlaena palustris* and *Lycopodium cernuum* are common in frequencies of up to 5%.

Zone TB-Z 1.80 m- 0.80 m. This uppermost section of the profile is characterised by very low frequencies of *Pandanus* pollen (less than 100%) and common Cyperaceae pollen (c. 10%). Pollen of riverside taxa is absent, with the exception of *Saurauia* comp. *Macaranga* comp and *Eugenia* comp are present in low frequencies (c. 5%). Other taxa characteristic of swamp-forest are rare or absent. Of herbs, *Utricularia flexuosa* pollen is present in low frequencies. Pteridophyte spores are represented by up to 15% *Dicranopteris* and undifferentiated types, up to 10% *Gleichenia* comp and the regular occurrence of low frequencies of *Lindsaya* comp, *Stenochlaena palustris* and *Lycopodium cernuum*.

2.2.5.4 *Zonation and description of dry-land pollen spectra* (Diagram B, Fig. 12). Two zones have been established on the basis of major changes in pollen assemblages. The upper of these has been further subdivided into three subzones on the basis of minor changes in composition.

Zone TB-1 8.65 m - 4.05 m. This zone is characterised by the presence of high percentages (generally above 75%) of pollen derived from arboreal forest taxa. Herb and shrub pollen is persistently present throughout in values ranging between 5% and 20%. Occurrences of arboreal non-forest pollen are small and irregular, frequencies rarely reaching more than 10%.

Dipterocarpaceae contribute the largest proportion of pollen derived from arboreal forest taxa (average values 10% to 20%), *Shorea* comp and *Hopea* comp (4 colpate) providing most of this. *Dipterocarpus* pollen is very rare. The second largest contributor of arboreal pollen is *Koompassia*. Other common types, probably derived from arboreal forest taxa, include *Anti-*

desma, *Garcinia merguensis* comp, *Lithocarpus* comp, *Nauclea* comp, Myrsinaceae, *Payena* comp and *Walsura* comp. Palmae pollen occur throughout in low frequencies. A significant feature of this zone is the virtual absence of pollen of Burseraceae. The shrub and herbaceous element is dominated here by *Elatostema* comp and *Pilea* comp. Other pollen in this category includes *Croton* comp and *Boea* comp. Pollen of Gramineae is absent and that of Compositae vere rare.

Spores of *Gleichenia* comp occur persistently throughout. Spores of *Cyathea*, *Cyclosorus* sect. *Sphaerostephanos* and *Taenitis blechnoides* occur sporadically.

Within this zone, several pollen types show distinct variations in their representation. Dipterocarpaceae pollen is more abundant above 5.15 m and below 6.80 m than in the interval between. *Koompassia* pollen is absent from the interval 5.50 m to 6.80 m. Other arboreal forest taxa which show variations include *Garcinia merguensis* comp, which is present only below 6.80 m, *Antidesma*, which shows a cluster between 6.15 m and 6.80 m and *Lithocarpus* comp, which is more common above 5.00 m. Myrsinaceae pollen and *Nauclea* comp are also more common in the upper part of the zone. *Livistona* shows a distinct cluster below 8.15 m. Of herbs and shrubs, *Boea* comp and *Croton* comp are confined to the lower part of the zone. *Gleichenia* is more common in the uppermost metre.

Among pollen and spore types which are derived from various habitats, but which may frequently occur in dry-land communities, *Dicranopteris* comp, Urticaceae/Moraceae (scabrate), *Macaranga* comp and *Elaeocarpus* comp are common. Changes in abundance occur within these taxa, which bear little or no relationship to changes of dry-land spectra but clearly follow changes in spectra of pollen types derived from taxa restricted to the hydrosere. It is therefore unlikely that, within this zone, these taxa contribute to any degree towards the dry-land vegetation.

Zone TB-2 4.05 m - 0.80 m. Within this zone, pollen of arboreal non-forest taxa is present in percentages ranging from 10% to 55%. Herb and shrub pollen is present in percentages of the same order. Pollen of arboreal forest taxa displays an irregular but distinct decline from 80% at the base of the zone to 20% - 30% in the uppermost samples examined. Several distinct changes are apparent. These are discussed below under heading of subzones.

Subzone TB-2A 4.05 m- 2.90 m. Pollen of arboreal non-forest and herbaceous taxa is present in percentages ranging from 20% to 55%. Pollen of arboreal forest taxa declines from 80% at 3.95 m to under 50% at 2.96 m depth.

This interval is characterised in particular by the absence of pollen of Dipterocarpaceae. Other arboreal forest taxa displaying a reduction in frequency include *Koompassia*, *Walsura* comp, *Nauclea* comp and *Payena* comp. Pollen of Fagaceae remains throughout, together with *Baeckia*. Burseraceae pollen is present in small frequencies. *Austrobuxus nitidus* pollen is practically restricted to this subzone.

Gramineae, *Elatostema* comp and *Pilea* comp occur irregularly. Pollen of *Trema* is abundant throughout. Regarding taxa derived from various habitats, spores of *Dicranopteris* display consistently high values throughout the interval. *Gleichenia* comp increases markedly in frequency toward the upper contact, a pattern shown also by *Macaranga* comp and Urticaceae/Moraceae (scabrate). Spores of *Stenochlaena palustris* occur sporadically throughout.

Subzone TB-2B 2.90 m - 2.15 m. High percentages of pollen of arboreal forest taxa characterise this section (values ranging from 55% to 75%). Arboreal non-forest pollen is present in percentages of 15% to 20% and herbaceous and shrub pollen 10% to 20%. Pollen of some arboreal forest taxa returns. Dipterocarpaceae is present in percentages of c. 10%. Fagaceae, *Walsura* comp and Myrisinaceae are present throughout. Pollen of *Trema* occurs in frequencies of c. 10% and Gramineae pollen in very low frequencies. In the uppermost part of the interval, pollen of *Myrica* and *Melastoma malabathricum* comp is present.

Spores of *Dicranopteris* are less well represented than in the underlying section. Spores of *Gleichenia* comp, *Stenochlaena palustris* and *Lindsaya* comp occur throughout. Pollen of Urticaceae/Moraceae (scabrate) is abundant in the lower part of the subzone, that of *Macaranga* comp and *Elaeocarpus* comp being well represented throughout the subzone.

Subzone TB-2C 2.15 m - 0.80 m. High frequencies of non-forest arboreal pollen characterise this subzone (20% to 50%), together with abundant herb and shrub pollen. Arboreal forest pollen is present in frequencies of 20% to 50%.

Pollen of Dipterocarpaceae and Fagaceae is common, that of *Payena* comp being present in low frequencies. Most arboreal forest taxa display a marked absence from this section.

Trema pollen is present in frequencies sometimes exceeding 50%. Pollen of Gramineae, *Myrica* and *Melastoma malabathricum* comp is regularly present. Spores of *Gleichenia* comp, *Dicranopteris* comp, *Lindsaya* comp and *Stenochlaena palustris* are frequent.

Urticaceae/Moraceae (scabrate), *Macaranga* comp and *Elaeocarpus* comp are present among pollen types derived from taxa of wide ecological tolerance, but in lower frequencies than in the underlying subzone.

2.2.5.5 *Hydrophytic and riparian vegetation history.* The rarity of pollen of obligate aquatic taxa from the lowermost zone (zone TB-V) of the profile indicates that, within this earliest recorded period, hydrophytic vegetation was of very limited extent in the area. Sediment characteristics (sands and fine silts) indicate that current speeds may have been sufficient to suppress any extensive hydrophytic development.

Within this zone, pollen of Cyperaceae, and *Utricularia* sp. were, with all probability, derived from local hydrophytic sources. The presence of percentages of *Pandanus* pollen in excess of 20% (calculated outside pollen sum) may also indicate local sources. The increase in frequency of both *Pandanus* and Cyperaceae pollen towards the upper part of the zone reflects a gradual, although limited, expansion of hydrophytic communities. The presence of pollen of *Utricularia* sp. in the uppermost samples also indicates this. The expansion of hydrophytic sources recorded here may be the result of reduced water currents, indicated by sediment characteristics, towards the top of the zone. The absence of pollen of Cyperaceae from the basal samples may indicate total absence of hydrophytic vegetation at the time of commencement of the zone. It is thought that a proportion of *Eugenia* comp pollen may be derived from non-hydrophytic sources. It is most probable, however, that the majority of this pollen is derived from swamp-forest sources, which may have occurred upstream from the Kuin area, the pollen being brought to the site of deposition by water transport. Today, *Eugenia* swamp-forest constitutes c. 67% of the total swamp area at Tasek Bera (Lim and Furtado 1974). The high frequencies of pollen of taxa particularly well represented in riparian environments form the most significant feature of pollen assemblages from this zone. *Canthium dicoccum* comp form the most abundant of these elements; other representatives include *Aglaia, Engelhardia, Gardenia, Lophopetalum, Pometia* and *Pterospermum.* Pollen of *Calophyllum, Barringtonia, Dillenia* and *Macaranga* comp were also well represented. These taxa frequent riparian habitats, but also characterise other communities. Their presence, therefore, is not of the same indicator value as the previous group of pollen types. They do, however, help to substantiate the nature of riparian communities during this time. The abundance of pollen of taxa of riparian habitats within this zone can be explained either by the much greater representation of riparian communities in the area, or by their occurrence much closer to the site of deposition than at present. It is likely that both these factors may have been the case. It is most feasible that, during this period, Tasek Bera was not flooded as it is today, but consisted of a normal river channel, the 'Proto-Tasek Bera'. If this were so, the river would have been much narrower than the present swamp, and therefore riparian communities would have been much closer to the site of deposition. Certain taxa represented in this zone are relatively rare in the Tasek Bera area today, e.g. *Canthium* spp. (Merton, pers. comm.). These communities must therefore also have undergone a considerable change in total representation in the area.

The community represented here is not easily explained in terms of present-day vegetation. *Canthium* spp. occur frequently in many riparian communities, but it is unlikely that anything but a gregarious occurrence of this taxon could account for the high frequencies of *Canthium* pollen encountered here. The only known occurrence today of gregarious *Canthium* spp. is that of *C. dicoccum*, which occurs abundantly along tidal rivers behind mangroves (Corner 1940). The presence of this pollen is not, however, sufficient evidence on which to suggest a former tidal regime for the Proto-Tasek Bera. Corner (1940) also notes *Pandanus helicopus* as characteristic of tidal rivers. This species forms the dominant *Pandanus* sp. at Tasek Bera today; there is no reason, therefore, to suspect that *Pandanus helicopus* is able to colonise both habitats, whereas *Canthium* cannot.

Certain phenomena at Tasek Bera do point to a former close connection to the sea, e.g. the occurrence of the puffer fish *Tetraodon palembangensis* (Furtado 1974). The close correspondence between the basal radiocarbon date from the pollen profile (4500 ± 90 BP) and basal radiocarbon dates from coastal peats from West Malaysia, e.g. peats from Sarawak yielded a basal date of 4270 BP ± 70* (Anderson and Muller 1975), may point toward some common causal event, possibly triggered by a eustatic change. It is not understood how a sea level change could affect the regime of an inland waterway of this nature. The closeness of these dates is, therefore, only brought to attention. Additional data is required to determine whether this correspondence is through chance or whether a common causal factor is involved.

The synchroneity of a change in sedimentation from fine silt to autochthonous peat, and a massive increase in frequency of *Pandanus* pollen (up to 500%, calculated outside pollen sum) at the base of zone TB-W, marks a change in the Tasek Bera environment from that of a river channel to a swamp dominated by *Pandanus* sp.

The complete disappearance or considerable reduction in frequency of pollen derived largely from riparian sources, e.g. *Canthium dicoccum* comp,

* Corrected for a Suess-effect of 230 years.

Aglaia, *Gardenia*, *Barringtonia*, *Dillenia* and *Macaranga* comp reflects a major reduction in the representation of riparian communities in the area.

A change of this nature could be the result of a substantial reduction in speed of the current, caused by a partial damming of the watercourse or increase in base level. This would facilitate the widespread expansion of swamp vegetation into the area. The change may also have rendered the area unsuitable for the continued regeneration of riparian taxa, perhaps owing to the reduced availability of nutrients in the slow-moving acid-rich water.

It is difficult at this stage to make any specific comment on the cause of this change, the likelihood being that it is due to more than one variable. In Section 2.2.3 of this chapter, the effects of river capture, tectonic movement and levée development on the origin of the flooding of Tasek Bera were discussed. A fourth possibility is that of the damming effect of *Pandanus* itself.

Pollen spectra of taxa thought to be of significance in the hydrophytic vegetation of this zone show a clear pattern, although the interpretation of this pattern in terms of vegetation is very difficult owing to problems of determination of certain pollen types present. The presence of up to 20% *Elaeocarpus* comp pollen in the lower part of the zone (subzone TB-WI) suggests that *Elaeocarpus* may rank considerably in the hydrophytic vegetation. *Elaeocarpus* comp pollen was found in abundance in surface samples from swamp-forest. An *Elaeocarpus* sp. is, however, known to occur frequently along watercourses bordering *Pandanus helicopus* swamp (Merton 1962). The presence of *Elaeocarpus* comp pollen may therefore indicate the presence of *Elaeocarpus* sp. in this latter habitat. This suggestion is further supported by the presence of only low frequencies of pollen of many taxa common to swamp-forest. The absence of *Lycopodium cernuum* spores also supports this; spores of this taxon were abundant in surface samples from swamp-forested areas.

The occurrence of up to 60% pollen similar to that of *Uncaria* within this interval remains unexplained. This pollen type is very small (c. 10 μ diameter) and may easily be confused with that of other taxa, e.g. *Weinmannia* and some Myrsinaceae. The pollen is, however, distinct from that of *Elaeocarpus*, *Gynotroches axillaris* and *Linociera*, all common swamp trees (Merton 1962, Soepadmo 1974) with pollen of similar size and shape to that of *Uncaria* comp.

Very high frequencies of *Pandanus* pollen in the upper part of this zone (subzone TB-W2) and the generally low frequencies of pollen of other hydrophytic taxa (excluding *Uncaria* comp) suggest a vegetation of almost pure *Pandanus* sp. during this period.

Zone TB-X marks a return of riparian taxa to the area, although the frequencies of pollen derived from this source differ from those present in zone TB-V. *Canthium dicoccum* comp and *Gardenia* pollen is present in low frequencies. *Saurauia* comp pollen is abundant. Pollen of *Macaranga, Barringtonia* and *Dillenia* may also be derived from a riparian source. Many spore types return within this zone; frequencies of *Dicranopteris* comp and *Gleichenia* comp are present in similar frequencies to those observed in zone TB-V. Gleicheniaceae may also, therefore, occur within riparian communities.

Conditions during this zone may have been similar to those present in the upper part of zone TB-V. The high frequencies of *Pandanus* pollen present, however, indicate that hydrophytic vegetation was still extremely widespread. High frequencies of *Eugenia* comp pollen here may indicate a more widespread *Eugenia* swamp vegetation than in the previous zone.

The very sudden change in many pollen spectra across the zone TB-W/TB-X boundary and the correspondence of this change with a lithological break, suggests that a period of non-deposition may have occurred between the two. Whether this is the case, and if so, the length of the hiatus involved, could only be tested with additional radiocarbon dates, which were unfortunately not available for this purpose.

Pollen spectra from zone TB-X show certain distinct changes from base to top. The increase in abundance of *Pandanus* pollen to over 800% at the upper contact (calculated outside the pollen sum) corresponds with a slight increase of Cyperaceae pollen. Pollen of certain riparian taxa, e.g. *Canthium* and *Barringtonia*, decrease in frequency, whereas others, e.g. *Dillenia* and *Macaranga* comp remain. These changes may be the result of a number of factors. The lithology of the zone indicates deposition in a largely open water environment (probably a water channel) in the lower part and accumulation in a closed swamp environment in the upper part (indicated by the ratio of inorganic to organic material (Fig. 10)). *Pandanus* pollen frequencies in the lower part correspond with those observed in surface sediment samples from channels in *Pandanus* swamp. Studies of pollen deposition of *Pandanus* inside a closed *Pandanus* swamp were not attempted in the surface sample study; however, the very high frequencies observed in samples from the upper part of this zone can best be explained by deposition beneath a closed vegetation of *Pandanus helicopus*. Pollen of riparian taxa, which show no decline toward the upper part of the zone, e.g. *Dillenia* and *Macaranga* comp, occur frequently in swamp and dry-land regrowth communities, whereas those not characteristic of other habitats, e.g. *Canthium dicoccum* comp, decline in frequency. Studies of dry-land vegetation history indicate that, at about this time, the indigenous population began extensive forest clearance. This may have extended to riparian communities, where those taxa tolerant of regrowth communities may continue, whereas those intolerent would decline. The slight increase of pollen of Cyperaceae towards the top of this zone may also be related to Man's activity. This will be discussed further below.

The rarity within zones TB-Y and TB-Z of pollen of riparian taxa suggests that, within these zones, riparian vegetation was of limited extent. Pollen of taxa common to both riparian and regrowth communities persists throughout, however, and it is considered that this pollen is derived primarily from regrowth vegetation sources. Within these zones, pollen spectra of *Dillenia* and *Macaranga* comp will be discussed under the heading of 'Dry-land vegetation history'.

The decrease in frequency of *Pandanus* pollen at the base of zone TB-Y is accompanied by an increase in frequency of pollen of Cyperaceae. No sediment change occurs across this boundary. The decrease in *Pandanus* pollen is attributed, therefore, to a general decrease in abundance of *Pandanus* sp. in the area. The increase of Cyperaceae pollen is thought to indicate an increase of Cyperaceae. Pollen rain studies indicate that Cyperaceae pollen is under-represented in pollen spectra, especially when compared with the very high pollen production of *Pandanus helicopus*. The increase of Cyperaceae in the area may therefore be of the same order as the decrease in area of *Pandanus* sp. The cause of this change may be due either to development of the hydrosere or the activity of the indigenous population.

Hydroseral development may be the result of an increase or lowering of water level, or an increase or decrease in speed of the water flow. An increase in Cyperaceae would most likely result from decreased water depth or current flow, since *Lepironia articulata*, the dominant species of Cyperaceae present at Tasek Bera, occurs more commonly in shallower, more protected areas than *Pandanus helicopus*. If the distribution of *P. helicopus* and *L. articulata* were closely related to water depths, one would expect a clear distributional relationship to occur between the two species, instead of the extremely haphazard distribution of both, which is actually the case.

A more satisfactory explanation is found by consideration of the activity of Man in the area. At the present time, *Pandanus helicopus* is being burned over extensive areas to maintain navigable channels through the swamp. These areas may then be colonised by *Lepironia*, which, with its growing points below the water surface, may be more fire-resistant than *Pandanus*. It is also most frequent in areas most commonly used by Man. The pollen record revealed here suggests that this practice may have continued for at least the last 600 years.

During the uppermost zone, TB-Z, *Pandanus* pollen is even less frequently represented and Cyperaceae more so, thus suggesting that this process has continued with particular vigour during this most recent phase of activity. Of particular interest in the uppermost part of the profile is the occurrence of tetrads of *Utricularia flexuosa*; this record may indicate that this species, which plays a very important part in the ecology of Tasek Bera's waterways, may have migrated to the area during historic time.

2.2.5.6 *Dry-land vegetation history*. Studies of dry land pollen spectra from surface samples from Tasek Bera (Morley 1976) indicate that the majority of pollen incorporated into sediments there, which is derived from dry-land sources, originates from vegetation within c. 2-3 km from the point of deposition. In the following section, therefore, the discussion relates only to the composition of vegetation in dry-land areas within a circle of radius c. 2-3 km from the point of collection of the pollen analysis core. It would require additional studies to determine whether the vegetation changes, shown here to have taken place within this local area, were of any regional significance.

The predominance of pollen derived from arboreal forest taxa, together with the virtual absence of pollen derived from non-forest taxa within zone TB-I, points to the conclusion that during the period represented by this zone, forests predominated in the area. The pollen record yields little to indicate the detailed components of this earliest recorded phase of non-hydrophytic vegetation. The predominance of pollen derived from Dipterocarpaceae and other taxa typical of lowland dipterocarp forest, e.g. *Antidesma* spp., *Garcinia* spp., *Lithocarpus*/*Castanopsis*, *Quercus* spp., Rubiaceae sect. Nauclea (*Nauclea* comp), Meliaceae and Sapotaceae, indicate that the general aspect of forest vegetation in the Tasek Bera area during this period may have varied little from that of the Meranti-Keruing forest type, which composes the majority of the present day surrounding natural vegetation. The absence of pollen of *Dipterocarpus* and Burseraceae is particularly noteworthy, however, during this zone. *Dipterocarpus* spp. account for a considerable proportion of the emergent taxa in Meranti-Keruing forest. This genus also is found frequently in Keruing forest, which occurs in many areas around Tasek Bera (Wyatt-Smith 1964). This forest type is reflected in place names in the area, e.g. Tanjong Keruing, near Kuin. It is not known whether the absence of this pollen type reflects the very low pollen production of *Dipterocarpus* or low frequencies of *Dipterocarpus* spp. in the dry-land vegetation. The absence of Burseraceae pollen is similarly unexplained, although pollen of this family occurs in low frequencies in other zones in the profile. Burseraceae constitutes the second most abundant family in Meranti-Keruing forest (Morley 1976, Poore 1968).

The alteration in frequency with depth of many pollen taxa also requires comment. Non-random distributions of arboreal forest pollen throughout this zone are shown by Dipterocarpaceae, *Koompassia, Garcinia merguensis, Lithocarpus* comp, Myrsinaceae, Nauclea comp and *Livistona* comp. These distributions may be explained in several ways. Peaks may be indicated by the close proximity of individual large trees to the site of deposition. Anderson and Muller (1975) explain such peaks, occurring in pollen spectra from a swamp-forest site in Sarawak, in this manner. In a swamp-forest, however, high over-representation could occur from individual trees growing actually on the site of deposition. These peaks could, on the other hand, indicate minor or widespread changes in the overall composition of the dry-land vegetation, as shown to occur by Poore (1968) by examination of the present distribution of taxa in the Jengka Forest Reserve. The pollen record indicates a dry-land ground flora consisting predominantly of Urticaceae, e.g. *Elatostema* spp. and *Pilea* spp. Understorey palms may also have been common, e.g. rattans and *Pinanga* spp. Many of these taxa may, however, have also occurred in the ground flora and understorey component of riparian vegetation.

The effects of widespread forest clearance within zone TB-2 are shown by the presence of high frequencies of pollen of *Trema, Myrica,* Gramineae and spores of *Gleichenia*. Pollen of many taxa typical of seral dry-land communities but which also occur in a variety of other habitats, including hydroseral and riparian communities, are also abundant. To determine whether these taxa have been derived from dry-land sources or otherwise, the scheme of subzones for the discussion of dry-land vegetation changes has been superimposed on spectra of many taxa of widespread ecology, e.g. *Elaeocarpus* comp, *Dicranopteris* comp, *Lindsaya* comp, *Macaranga* comp, *Stenochlaena palustris* and Urticaceae/Moraceae (scabrate). The presence of an apparent relationship between pollen spectra of taxa known to be derived from dry-land sources and spectra of pollen of taxa of wide-ranging ecology is taken to indicate the possibility of a common source of pollen of both ecologically restricted and ecologically widespread taxa. When the former dry-land vegetation is constructed from data presented from pollen spectra of both ecologically restricted taxa and those taxa in question, which are ecologically more widespread, but whose spectra suggest a probable dry-land origin, a much clearer picture of the possible pattern of dry-land seral vegetational succession emerges.

At the lowermost subzone, TB-2A, the abundance of pollen of *Trema* and *Baeckia*, together with the decrease in abundance of many pollen types derived from forest taxa (pollen of Dipterocarpaceae is absent from this interval), indicate widespread forest clearance followed by regrowth of, among trees, *Baeckia* sp. and *Trema* spp. Spores of *Dicranopteris* comp are very abundant throughout, suggesting that *Dicranopteris* spp. were very widespread. These thicket-forming ferns may therefore have constituted a considerable part of the dry-land vegetation. Towards the upper contact of this interval, the increase of pollen of Urticaceae/Moraceae (scabrate), *Macaranga* comp and *Gleichenia* comp spores may indicate a further stage in the seral succession, where Urticaceae/Moraceae (e.g. *Artocarpus* spp., *Poikilospermum* spp. etc.), *Macaranga* spp. and *Gleichenia* spp. play a prominent role.

The increase of pollen of arboreal forest taxa during subzone TB-2B may indicate some degree of recovery of the forest. This re-expansion of the forests may be at the expense of *Baeckia* sp., *Dicranopteris* spp. and *Gleichenia* spp. which display a marked reduction in frequency. The reduction toward the middle of this interval of Urticaceae/Moraceae pollen may also be related to this expansion. The fact that non-forest vegetation is still prominent in dry-land areas is indicated by the presence of up to 20% *Trema* pollen, also pollen of *Macaranga* comp, Gramineae, *Myrica* and *Melastoma malabathricum* comp. High frequencies of *Elaeocarpus* comp pollen may indicate dry-land areas as a source of this pollen type during this interval. The occurrence of regular frequencies of Gramineae pollen may mark the first expansion of Lalang (*Imperata cylindrica*) grassland. Within the uppermost subzone, TB-2C, a considerable reduction in frequency of pollen of arboreal forest taxa (although Dipterocarpaceae pollen is present in low frequencies), together with very high frequencies of *Trema* pollen (up to 50%) and high Gramineae frequencies (up to 20%), indicate considerable expansion of non-forest communities and, probably, of Lalang grassland, at the expense of primary forest.

The pollen record indicates that Man has been actively cutting natural forest on an appreciable scale in the northern part of the Tasek Bera area for only the last 600 years. There is no evidence to support a migration of Orang Asli to the area between 3500 and 4500 BP as a result of their being displaced by the Coastal Malays, during this period (Hodder 1968). The pollen record, however, may only indicate the time at which the indigenous population changed their way of life from a predominantly forest-orientated existence to one of shifting agriculture. The fact that the Semelai, the tribe which now inhabit the Tasek Bera area, have close relatives (the Temok) who are still hunters and gatherers (Carey 1976), lends considerable support to this suggestion.

2.2.6 *Conclusion*

The history of the water channel which presently forms the Tasek Bera (Fig. 13) has been long. Initially it formed a part of the ancestral Pahang drainage system, when the S. Pahang discharged into the Malacca Strait through the present S. Muar Valley. The S. Pahang may have flowed southward in this manner through a large part of the Late Cretaceous and Tertiary periods. Capture of the former S. Pahang headwaters by the eastward-flowing Lower Pahang, possibly during the Late Tertiary or Early Pleistocene, led to eventual discharge into the South China Sea by way of the present S. Pahang channel. The base level advantage of the eastern-flowing streams must then have been considerable, as the direction of flow in a large part of the ancestral Pahang channel below the elbow of capture was reversed, forming the present S. Bera. At one time, the catchment of this stream was extensive, but the headwaters have now largely been captured or recaptured by eastward or southward-flowing streams.

The most recent capture resulted in the southern discharge, through the S. Palong, of the former western headwaters of Tasek Bera. It is likely that this capture occurred in the recent past, as the 'wind gap' between the two systems, the S. Ayer Kuning channel, still holds water.

Sediments studied from Tasek Bera date only from the mid-Holocene, c. 4500 radiocarbon years BP. The oldest sediments examined were probably deposited in a moderately fast flowing river channel, the 'Proto-Tasek Bera'. This channel was bordered by riparian communities, which were probably dominated by *Canthium* sp. Hydrophytic vegetation within the channel was initially of limited extent, but as the flow of water through the channel reduced, as suggested by the reduction in size of sediment particles, hydrophytic communities became more widespread.

The change from a river channel to a swamp dominated by *Pandanus* was very sudden, being clearly marked by a change of both lithology and pollen content. It is thought that the sudden expansion of swamp communities may have been the result of a drainage impediment in the area caused possibly by slight crustal movement. It is possible, however, that it corresponded also with the capture of the former western headwaters of the Tasek Bera by the S. Palong.

Initially, the swamp vegetation consisted predominantly of *Pandanus* sp., with abundant *Elaeocarpus* sp. Later, *Elaeocarpus* became less widespread,

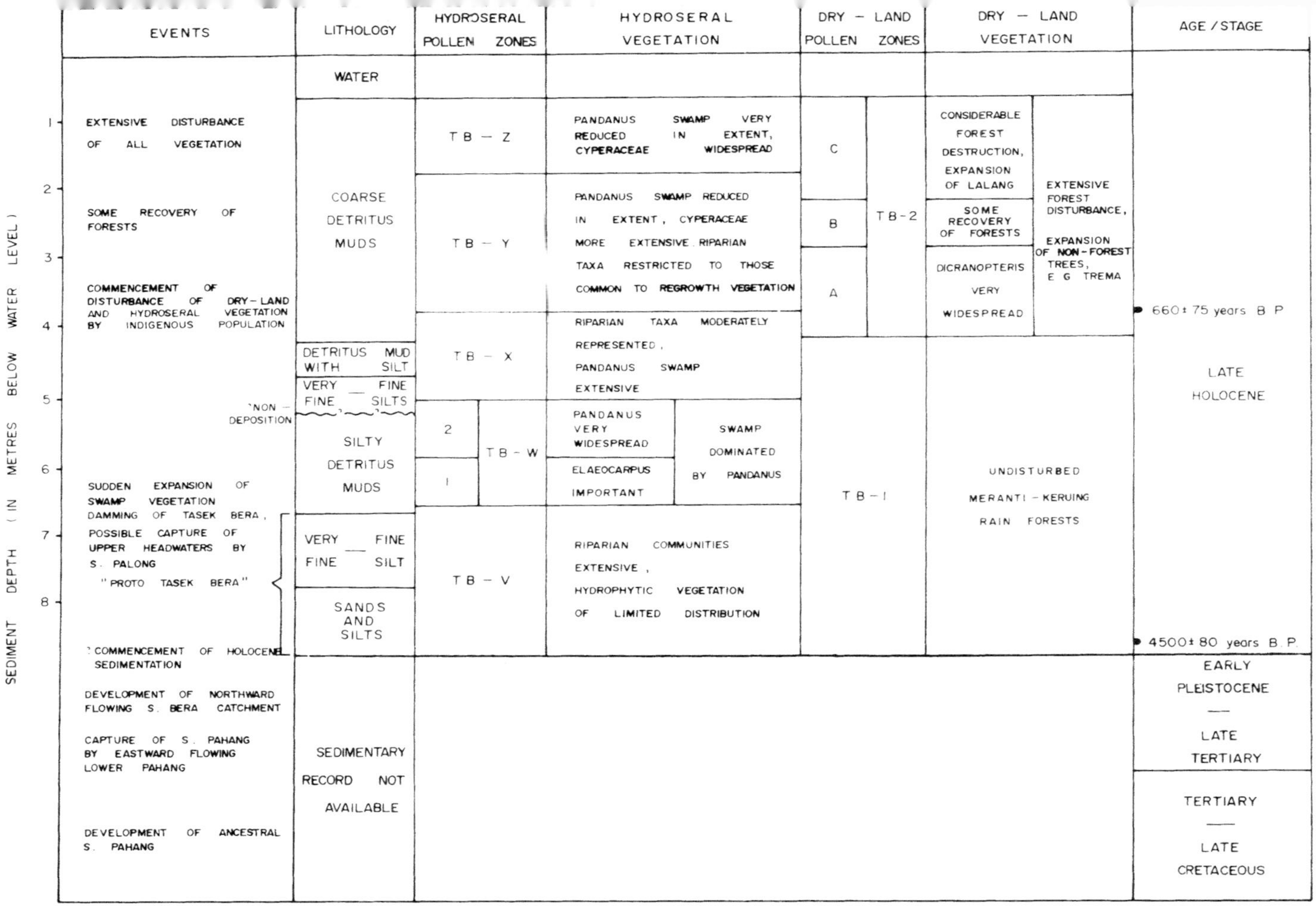

Fig. 13 Summary of the geomorphological and vegetational history of Tasek Bera.

the swamp vegetation probably consisting of almost pure *Pandanus* sp.

A sudden change in both pollen content and lithologies at 5 m depth suggests the possibility of a hiatus. Following this, certain riparian elements return to the area, at a time when the water flow must have increased, suggested by the presence of silt within the sedimentary sequence. *Pandanus* swamp, however, remained extensive.

The recent history of the swamp vegetation is thought to be largely controlled by human factors. *Pandanus* swamp shows a marked decline in representation, whereas Cyperaceae swamp increases in extent. It is probable that the *Pandanus* is burned in order to maintain navigable channels throughout the area. The initial stages of the *Pandanus* decline have been radiocarbon dated at 660 ± 75 radiocarbon years BP.

The pollen record suggests that, until this time, the dry-land areas within the Tasek Bera vicinity bore a vegetation similar to that of the Meranti-Keruing rain forests, which presently occur in the areas not affected by shifting agriculture. Since this date, however, destruction of the natural vegetation has been considerable, resulting in the widespread development of non-forest vegetation. It is suggested that this date marks the time at which the Semelai (the indigenous Orang Asli tribe) either migrated into the area, or changed their way of life from one orientated toward hunting and gathering in the forests to one of shifting agriculture.

2.3 Climate
(R. P. Lim and J. I. Furtado)

Tasek Bera is situated in the interior rainfall region of Peninsular Malaysia (Chia, Lin-Sien, pers. comm.), and experiences two monsoon and two dry seasons, each alternating with the other. It has a fairly high annual rainfall of over 5 inches (Dale 1959, 1960) and a fairly high air temperature which fluctuates withing 10°C annually (Dale 1963). It is situated in the humid tropics where the relative humidity is high (Dale 1963) and annual solar radiation also fairly high.

Rainfall data for Tasek Bera is based on readings at Pos Iskandar and at Dunlop Bahau Estate, Bahau. The former were obtained from the Drainage and Irrigation Department, Malaysia but were incomplete for 1971 with February 1971 unavailable.

The annual rainfall for Tasek Bera for 1971 measured at Pos Iskandar and Dunlop Estate, Bahau was 98.66 inches and 108.79 inches respectively (Fig. 14; Table 4). In 1970, the annual rainfall measured at Dunlop Estate, Bahau,

Table 4 Mean, minimum and maximum rainfall at Pos Iskandar, Tasek Bera and Dunlop Bahau Estate, Bahau, for 1970–1972. (R. P. Lim and J. I. Furtado)

Location and date	Rainfall (inches)		
	Mean	Minimum	Maximum
Pos Iskandar			
April 1970–March 1971	10.24	2.35 (July)	31.95 (January)
1971	8.97	1.67 (July)	31.95 (January)
June 1971–May 1972	6.31	1.10 (January)	15.44 (December)
Dunlop Bahau Estate			
1970	6.51	2.29 (July)	14.01 (May)
1971	9.06	2.91 (June)	20.10 (December)
July 1971–June 1972	7.65	2.06 (March)	20.10 (December)
Overall	8.12	1.10 (January 1972)	31.95 (January 1971)

was 78.23 inches which is more representative of the annual rainfall at Tasek Bera, conforming to most areas in Peninsular Malaysia which has about 80 inches per annum (Wycherley 1967). The high value in 1971 was due to abnormally heavy rains in January making it an atypical year for rainfall. The monthly mean rainfall for 1971 was 8.97 inches (minimum 1.67 inches in July, and maximum 31.95 inches in January) at Pos Iskandar, and 9.06 inches (minimum 2.91 inches in June, and maximum 20.10 inches in December) at Dunlop Estate, Bahau (Fig. 14; Table 4).

Seasonal rainfall fluctuations in 1970 showed a normal bimodal wet and dry cycle over a year, with wet periods corresponding to the lesser southwest monsoon (late March-May) and the greater northeast monsoon (late September-January) (Fig. 14A). The regularity of these seasons, and hence rainfall, was not periodic and definite so that in some years there may be only one extended wet season followed by an extended dry season. In 1971 this seasonal rainfall was not regular and strongly represented, although the southwest monsoon fell within April-May while the northeast monsoon extended from September to January. Thus, the annual rainfall consisted of:

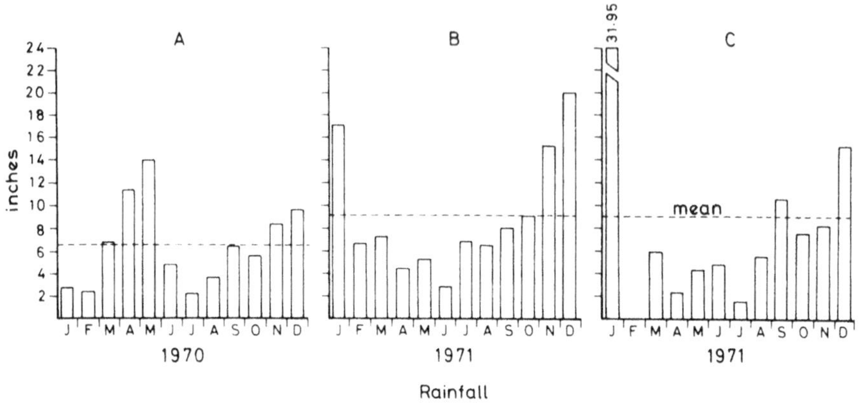

Fig. 14 Rainfall data at Tasek Bera (1970–71). Measured at Dunlop Bahau Estate, Bahau in 1970 (A) and 1971 (B), and at Pos Iskandar in 1971 (C). (N.B.: 'X' denotes no data available.)

(a) wet months, April-May and September-January; and (b) dry months, February-March and June-August. Available data and observations suggest that rainfall within the Tasek Bera watershed and its vicinity is patchy and sporadic being considerably influenced by microclimatic and local conditions.

2.4 Study area and stations (T. Mizuno, R. P. Lim, I. Ikusima and J. I. Furtado)

The study area was located at Pos Iskandar in the southwestern part of the swamp (Fig. 2). It was selected mainly because of its accessibility and its closeness to diverse habitat types, even though it is intensively used and highly disturbed by Semalai aboriginal inhabitants (see later).

The study area covers about 77,400 m^2. Stations representative of different habitat types in the swamp were selected in and within the proximity of this study area. Each station is described below in detail, and its location (Fig. 15) and representative profiles (Fig. 16) in the study area are indicated. The vegetation (Fig. 17) and bathymetric conditions (Fig. 18) of this area are also given. The study area basically consists of a man-influenced littoral region with *Lepironia* and *Pandanus* stands with the main open water (limnetic)*

* Limnetic region is not used in its usual context but is used to represent all the areas devoid of vegetation. Light penetrates through to the bottom for most of the open water regions.

channel running roughly from southeast to northwest, and with subsidiary limnetic channels traversing the littoral especially one running southwest to northeast, from the landing point to the main limnetic channel. The southwestern bay is fringed by swamp-forest trees with dense stands of swamp-forest to the southeast and northwest (Fig. 17). *Pandanus* occurs characteristically in clumps around the limnetic and the edges of the littoral region, and is also fairly abundant along the swamp-forest channels.

Station I. This station is located in a basin-like area of the main channel with interconnecting channels flowing in and out. This basin is clearly demarcated by the abrupt appearance of the littoral *Lepironia articulata* and *Pandanus helicopus* around it. Its surface area varies with the seasonal growth of *Utricularia flexuosa* along its margins. Its normal depth is three metres varying, as at all stations, with water level (see later). Its substrate is devoid of macrophytic vegetation, although temporary obstructions to free water flow encourages growth of macrophytic vegetation like *Utricularia*. Much of the water this area receives comes from the swamp-forest to the southeast, but there is also an exchange of water between it and the littoral region. The speed of the water is slow and flowing water occurs at all depths which is evidenced by the unstable nature of the benthos.

Stations II and VI. Limnetic-littoral transition, Pandanus-Utricularia. These stations are located along the edge of the limnetic main channel, where it meets clumps of *Pandanus helicopus*, and is characterised by an association of *P. helicopus* and *U. flexuosa*. The flow of water here is extremely slow, approaching stagnation. Its benthos is covered by a mat of *Cryptocoryne griffithii*. The sampling point lies in the shadow of tall *Pandanus* palms for part of the day. Its normal depth is two metres.

Stations III and XII. Limnetic-littoral transition, Utricularia. These stations are in the midst of a *Utricularia flexuosa* stand at the edge of the limnetic open channel. Isolated floating mats of *Utricularia* occur for most of the year but become almost continuous along the margins of the limnetic-littoral boundary at peak growth. Beneath these mats are dense stands of *Blyxa echinosperma* and *Hydrilla verticilata*, the former predominating. At low water levels, the tips of *Blyxa* and *Hydrilla* protrude through the *Utricularia* mats giving it the appearance of a mixed stand. The dispersion of this *Utricularia* stand is highly variable, because of its seasonal growth and dependence *inter alia* on fluctuations in water level, a function of rainfall. The

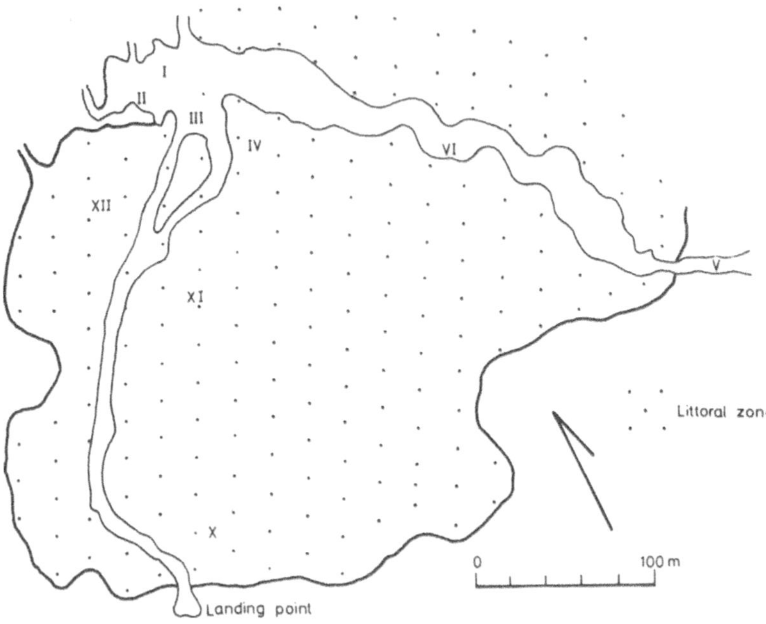

Fig. 15 Map of the study area showing sampling stations. (107)

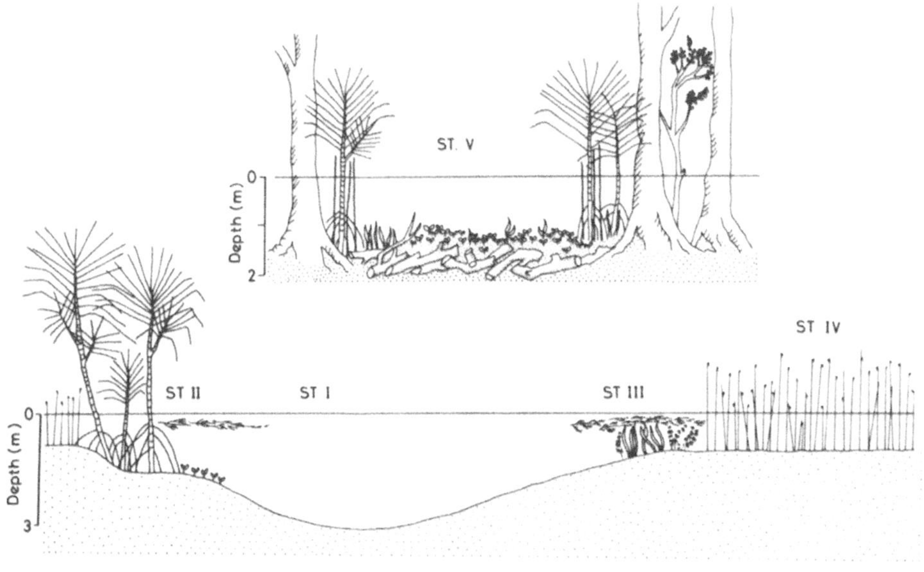

Fig. 16 Diagrammatic profile of the study area showing the structure of Sampling Stations I-V. (T. Mizuno)

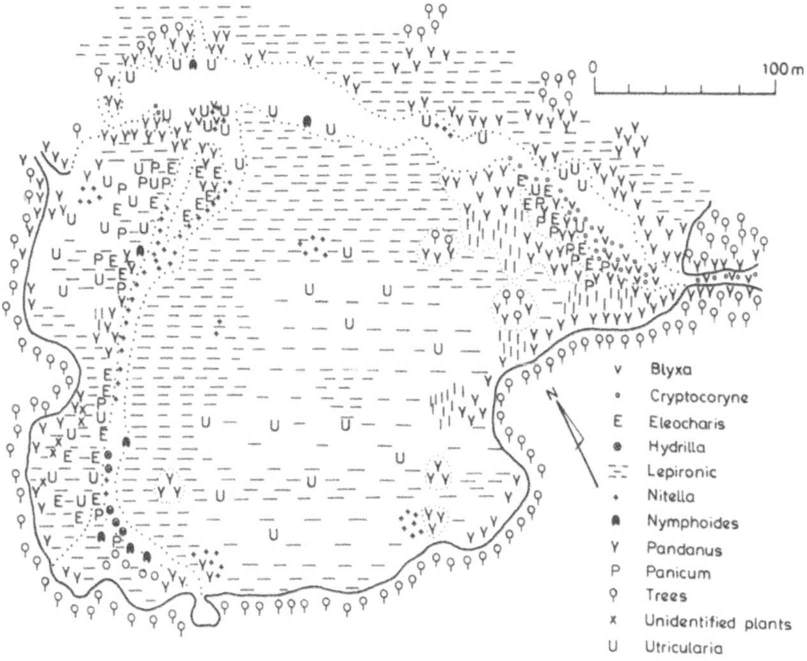

Fig. 17 Map of the study area showing vegetational distribution. (T. Mizuno)

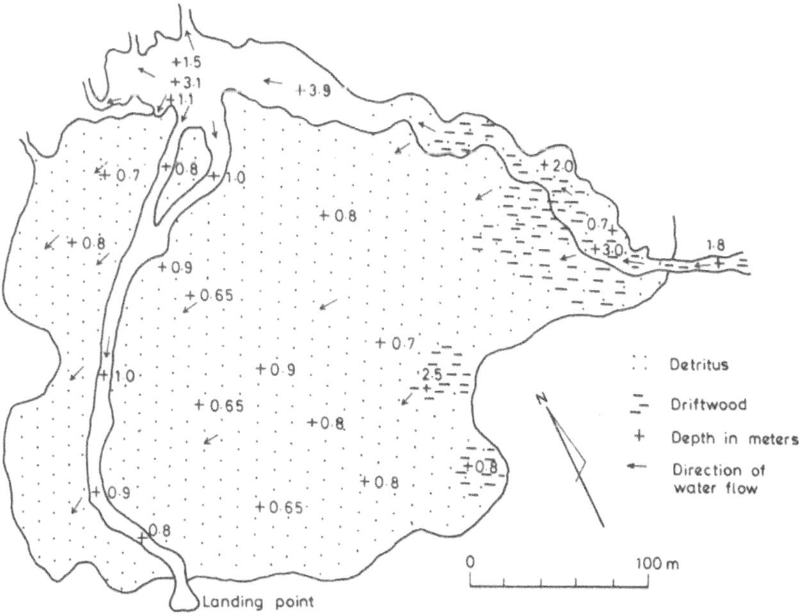

Fig. 18 Map of the study area showing the bathymetrical condition and direction of water flow. (T. Mizuno)

mat of *Utricularia* rises with rising water level but with a lowering of water level, parts of the plant either become entangled among the thorny *Pandanus* plants or are washed away in the water discharge. For plankton studies, a small opening was made in the centre of the mat, and left for two or three days before samples were taken. The normal depth of this station is one metre.

Stations IV, XI and XIII. Three stations were set up in the vast *Lepironia* field within the study area, and they represent the typical littoral of Tasek Bera. They are also exposed to the most vigorous changes in water level being dry for varying periods during the dry season. Within these fields occur patches of *Pandanus helicopus* and isolated mats of *Utricularia*. The reeds occur as dense stands (see later) but during the course of the study they were burnt several times, resulting in it being different from the less disturbed stands that occupy other parts of the swamp. Station XI was originally set up for experimental studies but was burnt in August 1971, and this was replaced with Station XIII. Experiments and regular measurements were carried out at these three locations. Its normal depth is 0.5 metres.

Station V. This station is located in the *Eugenia* swamp-forest. It characterises a lotic environment within the swamp. The vegetation is mixed *Eugenia* spp. and *Pandanus helicopus*. *Thoracostachyum baucanum* appears together with the *Pandanus* along the edges of the stream channels. Rich cultures of *Batrachospermum* spp. are often found attached to the submerged leaves of macrophytes. *C. griffithii* and *E. echinospermum* are the only submerged plants growing on the bottom of the stream channels, *C. griffithii* being the dominant species. The normal depth is 2.5 metres.

2.5 Land utilisation
(R. P. Lim and J. I. Furtado)

A population of about 700 Semalai aborigines have inhabited Tasek Bera and its watershed for several hundreds of years. Like the remainder of Malaysia, agriculture including fisheries and forestry forms the resource base of the subsistence economy of the Semalai inhabitants. In this economic system, the Semalai have exploited the swamp and watershed resources while maintaining some degree of balance with their ecosystems. For subsistence, the Semalai gather forest products both from the swamp and watershed for their housing, domestic and industrial implements; they fell and burn forests in the watershed for their settlements as well as for the cultivation of upland rice, other staples, vegetables and fruits on some form of shifting cultivation; they

Plate 2 Aerial view of abandoned plots of shifting cultivation.

capture fish from the swamp for their food; and they hunt for game both in the swamp and watershed for additional food.

Extensive areas on the eastern and western shores of the swamp, in the northern part of the watershed, have been managed thus on a shifting cultivation basis and are now regenerating secondary forests ('belukar' in Malay) (Plate 2). The widely dispersed settlements of the Semalai were concentrated around Pos Iskandar during the 1948 emergency (the Communist Insurgency), in six villages ('Kampungs' in Malay), four on the western and two on the eastern shores. Consequently, there is a concentration of cultivated land and shifting cultivation in the vicinity of Pos Iskandar, and the study area is intensely utilised by the Semalai. In the past 10-15 years, the Semalai have been introduced to plantation agriculture by the allocation of five acres of rubber smallholding to each family.

About 32% of the swamp-forest was cut in historical times, and is replaced by *Lepironia* and *Pandanus* stands. This man-modified littoral is exploited intensively for various plant products such as the *Lepironia* reeds and *Pandanus* leaves which are used for household implements. Fishing is carried out in this part of the littoral and in the limnetic region. Fish are captured by baited hooks and lures, by gill nets and bamboo or rattan traps, and by harpoon using a light at night, especially in or near *Pandanus* stands. During the dry season when the littoral may be dry, especially between June and August, the *Pandanus* clumps and *Lepironia* reeds are burnt and the former also cut, particularly in some arms of the swamp. This cutting and/or burning helps to keep navigation channels open, although it encourages the vegetative regeneration of both *Lepironia* and *Pandanus*. It also aids in the harvesting of fish isolated in patches in the arms of the swamp.

The watershed forests have been used for settlements and shifting cultivation by the Semalai over hundreds of years: and have been felled in the past six years for commercial timbers on the western part of the swamp. Much of this logging has been close to the swamp.

3. Environmental conditions

I. Ikusima, R. P. Lim and J. I. Furtado

3.1 Physical properties
(I. Ikusima and R. P. Lim)

3.1.1 *Light*

3.1.1.1 *Solar radiation* (I. Ikusima). Solar radiation was measured at two localities, namely Tasek Bera and Kuala Pilah (the Japanese IBP headquarters) which is about 50 km from Tasek Bera. An Ishikawa-Bellani pyranometer (daily radiation integrator) was used to measure both circumglobal and horizontal radiation (Fig. 19), while diurnal changes of horizontal radiation were measured with a Gorczynski radiometer (EKO, MS-12 type, Tokyo).

Circumglobal radiation flux was small as is normal of all tropical latitudes (Gates 1965) (Table 5). There were also very small variations in solar radiation between Tasek Bera and Kuala Pilah indicating that it does not vary over large areas (Table 6).

The daily mean horizontal radiation over 65 days was 325 cal $\cdot$ cm$^{-2}\cdot$day^{-1} with a maximum of 489 cal $\cdot$ cm$^{-2}\cdot$day^{-1} on a clear day in June (Table 5). On rainy days the value was approximately 170 cal $\cdot$ cm$^{-2}\cdot$day^{-1}.

Seasonal variation in radiation are small but are distinct between the wet and dry seasons (Fig. 19). The mean horizontal radiation is 392 cal $\cdot$ cm$^{-2}\cdot$day^{-1} in the dry season (June-September) and is 275 cal $\cdot$ cm$^{-2}\cdot$day^{-1} in the wet season (November-January), which is typical of the radiation values obtained by Wycherly (1967) for Peninsular Malaysia.

Diurnal patterns in solar radiation measured over three months indicate a normal distribution from sunrise (0630 hours) to sunset (1830 hours) (Fig. 20). The meridian time (i.e. when the sun is directly overhead) was 1245 hours. (Fig. 20). The average daily radiation was 458.6 cal $\cdot$ cm^{-2}.

Furtado, J. I. and Mori, S., eds., Tasek Bera. ISBN 90 6193 100 2.

Table 5 Mean, minimum and maximum horizontal solar radiation (cal · cm^{-2} · day^{-1}) at Pos Iskandar from June 1972 to March 1973. (I. Ikusima)

Month		Mean	Minimum	Maximum	Number of days recorded
1972	June	378	242	489	4
	July	351	339	358	3
	August	408	325	488	8
	September	360	139	460	6
	October	313	238	412	6
	November	273	146	395	7
	December	261	109	364	7
1973	January	275	111	370	9
	February	303	102	432	8
	March	327	248	415	7
Overall mean		325	200	418	Total 65

Table 6 Circumglobal solar radiation measured at Pos Iskandar and Kuala Pilah. Figure in parentheses indicates the number of days recorded. (I. Ikusima)

Date		Circumglobal radiation (cal · cm^{-2} · day^{-1})	
		Pos Iskandar	Kuala Pilah
1972	June	205 (4)	190 (16)
	July	191 (3)	202 (29)
	August	216 (8)	196 (28)
	September	192 (6)	185 (80)
	October	160 (6)	218 (31)
	November	148 (7)	198 (27)
	December	173 (3)	162 (18)

3.1.1.2 *Transparency* (I. Ikusima, R. P. Lim and T. Mizuno). Transparency was measured using a Secchi disc at Station I around noon on clear days when the sun was directly overhead. Measurements for three years are summarized in Fig. 21. Transparency ranged from 1.5 to 2.5 m with a maximum of 2.67 m in September 1971, just after the dry season with rising water level and a minimum of 1.08 m in April 1971. During the study period, transparency at Tasek Bera fluctuated daily with the fluctuation of water level which is related to the amount of rainfall. There was a positive correlation between water level and transparency (coefficient of correlation = 0.60).

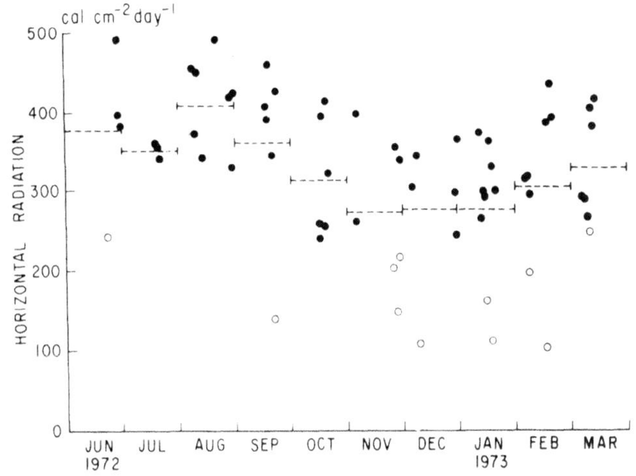

Fig. 19 Horizontal radiation measured with an Ishikawa's Bellani pyranometer at Pos Iskandar from June 1972 to March 1973.
Solid circles: clear, fine, cloudy days.
Open circles: rainy (daytime) days.
Broken lines: monthly mean daily radiation. (I. Ikusima)

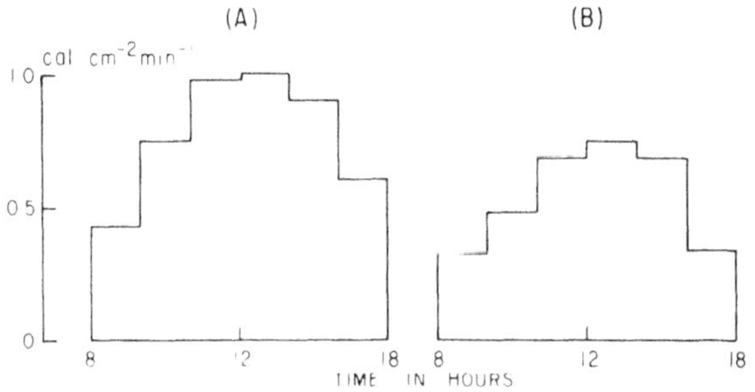

Fig. 20 Diurnal marches of (A) maximum horizontal radiation and (B) mean horizontal radiation, at Pos Iskandar, from June to September 1972. (I. Ikusima)

In general, low transparency was observed at low water level during the dry season or at extreme high water levels during the wet months such as the flood in January 1971. Increased turbidity was caused (1) by an increase in density of plankton and a decrease in a water renewal due to a reduction in water velocity of the streams, (2) by an increase of particulate materials especially inorganic particles such as clay, in the run off water from the forest floor of the watershed, and (3) by an agitation of organic particles and debris attached to the macrophytes by the strong water currents and the action of

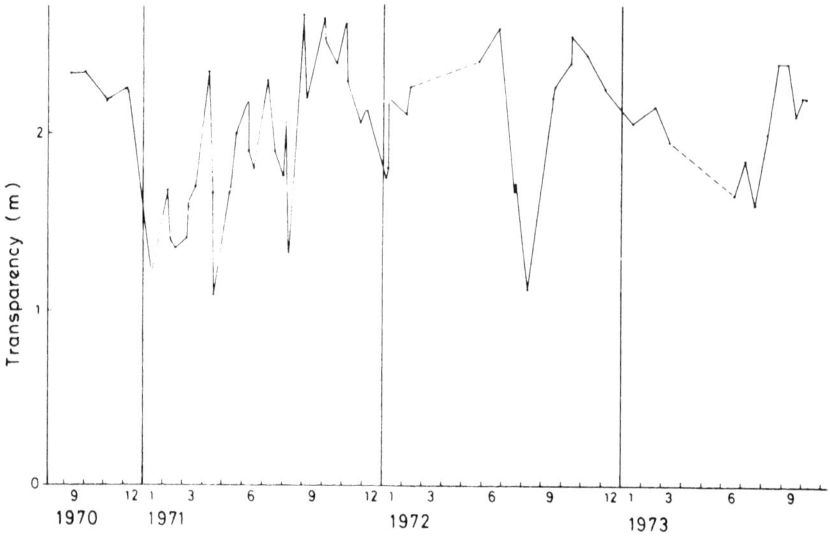

Fig. 21 The Secchi disc transparency readings at Station I, Tasek Bera from 1970 to 1973. (Ikusima, Lim, Furtado, Nagai, Kumano, N. Mizuno, Watanabe, T. Mizuno, Tezuka and Okino)

rain drops. Low transparency values of 1.23 at Station I and 0.87 near the swamp shore were recorded during the great floods in January 1971. The colour of the water was yellowish brown due to clay particles especially near the shore.

3.1.1.3 *Underwater light* (I. Ikusima. R. P. Lim and J. I. Furtado).
Underwater light intensity was measured with a waterproof photometer and an electric recorder. Measurements were made around noon on clear and cloudy days after or before the measurement of transparency. Opal filters were used to measure neutral light. The coefficient of extinction based on an exponential rate of transparency measurements is given in table 7. The mean value of the extinction coefficient for neutral light at Station I taken over a period of 4 years was $2.43 \cdot m^{-1}$. A maximum value of $3.58 \cdot m^{-1}$ was observed at Station I ($4.65 \cdot m^{-1}$ near the shore) at the time of the great flood in January 1971. High values of the coefficient of extinction were observed at low water level in dry season and during the floods in a wet season.

In the study area, the coefficient of extinction varied between stations. The values at the open water and at the open channels were similar. The values at the *Utricularia* community and at the *Lepironia* community were slightly

Table 7 Coefficients of extinction and transparency at Station I, Tasek Bera. (I. Ikusima, R. P. Lim, J. I. Furtado and T. Okino)

Date		Coefficient of extinction (l/m)	Transparency (m)	Observer(s)
1970	June 6	2.35		R. P. Lim and J. I. Furtado
	September 10	2.88	2.35	I. Ikusima
	October 4	2.49	2.35	I. Ikusima
	November 4	3.18	2.20	I. Ikusima
	December 5	2.83	2.27	I. Ikusima
1971	January 13	3.58	1.23	I. Ikusima
		4.65[a]	0.87[a]	I. Ikusima
	February 5	3.44	1.68	I. Ikusima
	September 5	2.35	2.20	R. P. Lim and J. I. Furtado
	October 4	2.31	2.52	R. P. Lim and J. I. Furtado
	November 5	2.43	2.65	R. P. Lim and J. I. Furtado
	December 5	2.35	2.14	R. P. Lim and J. I. Furtado
1972	April 2	2.38		R. P. Lim and J. I. Furtado
	June 27	1.82	2.60	I. Ikusima
	July 20	3.24	1.80	I. Ikusima
	August 8	4.11	1.12	I. Ikusima
1973	September 7	0.59[b]	2.40	T. Okino
	September 19	1.71	1.90	T. Okino
	September 29	1.30	2.20	T. Okino
	October 4	0.89	2.20	T. Okino

[a] Near a lake shore.
[b] Water layer from the surface to 1 m deep.

higher than that of the former stations. The extinction coefficients for neutral light at Stations I, II, III, IV and V are given in Table 8. These differences in extinction coefficients between stations were due to differences in the amount of plankton, debris attached to the macrophytes and fragments of the macrophytes in the water. Little photosynthesis can occur at depths below two metres, because light conditions were not sufficient.

An empirical formula that represents the inverse proportionality between coefficient of extinction of neutral light (K) and transparency (Tr), i.e. $K \cdot Tr = 1.9$ does not satisfy the experimental values, probably because of a nature of the black water (Fig. 22).

The vertical distribution of light intensity for different wavelengths was measured with the photometer using red, yellow, green, blue, and violet glass filters. This is to determine the spectral distribution of transmissibility. A

Table 8 Coefficients of extinction (m^{-1}) of neutral light at Stations I, II, III, IV and V in Tasek Bera, taken on 6 June 1970 at 1045 hours. (R. P. Lim and J. I. Furtado)

Station	Coefficient of extinction
I	2.43
II	2.41
III	3.00
IV	4.65
V	2.34

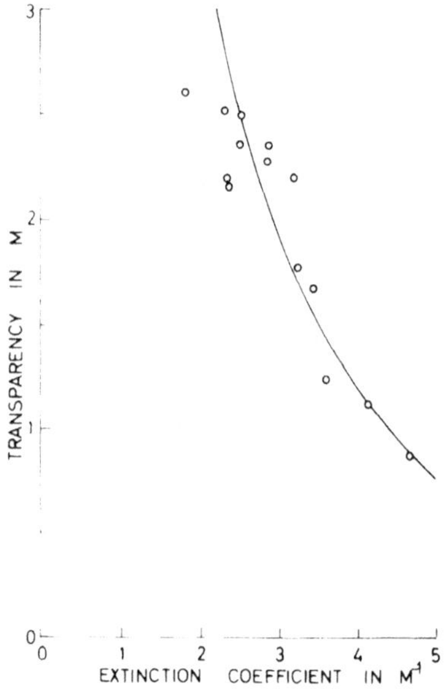

Fig. 22 Relationship between the transparency (Tr) and the coefficient of extinction (K). Equation, $Tr = 8.86/K - 1.023$ can be fitted to the data as shown by the curve. (I. Ikusima)

linear relationship exists between the logarithm of the relative light intensity and depth in different spectral regions at a certain depth. The coefficient of extinction for each wavelength is given in Table 9. Light transmission in swamp water at different wavelengths was calculated using the constants given in Table 9. In Tasek Bera, the higher extinction coefficient in the blue-green spectral region (400-500 nm in wavelength) indicates that light attenuation in the blue-green region is smaller than for red or yellow wavelength (Fig. 23). This is principally the result of the brown colour of the water.

Table 9 Extinction coefficients (m^{-1}) of selected wavelengths of light (nm) under different transparencies (m) observed on the different date at Station I, Tasek Bera. (I. Ikusima)

Toshiba's glass filter	Neutral	V-R63	V-R60	V-G54	V-Y48	V-B46	V-V40
Wavelength at max. transmission (nm)	–	630	600	540	480	460	400
Sensitivity center of spectral response of the filter-photocell combination (nm)	–	640	610	540	560	520	450
Extinction coefficient (m^{-1}) (Transparency, m)							
22 June 1972 (2.75)	2.94	1.43	1.41	2.09	2.30	4.60	3.43
27 June 1972 (2.60)	1.82	1.47	1.47	2.24	2.26	3.07	6.40
8 Aug. 1972 (1.12)	4.12	2.73	4.01	5.21	5.25	7.89	9.06

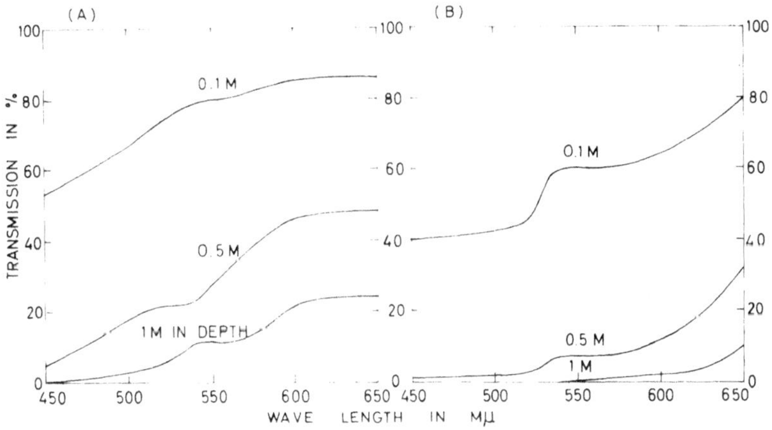

Fig. 23 Percentage transmission of different wavelengths of light at 0.1, 0.5 and 1 m depths at Station I, in June 1972 (A) and in August 1972 (B). (I. Ikusima)

Ho (1970) showed a reverse phenomenon at Ampang Reservoir, Kuala Lumpur, which is mesotrophic and has extinction coefficients of $0.7 \cdot m^{-1}$ at $\lambda = 430$ nm, $1.37 \cdot m^{-1}$ at $\lambda = 530$ nm and $1.66 \cdot m^{-1}$ for neutral light; this is related to the green colour of the phytoplankton rich water.

Correct interpretation and application of the results for characterising the optical property of water bodies as well as the relation between transparency and primary productivity studies has been discussed by several authors (e.g. Poole and Atkins 1929, Clarke 1941, Ichimura 1956, Becton 1957, Vollenweider 1960, Ikusima 1967, 1970, and Schindler 1971). In eutrophic and mesotrophic waters, a transparency reading gives an approximate indication of the quantity of allochthonous and autochthonous material in the water (Hutchinson 1957). Yoshimura (1938) has computed the compensation point by multiplying Secchi disc readings by 1.2. Where Secchi disc depths have been found to give a good indication of turbidity due to plankton they have been used, with limitations, for primary productivity studies (Vollenweider 1960). Ikusima (1967), using empirical information has shown that the compensation depth for aquatic green plants can be expressed as a linear function of transparency. It is also noted that Secchi disc readings vary with conditions under which the readings are taken (Vollenweider 1969), and thus readings cannot be used to compare different water bodies unless they are very similar.

3.1.2 *Air, water and land temperature*
(R. P. Lim, J. I. Furtado and I. Ikusima)

Air temperature above the water was fairly high averaging 29.5°C (minimum 25.2°C and maximum 36.8°C) (Table 10). The limnetic region and *Utricularia* sublittoral (Stations I and III) were cooler than the littoral region (Station IV) or the sublittoral *Pandanus* (Station II) with the forested limnetic channel (Station V) being the coolest. The slightly higher air temperatures at the *Lepironia* littoral and the *Pandanus* sublittoral were due to a partial obstruction of wind action by the emergent littoral vegetation (*Lepironia* and *Pandanus* respectively), thus hampering air circulation above the water and making it warmer. This slightly higher temperature above the water surface and the vegetated littoral probably causes less heat exchange between air and water. In comparison the air above the limnetic region in the main channel, at the *Utricularia* and in the forest channel is exposed to wind action, thus making it cooler; furthermore the swamp-forest canopy and stand completely shelters the air above the water from radiation and partly from wind action thus making its temperature much lower. Rainfall also markedly reduces the air temperature of the littoral (Table 10). The range in air temperatures above the water also varied being lowest (5.8°C) at the limnetic forest channel, highest (11.5°C) at the *Lepironia* littoral, and somewhat in between (6.2-8.3°C) in the limnetic and sublittoral regions. The differences in temperature range are also related to the influence of shade, density of vegetation and air movement, all of which are important in heat exchange.

Water temperatures were high averaging 26.3°C at the surface (minimum

Table 10 Mean, minimum and maximum air temperatures at Stations I-V at Tasek Bera (1970–1972). (R. P. Lim and J. I. Furtado)

Stations	Air temperature (°C)		
	Mean	Minimum	Maximum
Limnetic, man-modified:			
Station I	29.6	25.2	33.5
(Main channel)		(Jan. 1972)	(Sept. 1970)
Station II	30.1	27.0	34.3
(Pandanus-Utricularia)		(Jan. 1971)	(Sept. 1970)
Station III	29.8	26.1	32.1
(Utricularia)		(Jan. 1972)	(April 1971)
Littoral, man-modified:			
Station IV	30.4	25.5	36.8
(Lepironia)		(Dec. 1971)	(Aug. 1970)
Limnetic, swamp-forest:			
Station V	27.9	25.2	31.0
(Main channel)		(Jan. 1971)	(Sept. 1970)
Overall	29.5	25.2	36.8

23.3°C and maximum 31.2°C and 25.4°C at the bottom (minimum 23.2°C and maximum 26.6°C) (Fig. 24). They varied less between Stations I-V than air temperatures. The *Lepironia* littoral (Station IV) had a slightly higher surface water temperature than the other stations, but the bottom temperatures were comparable between stations (Table 11). This difference may be due to less heat exchange at the water-air interface in the *Lepironia* littoral because of prevailing higher air temperatures.

Water temperatures showed distinct seasonal fluctuations related to the wet and dry seasons (Fig. 24). Low water temperatures occurred in the wet monsoon seasons (late September-January and April-May), while higher temperatures occurred in the dry seasons (February-March and June-September). The lowest water temperature was recorded in January 1971 (north-east monsoon) while the highest temperature was recorded in September 1971. There was a slight change in the duration and timing of these seasons from year to year which accounted for September 1971 being a dry season.

The range in water temperature fluctuation also varied between stations being highest in the *Lepironia* littoral (Station IV) both at the surface (7.7°C) and bottom (4.9°C), low in the limnetic main channels both in the modified area (Station I: surface 3.5°C, bottom 3.2°C) and the swamp-forest (Station V:

Table 11 Mean, minimum and maximum water temperatures at surface and bottom depths at Stations I-V in Tasek Bera (1970–1972). (R. P. Lim and J. I. Furtado)

Stations		Water temperature (°C)		
		Mean	Minimum	Maximum
Limnetic, man-modified:				
Station I	Surface	26.0	23.7	27.2
(Main channel)			(Jan. 1971)	(June 1970)
	Bottom	25.4	23.4	26.6
			(Jan. 1971)	(May 1970)
Sub-littoral; man-modified:				
Station II	Surface	26.1	23.7	29.4
(Pandanus-Utricularia)			(Jan. 1971)	(Sept. 1970)
	Bottom		23.3	26.6
			(Jan. 1971)	(May 1970)
Station III	Surface	26.6	23.7	29.0
(Utricularia)			(Jan. 1971)	(June 1970)
	Bottom	25.6	23.3	26.7
			(Jan. 1971)	(Aug. 1971)
Littoral, man-modified:				
Station IV	Surface	27.3	23.5	31.2
(Lepironia)			(Jan. 1971)	(April 1971)
	Bottom	25.6	23.3	28.2
			(Jan. 1971)	(Aug. 1971)
Limnetic, swamp-forest:				
Station V	Surface	25.3	23.3	26.6
(Main channel)			(Jan. 1971)	(May 1970)
	Bottom	25.3	23.2	26.6
			(Jan. 1971)	(May 1970)
Overall	Surface	26.3	23.3	31.2
			(Jan. 1971)	(April 1971)
	Bottom	25.5	23.2	26.6
			(Jan. 1971)	(May 1970)

surface 3.2°C, bottom 3.3°C; Station III: surface 5.3°C, bottom 3.4°C) (Table 11). Smaller fluctuations in water temperature in the limnetic region was due to air and water movements facilitating heat exchange; whereas, in the *Lepironia* littoral, the dense reed stands hinder air and water movement, thus reducing heat exchange. The higher bottom temperature fluctuation in the limnetic channel in the swamp-forest (Station V) may be due to water influx from a man modified area upstream and to insufficient dissipation of its heat.

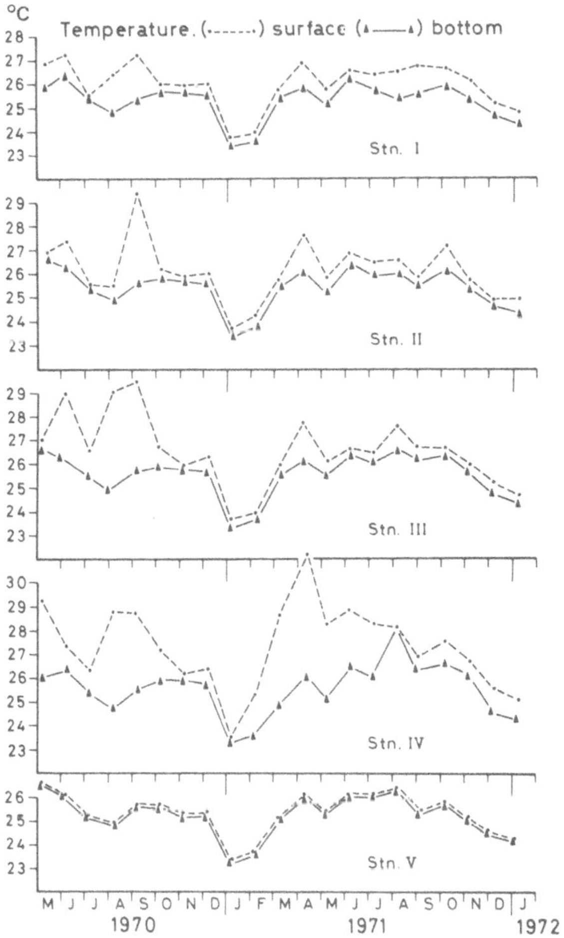

Fig. 24 Water temperature (°C) at Stations I-V in Tasek Bera (1970–1972), at surface and bottom depths. (R. P. Lim and J. I. Furtado)

The *Lepironia* littoral (Station IV) also showed a greater water temperature differential between the surface and bottom (1.7°C) than the *Utricularia* sublittoral (Station III: 1.0°C) or the other limnetic and sublittoral stations (Table 11). This was due to aquatic macrophytes (*Lepironia* and *Utricularia*) preventing water circulation thus hindering heat dissipation.

Water temperatures were high as in other tropical regions because of strong radiation throughout most of the year: cf. Talling (1969) and Carter (1955) in Africa; Sreenivasan (1964 and 1966) in India; Marlier (1967) in the Amazon Basin; Ruttner (1931) in Indonesia and Johnson (1967, 1968), Ho (1970), and

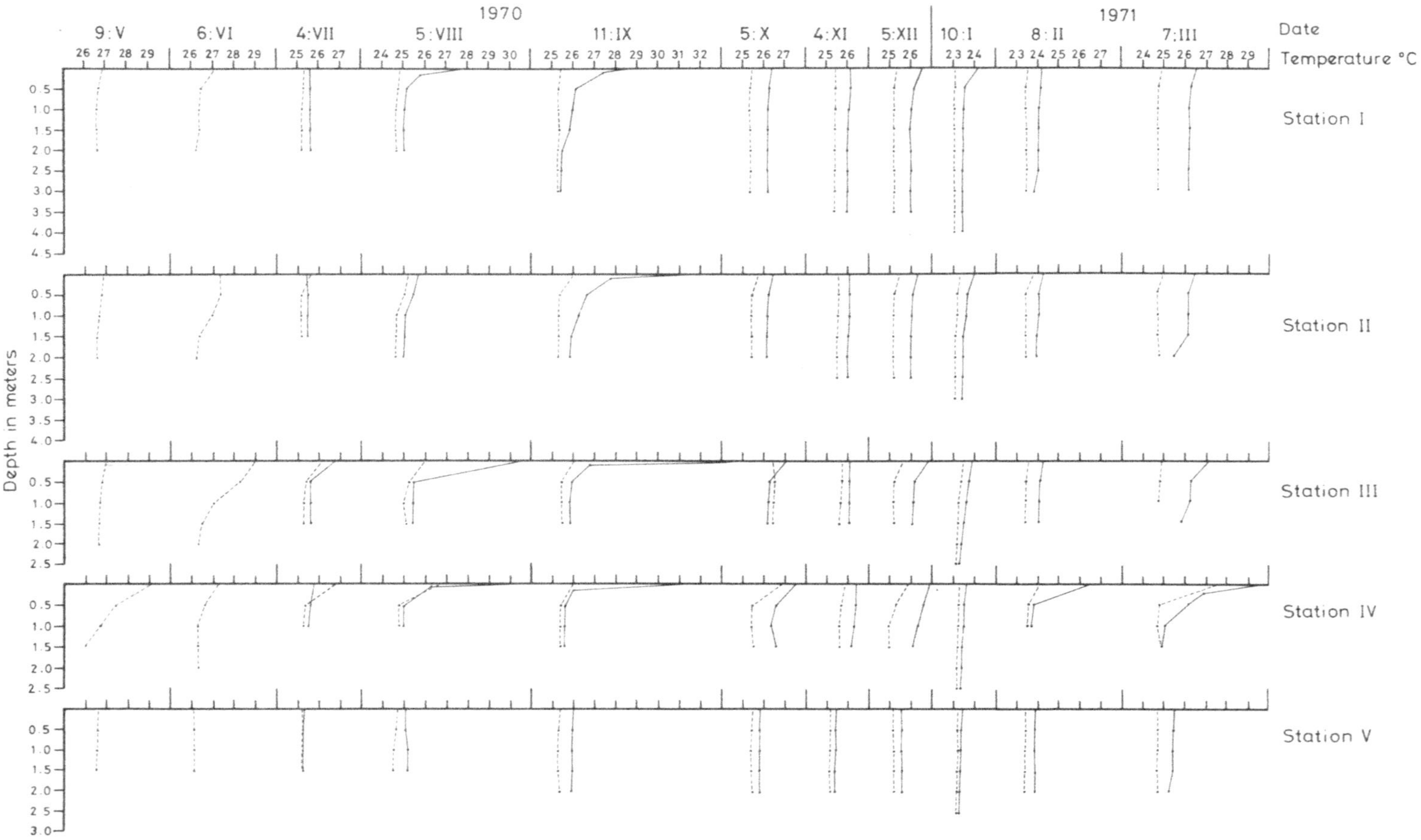
1970
1971
9:V
6:VI
4:VII
5:VIII
11:IX
5:X
4:XI
5:XII
10:I
8:II
7:III
Date
Temperature °C
Station I
Station II
Station III
Station IV
Station V
Depth in meters

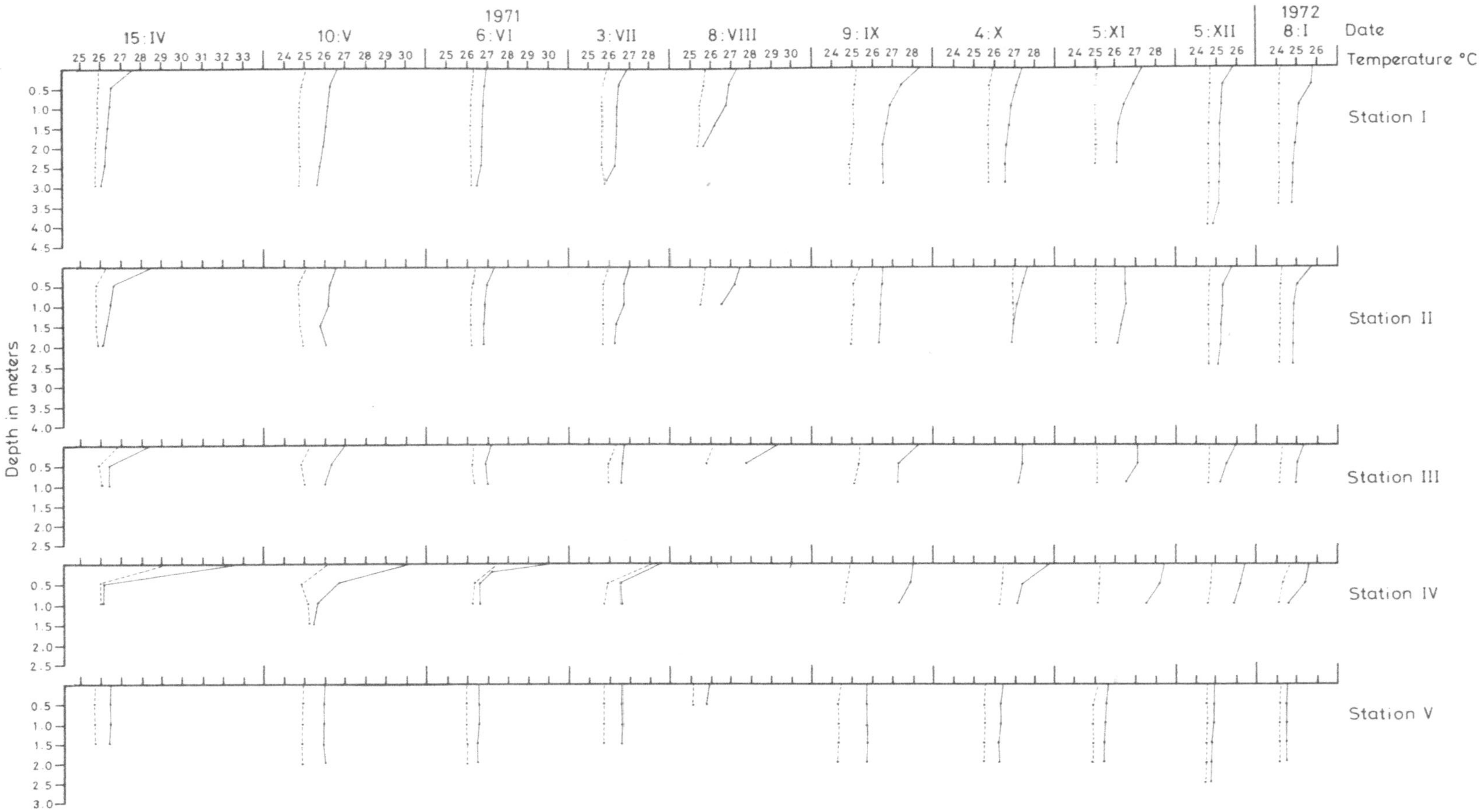

Fig. 25 Water temperature – depth profiles for Stations I-V at Tasek Bera (1970–1972), in the morning (0900 hours) and afternoon (1500 hours). (R. P. Lim and J. I. Furtado)

Arumugam (1971) in Peninsular Malaysia. Small monthly fluctuations in water temperature are due to weak variations in radiation flux.

The water temperature profile of Stations I-V over 21 months (Fig. 25) showed that the limnetic region without dense aquatic macrophytes (Station I, II and V) differed from the vegetated littoral (Station II and IV) in that the latter had a steeper temperature gradient, especially during the dry seasons (July-September 1970, February-April 1971 and July-September 1971). This is related to the water temperature differential between the surface and bottom, discussed above. Thermal gradients were otherwise small throughout the study period. The marked water temperature gradient in the *Utricularia* sublittoral and the *Lepironia* littoral tended to be prolonged into the wet seasons, and is due mainly to poor water circulation caused by dense aquatic macrophytes. The water temperature increased during the day from 0900 to 1500 hours due to heating by radiation, except at Stations III and V on October 1971 and 4 July 1971 respectively (Fig. 25). These exceptions were due to sporadic rains on these days. Particularly low temperature differences between 0900 and 1500 hours occurred in January 1971 at Stations I-V again due to rains.

The difference in water temperature profiles between the limnetic and the *Utricularia* sublittoral and *Lepironia* littoral was apparent. In the limnetic the gentle curves were typical of shallow tropical standing waters (Ho 1970; Arumugam 1971); whereas in the *Utricularia* and *Lepironia* zones there was definite stratification, especially during the dry seasons in spite of the shallow depth. This marked stratification in the littoral and sublittoral of the modified areas was due to differences in the amount of mixing between the layers of the water column, caused by (1) the absence of a strong current in the littoral and sublittoral in comparison to the limnetic main channel (see earlier); (2) the prevention of wind action on the water surface by surface and emergent aquatic macrophytes, especially *Utricularia* and *Lepironia*, when they occurred in dense stands; and (3) the prevention of water circulation by surface and rooted aquatic macrophytes such as *Utricularia* and *Lepironia* especially in dense stands. Thus, a thermocline forms in the *Utricularia* sublittoral and the *Lepironia* littoral at about 25 cm depth. Similar observations have been reported by Carter (1955) in the lily zone of an Ugandan swamp, and by Carter and Beadle (1930) in the swamps of the Paraguayan Chaco.

Monthly isotherms based on daily peak temperatures (i.e. at 1500 hours) indicate that, despite a daily turnover (see later), the temperature characteristics of a water body at a given time of day were fairly similar over a period of several months (Fig. 26). There was complete mixing daily at the limnetic

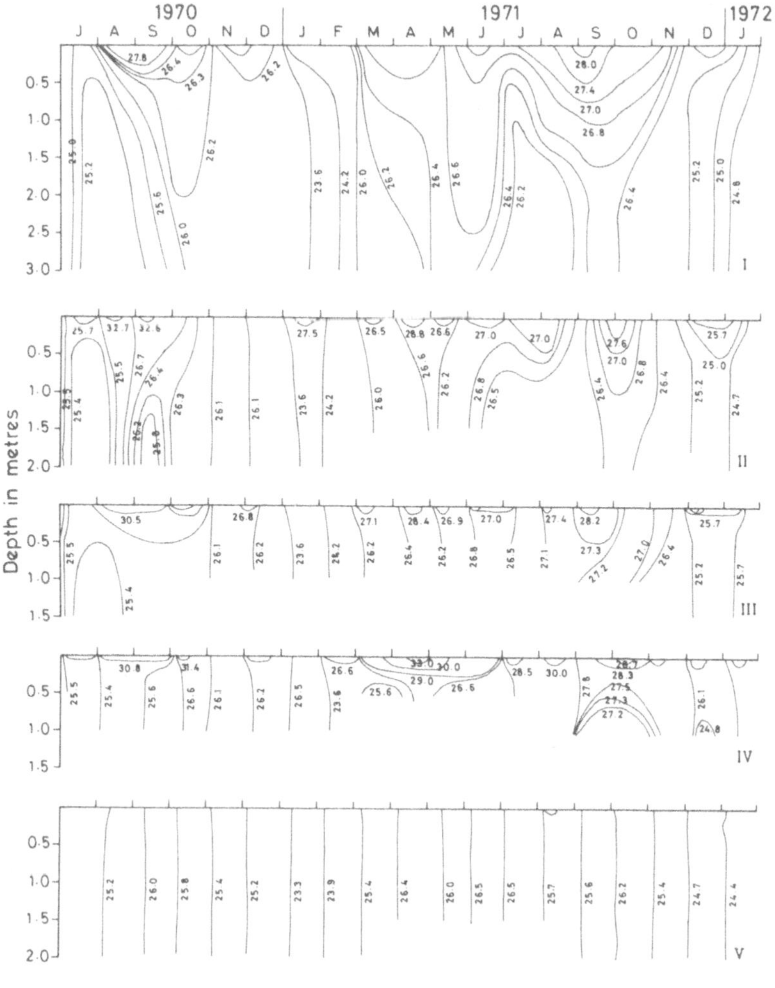

Fig. 26 Monthly isotherms at Stations I-V in Tasek Bera (1970–1972). (R. P. Lim and J. I. Furtado)

(Station I) in the modified area towards the end of the wet seasons (December-early February, and May-June), while persistent stratification occured in the drier seasons and extended into the wet seasons (July-November, and March-May). In the *Pandanus* and *Utricularia* sublittoral (Station II and III) and in the *Lepironia* littoral (Station IV) superficial stratification was common, while the limnetic in the swamp-forest (Station V) showed a complete homothermy as is expected of lotic and completely shaded waters. It is evident from these observations (1) that radiation contributes to greater heating of the water during the dry season and, hence, the stratification; (2) that the *Utricularia* and *Lepironia* vegetations prevent water circulation and thus enhance superficial stratification in the water columns; and (3) that the swamp-forest prevents heating and stratification of its limnetic channel. Thermal stratification is also dependent on other factors such as water currents, rainfall, and changing water levels, which are discussed elsewhere.

3.1.3 *Water velocity and flux within the study area*
(I. Ikusima, T. Mizuno, S. Mori and O. Sato)

The direction of water flow and its velocity within the study area was measured at several localities (Fig. 27). Velocity-depth profiles along the main channels were also measured. From the flow rates obtained together with corresponding depth measurements the flux of water flowing through the various regions of the swamp was computed.

The general direction of flow of water is northerly. The water enters the study area from the swamp forest (Station V) and flows along a channel into

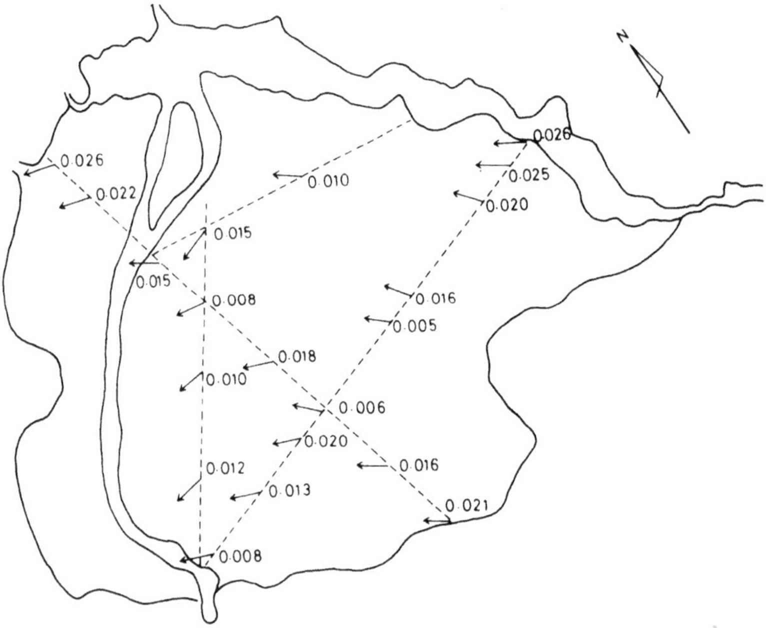

Fig. 27 Velocity and direction of water flow in the *Lepironia* littoral region measured on 5 and 6 November 1972. Arrows on the map show the direction of water flow and figures indicate the velocity of water in metres per second. (S. Mori and T. Mizuno)

the open water region (Station I) where it continues out through three main channels to the rest of the swamp. However, there is also a subsidiary flow from the main channel into the *Lepironia* littoral region, which generates slight currents within this region. The average velocity within the main channel is 0.25 $m \cdot sec^{-1}$, while the velocity within the *Lepironia* littoral region is 0.016 $m \cdot sec^{-1}$. This indicates that there is a fair amount of flow in the main channel approaching that of a lotic situation.

Velocity of the water varies greatly horizontally and vertically in the same stream (Table 12). It is a function of the amount and time distribution of rainfall affecting the depth and width of the channel. The type of substrate also affects velocity profiles (Fig. 28) which approaches zero near the bottom. The rate of flow varies greatly with water level. Measurements of the rate of flow made at Station V in July and August were ca. $3 \times 10^4\ m^3 \cdot day^{-1}$ and $8 \times 10^3\ m^3 \cdot day^{-1}$ respectively. Although the water level in August was only 40 cm lower than that in July the cross-sectional area was 0.8 times as large as that in July. There was a large increase in velocity with a rise in water level measured at the inlet (Station V) and the open water region (Station I) respectively (Fig. 29).

Water balance in the study area is important in estimating production based on oxygen dynamics. Computations of the flux of water show that the main source of water into the study area is derived from the swamp-forest (Station V). Of the $4.07 \times 10^4\ m^3 \cdot day^{-1}$ that flows from the swamp-forest (Station V), $3.02 \times 10^4\ m^3 \cdot day^{-1}$ flows along the main channel while $1.05 \times 10^4\ m^3 \cdot day^{-1}$ filters through the *Lepironia* littoral region as the water passes along the main channel. $4.04 \times 10^4\ m^3 \cdot day^{-1}$ leaves the study area through the three outlet channels. The daily loss of water mainly through evapotranspiration from the study area is estimated at $3 \times 10^2\ m^3 \cdot day^{-1}$ at mean water level in July 1972. Although the flux has been conveniently compartmentalised the situation is more complex in reality. This exercise however, does indicate a heterogeneous retention time within various habitats in the swamp. The high flow rates in the main channel may explain several physical, chemical and biological phenomena within the swamp (see later).

3.1.4 *Water level fluctuations*

(T. Mizuno and R. P. Lim)

Water level in the swamp was closely associated with rainfall (Fig. 30). The mean, minimum and maximum water level of the swamp throughout the study period were 1.53 m, 0.66 m (9 August 1971), and 5.72 m (2 January 1971)

Table 12 Mean water velocity, rate of flow and area of the cross-section at the main inlet (Station V) and outlets (Stations VI, VII and VIII). (I. Ikusima)

Date	Station V			Station VI			Station VII			Station VIII			Relative value of water level (m)
	Mean water velocity (m/sec)	Rate of flow (m^3/day)	Area of cross-section (m^2)	Mean water velocity (m/sec)	Rate of flow (m^3/day)	Area of cross-section (m^2)	Mean water velocity (m/sec)	Rate of flow (m^3/day)	Area of cross-section (m^2)	Mean water velocity (m/sec)	Rate of flow (m^3/day)	Area of cross-section (m^2)	
1970 October	0.253*			0.169*			0.177*			0.083*			+1.39
1972 June	0.205												+1.09
1972 July	0.071	30,181	5.08	0.067	9,642	1.60	0.154	27,015	2.32	0.028	3,694	1.50	+0.40
1972 August	0.038	7,934	3.91	0.025	953	0.51	0.084	6,226	0.97		dried up		+0.00

* Water velocity at the surface.

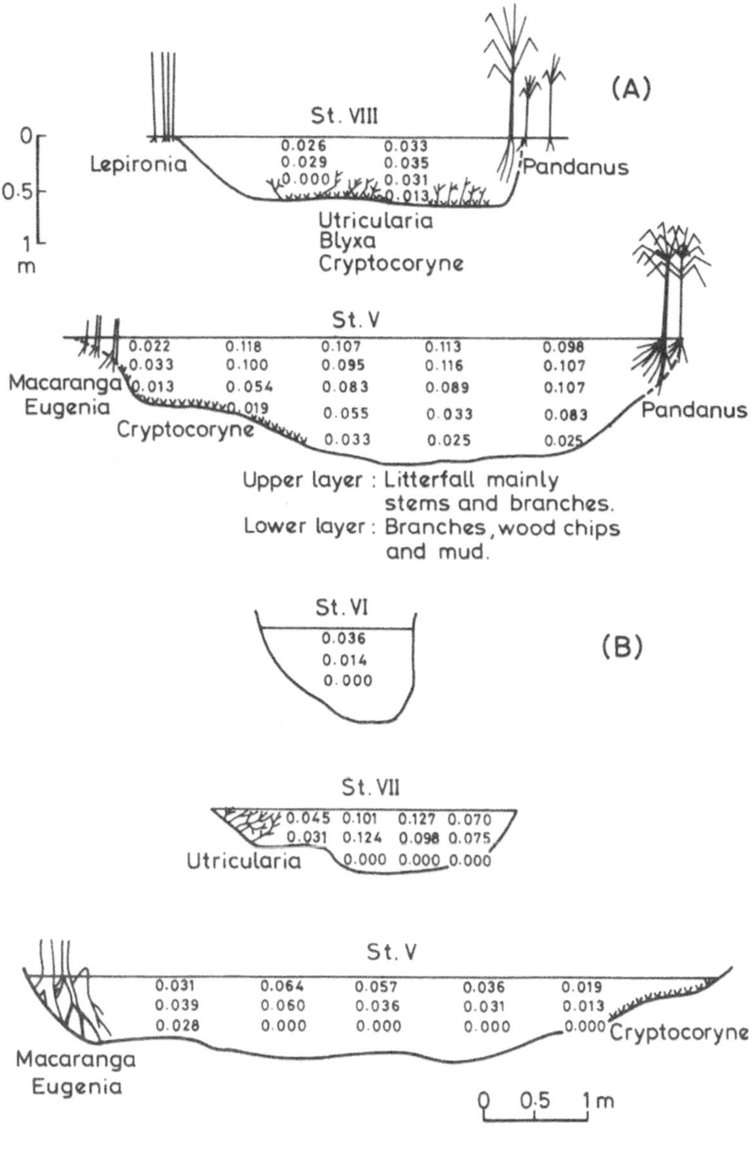

Fig. 28 Cross-section of the main inlet (Station V) and outlets (Stations VI, VII and VIII), showing the distribution of velocity of water (metres per second) measured on 18 July (A) and 8 August (B), 1972. (I. Ikusima)

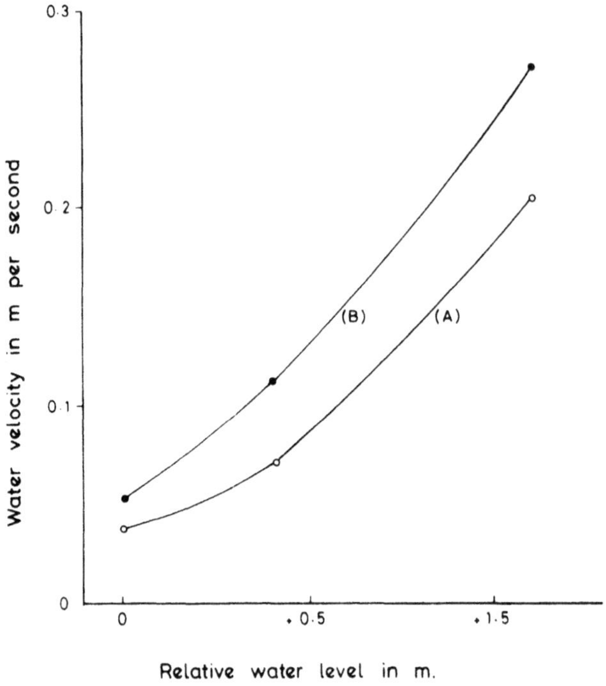

Fig. 29 Relation between the water level (relative value in metres) at Station I and the velocity of water (metres per second) at the inlet, Station V. Curve A indicates the mean velocity at the inlet; Curve B, the maximum velocity at the inlet. (I. Ikusima)

respectively. Water level readings were taken on each field sampling day and were not representative of the month since they were recorded at the beginning of each month. Hence it could result in some artefact when associated with rainfall. A more intensive recording of water level and rainfall is needed to verify the above relationship.

3.2 Chemical properties
(R. P. Lim, J. I. Furtado and O. Sato)

Water samples for alkalinity, conductivity and chemical analyses were taken at specified depths with a Kemmerer sampler, at the same time as samples were taken for dissolved oxygen and pH. For chemical analysis, samples were taken at 5 stations (I-V) from surface, mid- and bottom levels, the actual depth varying with the water level of the swamp. Conductivity was measured with a 'Dionic' conductivity meter, Series 3, which was calibrated with a

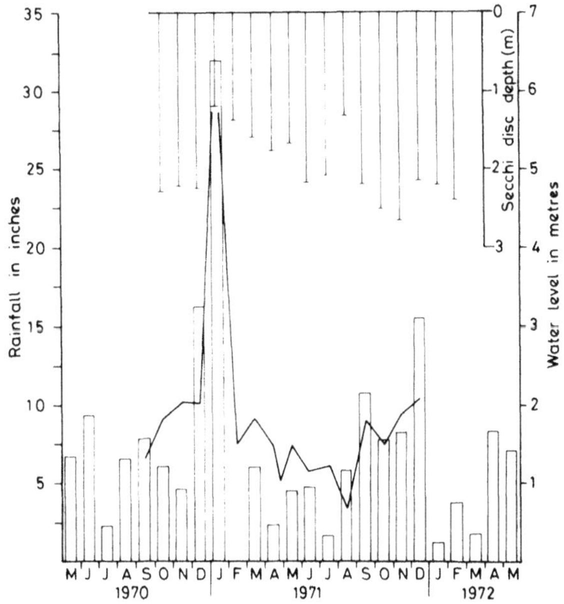

Fig. 30 Rainfall (histogram), water level (frequency polygon) and Secchi disc readings (linear projections) at Tasek Bera (1970–1971). (R. P. Lim and J. I. Furtado)

standard KCl dilution series (Golterman and Clymo 1969). Alkalinity was determined by titration with 0.01 N HCl using a B.D.H. 4.5 indicator. The natural brown colouration of the samples tended to mask the end-point, making accurate determination difficult. At pH 4.5 (determined potentiometrically), there is a change to a slightly greenish-yellow colour, which was considered subsequently as the end-point. (Discolouration with activated charcoal is laborious and of long duration, leading to inaccurate results because alkalinity varies rapidly with time.) Between March 1971 and January 1972, a more extensive study was carried out to determine the change in alkalinity with depth and over a period of six hours (i. e. from 0900 hours to 1500 hours). Water samples for chemical analysis were collected in 1.5 litre acid-washed polyethylene bottles and taken back to the laboratory. A portion was filtered and stored at 5°C until analysed. Unstable elements (PO_4-P, NO_3-N, NH_4-N, and NO_2-N) were analysed on the day of collection while SO_4, Cl, Ca, Mg, Fe, Al, Na, K and organic-N were analysed later. Deionised distilled water was used for the preparation of standard solutions, and for final washing of the acid-washed glassware. Chemical constituents were determined according to methods in Anonymous (1965), Golterman and Clymo (1969), Mackereth (1963) and Swingle (1963) as follows:

Sodium and potassium: E.E.L. filter flame photometer*
Calcium: EDTA titration*
Magnesium: by total hardness method using Eriochrome T-Black-EDTA titration*. Magnesium values were obtained as described in Mackereth (1963)*.

Determinations of calcium and magnesium by the EDTA titrations were found to be imprecise and a gross overestimate, because of the poorly defined end-point. Spectrophotometric estimations were more accurate, although Likens et al. (1967) considers that it provides an under-estimate for calcium, and suggests a correction factor of 1.6. The range for magnesium was rather narrow, and for values above one ppm, the calibration was not linear.
Aluminium: spectrophotometrically using aluminon
Iron: o-phenanthroline method*
Sulphate: ion exchange columns (Mackereth 1963)
Chloride: titrimetric method using 0.01 N $HgNO_3$
Orthophosphate-phosphorus: $SnCl_2$-molybdate method
Nitrate-nitrogen: phenoldisulphonic acid method
Ammonium-nitrogen: nesslerisation of the distillate of the sample (after Swingle 1963)
Nitrite-nitrogen: sulphanilic acid method
Organic nitrogen: Kjeldahl digestion
Reactive silicate: yellow silicate-molybdate method (Golterman and Clymo 1969)

3.2.1 *Dissolved gases: oxygen and carbon dioxide*
(I. Ikusima, R. P. Lim, J. I. Furtado and O. Sato)

Dissolved oxygen was sampled at 0.25, 0.5 or 1 metre depth intervals from the surface to the bottom, and was determined by the unmodified Winkler method or with an oxygen meter (oxygen meter with thermistor: model 1520 with electrode model 1521, Electronic Instruments Co. U.K.). The measurements were made at 0900 and 1500 hours, or at dawn, around noon and at dusk, as the dissolved oxygen content of Tasek Bera has a daily periodic cycle with a maximum value at around 1500 hours and a minimum value just before the sunrise. In general, the dissolved oxygen content of Tasek Bera near Pos

* Calcium, magnesium, sodium, potassium and iron were determined by a Techtron Atomic Absorption Spectrophotometer, Model type AA^4, when this was made available.

Iskandar was low averaging 2.09 mg $O_2 \cdot l^{-1}$ but varied at different localities in the study area (Tables 13 and 14, and Figs. 31 and 32).

Dissolved oxygen was higher in the littoral *Lepironia* and *Utricularia* (Stations III, XI and XII) than the limnetic and sublittoral stations in both the forested and man-modified areas (Stations I, II and V). This was due to photosynthetic activity of the macrophytes and its attached flora (see later). Nevertheless, the low oxygen content is comparable to the Paraguayan swamps (Carter and Beadle 1930), and may be due to (1) low primary production as in the Amazon basin (Klinge and Ohle 1964, Marlier 1967, and Sioli 1967); (2) oxygen uptake by humic colouring matter undergoing oxidation (Hutchinson 1957, Lonnerblad 1930, and Birge and Juday quoted by Hutchinson 1957); and (3) bleaching reactions, presumably by ultraviolet radiation, resulting in photoxidation (Hutchinson 1957). The low oxygen may be also due to a high C.O.D. by suspended organic matter which was abundant (see later). Bacterial action is the unlikely cause of this oxygen since B.O.D. was low (see later), and since the population size and species diversity of bacterial heterotrophs was low (see later).

Seasonal trends of oxygen concentration were observed, a high concentration occurred in September-November (monsoon season) and a low concentration was measured in January-April and August (except in the year with high water-level). Carter (1955) observed an increase in oxygen with a rise in water level in the *Papyrus* swamp of Uganda. In Tasek Bera, the same occurred except during the great floods in the wet season because of (1) increased turbidity, (2) decrease in the amount of submerged macrophytes and the density of attached algae, and (3) lower radiation levels.

A distinct positive correlation between oxygen concentration and transparency was found throughout the year. The intensity of oxygen uptake may be accelerated by the increase of brown organic suspended materials at a time of low transparency.

Daily changes of concentration of dissolved oxygen are shown in Fig. 32. The dissolved oxygen increased gradually after dawn and reached its maximum around 1500 hours then gradually decreased. An oxygen increment over six hours, from 0900 to 1500 hours was significant on clear days in the *Lepironia* littoral and *Utricularia* sublittoral (Station III, IV, XI and XII). In comparison, the forested (Station V) and man-modified limnetic (Station I) and the *Pandanus* sublittoral (Station II) showed only slight changes in dissolved oxygen, because of poor benthic and planktonic flora, and poor attached flora and fauna. However, these daily fluctuations in oxygen concentration was insignificant at all stations during the January 1971 floods, and this trend was maintained until February.

Table 13 Average dissolved oxygen (p.p.m. and percentage saturation) at Stations I, II, III, IV and V at Tasek Bera from July 1970 to January 1972. (R. P. Lim and J. I. Furtado)

Date	Station I		Station II		Station III		Station IV		Station V		Averages		Minimum		Maximum	
	p.p.m.	% Sat.	p.p.m.	% Sat.	p.p.m.	% Sat.	p.p.m.	% Sat.	p.p.m.	% Sat.	p.p.m.	% Sat.	p.p.m.	% Sat.	p.p.m.	% Sat.
4 July '70	1.95	24.2	1.92	23.8	3.10	38.7	2.88	35.9	2.08	25.7	2.39	29.6	1.92	23.8	3.10	38.7
5 Aug. '70	2.16	26.7	1.73	21.4	4.06	50.8	1.89	23.7	2.03	25.0	2.37	29.5	1.73	21.4	4.06	50.8
4 Sept. '70	2.06	25.7	2.00	25.3	2.68	33.9	1.18	14.8	2.10	26.2	2.00	25.2	1.18	14.8	2.68	33.9
5 Oct. '70	1.65	13.3	1.75	21.9	2.08	26.0	2.56	32.2	1.54	19.1	1.92	22.5	1.54	13.3	2.56	32.2
4 Nov. '70	1.61	20.1	1.67	20.8	1.77	22.1	2.32	29.0	1.38	17.2	1.75	21.8	1.38	17.2	2.32	29.0
5 Dec. '70	1.88	23.5	1.91	23.8	2.27	28.3	2.82	35.0	1.54	19.1	2.08	25.9	1.54	19.1	2.82	35.0
10 Jan. '71	1.21	14.5	1.29	15.5	1.33	16.0	1.34	16.1	1.19	14.3	1.27	15.3	1.19	14.3	1.34	16.1
8 Feb. '71	0.99	12.0	1.02	12.4	1.21	14.6	1.38	16.8	1.01	12.2	1.12	13.6	0.99	12.0	1.38	16.8
7 Mar. '71	1.40	17.4	1.46	18.1	1.71	21.3	1.51	18.9	1.42	17.5	1.50	18.6	1.40	17.4	1.71	21.3
15 Apr. '71	1.03	12.9	1.13	14.2	1.24	15.7	0.84	10.8	1.37	17.2	1.12	14.2	0.84	10.8	1.37	17.2
10 May '71	2.10	26.1	2.10	26.1	2.23	27.8	2.09	26.2	2.08	25.8	2.12	26.4	2.08	25.8	2.23	27.8
6 June '71	1.41	17.8	1.52	19.2	1.83	23.1	1.32	16.8	1.64	20.6	1.54	19.5	1.32	16.8	1.83	23.1
3 July '71	1.95	24.4	2.12	26.6	2.33	29.3	2.00	25.4	2.33	29.2	2.15	27.0	1.95	24.4	2.33	29.3
2 Aug. '71	2.68	33.5	2.72	34.2	4.43	56.5	1.83	23.9	3.15	39.3	2.96	37.5	1.83	23.9	4.43	56.5
5 Sept. '71	2.36	29.5	2.60	32.4	2.89	36.4	2.06	26.1	1.19	14.8	2.22	27.8	1.19	14.8	2.89	36.4
4 Oct. '71	1.42	17.8	1.78	22.5	2.23	28.2	2.07	26.3	1.42	17.7	1.78	22.5	1.42	17.7	2.23	28.2
5 Nov. '71	1.80	22.4	1.79	22.3	2.41	30.1	2.84	35.3	1.45	17.9	2.06	25.6	1.45	17.9	2.84	35.3
5 Dec. '71	1.54	18.9	1.83	22.5	1.91	23.6	2.36	29.2	1.28	15.7	1.78	22.0	1.20	15.7	2.36	29.2
8 Jan. '72	1.87	22.9	1.76	21.5	2.08	25.4	2.53	31.1	1.44	17.5	1.94	23.7	1.44	17.5	2.53	31.1
Average	1.74	21.2	1.79	22.3	2.31	28.8	1.99	24.9	1.67	20.6	1.90	23.6	–	–	–	–
Minimum	0.99	12.0	1.02	12.4	1.21	14.6	0.84	10.8	1.01	12.2	–	–	0.84 (Apr. '71)	10.8 (Apr. '71)	–	–
Maximum	2.68	33.5	2.72	34.2	4.43	56.5	2.88	35.9	3.15	39.3	–	–			4.43 (Aug. '71)	56.5 (Aug. '71)

Table 14 Yearly mean, minimum and maximum dissolved oxygen values (mg·litre^{-1}) and percentage saturation of oxygen in parentheses at Stations I, V, XI and XII at Tasek Bera from September 1970 to August 1973. (I. Ikusima, M. Nagai, S. Kumano, T. Watanabe, O. Sato, Y. Tezuka and T. Koyama)

Station	Time	1970 (Sept.-Dec.)			1971 (Jan.-Dec.)			1972 (Jan.-Dec.)			1973 (Jan., July, Aug.)	1970–1973 Overall
		Mean	Min.	Max.	Mean	Min.	Max.	Mean	Min.	Max.	Mean	Mean
I	M[a]	1.60(20)	1.42(17)	1.98(24)	1.64(20)	0.95(13)	2.54(31)	1.28(16)	0.81(10)	1.92(24)	1.05(15)	1.46(18)
	N[a]	2.15(27)	1.95(24)	2.26(28)	1.93(22)	1.11(13)	2.80(35)	2.03(27)	0.86(11)	2.91(37)	2.17(34)	2.02(26)
	E[a]	2.53(32)	2.10(26)	2.84(36)	2.12(24)	1.46(18)	2.94(26)	2.48(33)	0.93(12)	3.72(47)	1.99(30)	2.30(29)
V	M	1.40(18)	1.30(16)	1.50(20)	1.69(17)	0.90(11)	2.60(31)	1.34(17)	0.80(10)	1.63(20)	0.88(12)	1.46(17)
	N	1.74(22)	1.62(20)	1.82(23)	1.88(23)	0.90(11)	3.05(37)	1.73(23)	1.16(15)	2.72(34)	1.39(21)	1.76(23)
	E	2.17(27)	1.75(22)	3.01(37)	1.87(23)	0.89(11)	3.25(40)	1.84(24)	0.97(13)	3.08(39)	1.44(21)	1.87(23)
XI	M	1.54(19)[b]			1.35(17)	0.76(9)	2.15(26)	1.40(18)	0.76(9)	2.95(38)	1.07(15)	1.36(17)
	N	4.16(53)[b]			2.88(41)	0.78(10)	5.31(69)	4.19(59)	1.87(22)	7.20(92)	2.94(48)	3.46(49)
	E	3.52(44)[b]			2.98(38)	0.65(8)	7.41(96)	5.67(81)	2.05(24)	9.02(119)	2.58(41)	4.00(55)
XIII	M				1.53(18)	1.03(13)	2.34(28)	1.21(15)	0.55(9)	1.89(24)	1.08(15)	1.35(16)
	N				3.53(45)	1.36(17)	6.89(92)	3.26(46)	2.14(28)	4.58(29)	3.14(50)	3.36(46)
	E				3.05(38)	1.54(20)	5.21(66)	3.55(50)	1.9(22)	5.03(66)	2.17(32)	3.17(43)

[a] M = early morning; N = around noon; E = evening just before the sunset.
[b] November and December, 1970.

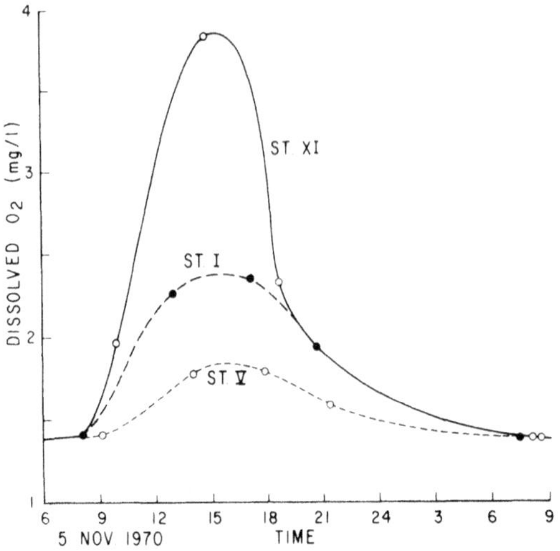

Fig. 31 Diurnal changes in dissolved oxygen at Stations I, V and XI. (I. Ikusima)

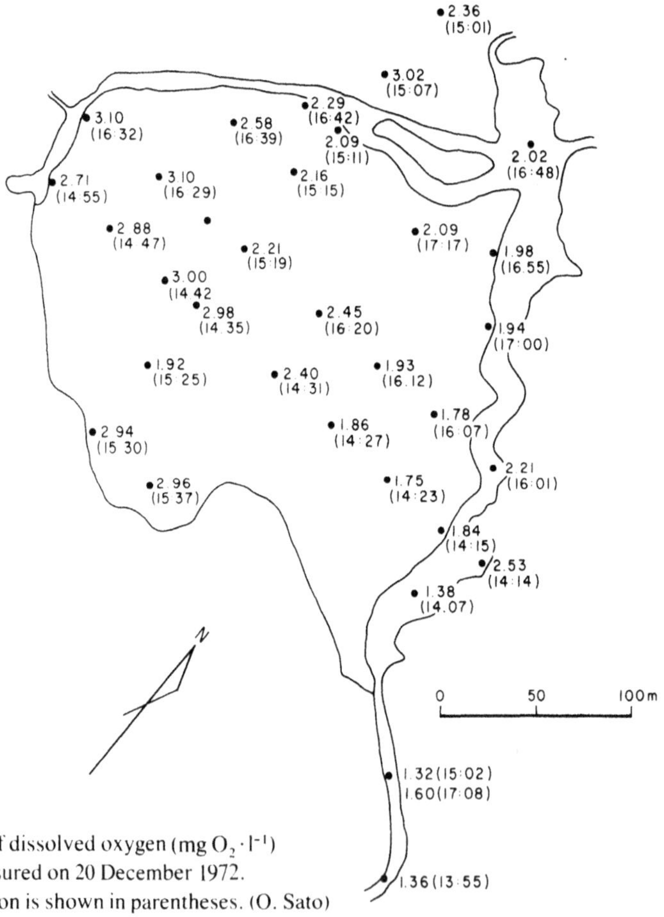

Fig. 32 Variation of dissolved oxygen (mg $O_2 \cdot l^{-1}$) in the study area measured on 20 December 1972. The time of observation is shown in parentheses. (O. Sato)

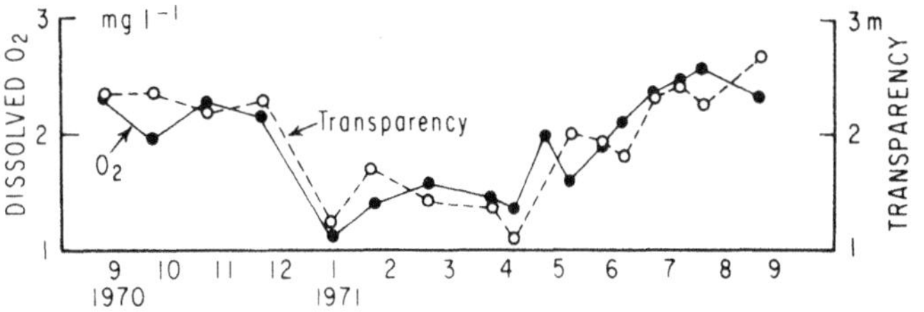

Fig. 33 Annual changes in dissolved oxygen (mg $O_2 \cdot l^{-1}$) and transparency (metres) at Station I. (I. Ikusima, S. Kumano and N. Nagai)

Over six hours, from 0900 to 1500 hours, changes in dissolved oxygen indicating overall net primary production of the aquatic ecosystem showed seasonal trends, with peak changes occurring in February-March and July-August (both dry seasons) and in October-November (monsoon season) (Fig. 33). On the basis of dissolved oxygen changes, primary productivity was greater in the littoral and sublittoral than in the limnetic, ranking decreasingly from Stations XI, XII, IV, III, II, I, V.

Carter (1955) observed an increase in oxygen with a rise in water level in the *Papyrus* swamp of Uganda. In Tasek Bera, the same occurs except for a great flood in the wet season because of (1) increased turbidity, (2) a decrease in the amount of submerged macrophytes and density of attached algae, and (3) lower radiation levels.

Total carbon dioxide measurements were made by the modified Koyama method (Koyama 1953). Free carbon dioxide measurements were determined titrimetrically against N/22 sodium carbonate using phenolphthalein as indicator. The maximum value of the total carbon dioxide in water was observed in the early morning and averaged 14.9 $mg \cdot l^{-1}$ (Table 15) at Stations I, V, XI and XII. The total carbon dioxide concentration during the January 1971 floods is shown in Table 16. There was no significant differences between the values at the time of the flood when compared normal conditions. These results differ from those of dissolved oxygen.

A total carbon dioxide decrease of 9.8 $mg \cdot l^{-1}$ occurred from morning to evening at Station XI between March and October 1972. This large decrease may be due to the uptake of CO_2 as a carbon source during photosynthetic activity of attached algae and macrophytes. However, it is replenished by respiratory activity during the dark period. Since the waters of Tasek Bera have low pH values, the proportion by weight of free carbon dioxide to the total carbon dioxide is high.

The concentration of free carbon dioxide falls within the range of the

Table 15 Total carbon dioxide and pH for Stations I, V, XI and XII at Tasek Bera, on 22 June 1972 at 0645 hours. (O. Sato)

Station	Depth (m)	pH	Total CO_2 ($mg \cdot l^{-1}$)
I	0	5.20	12.00
	1	5.20	13.49
	2	5.20	13.53
V	0	5.20	14.30
	1	5.20	23.30
XI	0	5.26	12.36
XII	0	5.20	15.07

Table 16 Total carbon dioxide and pH for Stations I and XI on 13 January 1971. (I. Ikusima)

Station	Time	Depth (m)	pH	Total CO_2 ($mg \cdot l^{-1}$)
I	0718	1	5.4	14.3
		3	5.4	14.8
	1515	1	5.4	14.3
		3	5.4	9.7
XI	0741	1	5.4	22.3
		2	5.4	8.0
	1553	1	5.4	11.9
		2	5.4	17.4

concentrations in the waters of the Ugandan swamps (Carter 1955). It is best developed in the limnetic and in the morning, and decreases over six hours from 0900 to 1500 hours when photosynthetic activity occurs. The amount of carbon dioxide as a source of carbon for photosynthesis by the primary producers may be sufficient in Tasek Bera.

A uniform water temperature from the surface to the bottom was recorded, with a low value from midnight up to early morning just before sunrise, causing the whole water column to mix. This resulted in the uniform distribution of dissolved oxygen from the surface to the bottom at all stations in the study area. The water was heated by radiation from the surface to the bottom during the day with concurrent production of oxygen throughout the water column as a photosynthetic by-product of algae and macrophytes. The highest oxygen concentration occurred near the surface, and decreased with depth during the day. An increase in oxygen near the bottom, however, occurred on fine days where the attached algae on the bottom mud, the filamentous algae and the benthic macrophytes grew densely. There was a lack of oxygen near the bottom or near zero oxygen concentrations at the water-mud interface, because of strong reducing processes in the bottom mud.

There was a close correlation between oxygen saturation and oxygen concentration (Table 14). Percentage saturation of oxygen is never very high except at Stations XI and XII, reaching a mean value of 26% over the four year survey. A maximum value of 119% was reached at Station XI in the afternoon on clear days in June, and 92% at Station XII in July, respectively. These high values were due to oxygen production through photosynthesis of submerged macrophytes and attached algae. A minimum value of 8% was observed during the floods of January 1971.

3.2.2 *pH and RpH*

(I. Ikusima, R. P. Lim and T. Mizuno)

The changes of pH with seasons, stations and depth were followed over a four-year period from 1970 to 1973. The pH measured with a glass electrode was always higher than that measured by the colorimetric method by 0.2 - 0.5 pH units. The hydrogen ion concentration at Station I is summarised in Fig. 34. The pH at Tasek Bera is low averaging 5.33 (minimum 4.57 and maximum 6.83) measured with a glass electrode, and 4.8 (minimum 4.6 and maximum

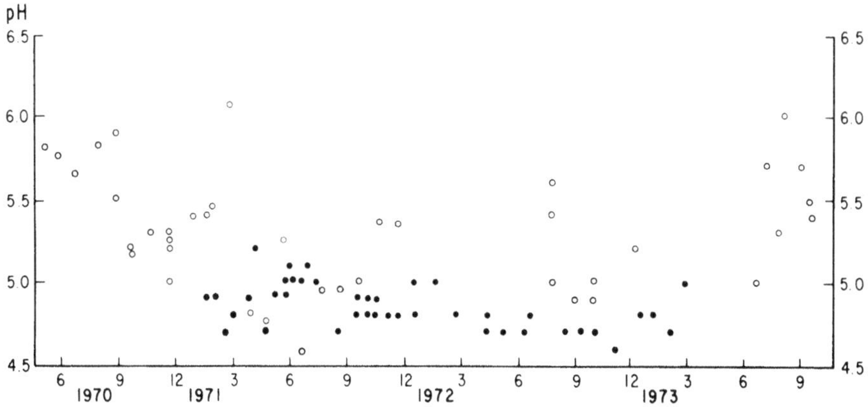

Fig. 34 Seasonal changes in pH measured at Station I. ○: glass electrode; ●: colorimeter. (1970–1973). (I. Ikusima, T. Mizuno, R. P. Lim, J. I. Furtado, O. Sato, S. Kumano, N. Mizuno and T. Watanabe)

5.2) by the colorimetric method. This corresponds to the blackwaters in the Amazon basin (Sioli 1967, 1968; Marlier 1967) and is slightly higher than that recorded by Johnson (1967a) for southern Malaya.

Changes in pH with stations are shown in Tables 17 and 18. There appears to be no significant variation of pH with depth, time in a day or between stations (see later). Consistently higher pH of about 6 was observed in the littoral stand near Stations XI and XIII, between 1300 and 1500 hours. This

Table 17 pH measured with a glass electrode, at Stations I, V, XI and XII, from Oct. 1972 to Jan. 1973. (O. Sato and I. Ikusima)

Stations		pH		
		Mean	Minimum	Maximum
I	Limnetic, open water	5.14	4.90	5.40
V	Limnetic, main channel in swamp-forest	5.06	4.95	5.21
XI	Littoral, *Lepironia* swamp	5.09	4.70	5.20
XII	Littoral, *Utricularia* stand	5.15	5.07	5.23

Table 18 Mean, minimum and maximum pH at Stations I-V in Tasek Bera, measured with a glass electrode (1970–1972). (R. P. Lim and J. I. Furtado)

Stations	pH		
	Mean	Minimum	Maximum
Limnetic, man-modified			
Station I	5.33	4.57	6.83
(Main channel)		(July 1971)	(Aug. 1970)
Sublittoral, man-modified			
Station II	5.30	4.48	6.37
(Pandanus-Utricularia)		(July 1971)	(Aug. 1970)
Station III	5.28	4.50	6.04
(Utricularia)		(July 1971)	(March 1971)
Littoral, man-modified			
Station IV	5.27	4.60	6.05
(Lepironia)		(July 1971)	(March 1971)
Limnetic, swamp-forest			
Station V	5.24	4.45	6.10
(Main channel)		(May 1971)	(March 1971)
Overall	5.28		

may be due to high photosynthetic activity of attached and filamentous algae.

In bicarbonate waters there is a relationship between pH and bicarbonates between pH 6.0 and 8.0 (Rhode 1949, Hutchinson 1957). Such a relationship does not exist at Tasek Bera because of the low pH which induces a shift in the HCO_3 equilibrium to CO_2. There was, however, a slight positive correlation between pH and alkalinity ($r = 0.62$; $p < 0.05$).

The pH at Tasek Bera is mainly affected by organic solutes derived from allochthonous and autochthonous sources.

Water samples for RpH measurements were bubbled with air for at least 30 minutes before measurements were made. The pH and RpH at Stations I, V and XI measured on 9 Sept. 1970, 5 Nov. 1970 and 10 Aug. 1972 are as follows:

pH			RpH		
Mean	Minimum	Maximum	Mean	Minimum	Maximum
5.27	4.90	5.60	6.40	5.80	6.80

This indicates that carbon dioxide is present quite abundantly in water and does not limit photosynthesis of aquatic plants.

3.2.3 *Alkalinity*
(R. P. Lim and J. I. Furtado)

The alkalinity values at Tasek Bera (Table 19) were low, averaging 3.07 ppm $CaCO_3$ (minimum 1.10 and maximum 1.18). It was fairly similar at Stations I-V, and those of 'blackwaters' in southern Peninsular Malaysia (Johnson 1967a), the Amazon (Sioli 1967, Marlier 1967) and the Paraguayan swamps (Carter and Beadle 1930). Alkalinity fluctuated seasonally (Fig. 35) with values increasing during the monsoons and reaching their peak at the beginning of the following dry season. This is probably due to decomposition of litter and its mineralisation, and to the concentration of salts through evaporation. The unusually high alkalinity in January 1971 (northeast monsoon) is presumably due to import of colloidal and decomposing matter from the watershed during the intense floods in this season.

Alkalinity is a measure of weak acid and weak acid salts present in the water, and its chemical mechanics is discussed by Hutchinson (1957). In neutral waters the system is basically a CO_2-HCO_3 equilibrium while, in acid

Table 19 Mean, minimum and maximum alkalinity at Stations I-V in Tasek Bera (1970–1972).

Stations	Alkalinity (ppm $CaCO_3$)		
	Mean	Minimum	Maximum
Limnetic, man-modified			
Station I	3.20	2.27	6.18
(Main channel)		(July 1971)	(Febr. 1971)
Sublittoral, man-modified			
Station II	3.10	2.20	6.10
(Pandanus-Utricularia)		(Sept. 1971)	(Febr. 1971)
Station III	3.00	1.10	5.85
(Utricularia)		(Jan. 1972)	(Febr. 1971)
Littoral, man-modified			
Station IV	3.07	2.00	5.65
(Lepironia)		(Jan. 1972)	(Febr. 1971)
Limnetic, swamp-forest			
Station V	2.96	2.10	5.80
(Main channel)		(Sept. 1971)	(Febr. 1971)
Overall	3.07	1.10	6.18
		(Jan. 1972)	(Febr. 1971)

waters below pH 5, only total free CO_2 is of any quantitative importance. Although alkalinity has often been expressed in meq/l or ppm $CaCO_3$, Johnson (1967a) has found no relationship between calcium and alkalinity. This is true also for the Tasek Bera waters which are very soft, following the classification of Brookes and Deevey (1963). Variations in alkalinity may be due to flora and fauna incorporating bicarbonates in their tissues, to community respiration, chemical oxidation, mineral turnover at night and CO_2 uptake for photosynthesis in the day. Diel and daily changes in alkalinity are discussed later.

3.2.4 *Conductivity*
(R. P. Lim and J. I. Furtado)

The conductivity of the water at Stations I-V was low, averaging 14.2 μmhos$\cdot$cm^{-1} (minimum 10.5 and maximum 23.0). There was no marked difference in conductivity measurements between the various stations (Table 20). Nevertheless, these measurements were low, and comparable with the 'blackwaters' of the Amazon basin (Sioli 1968) and Peninsular Malaysia

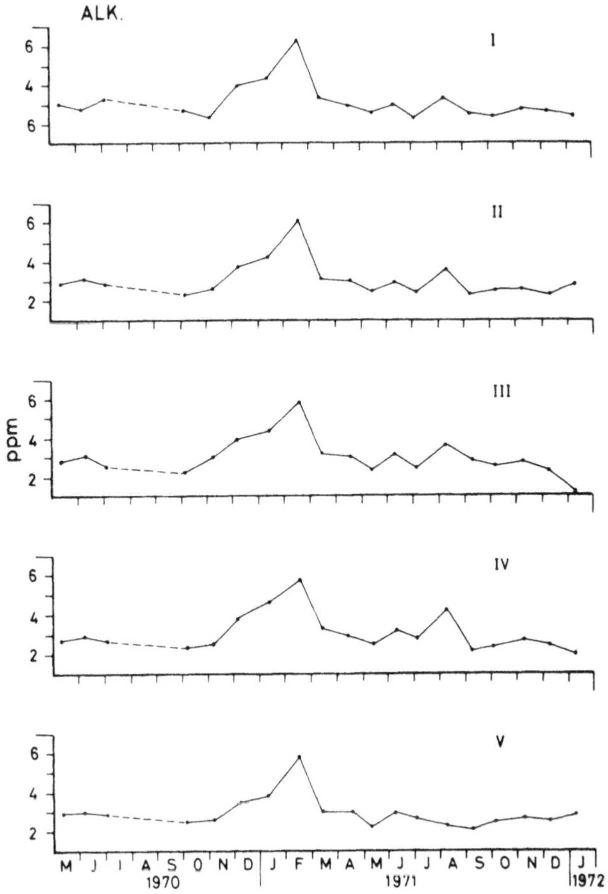

Fig. 35 Alkalinity (ppm $CaCO_3$) at Stations I-V at Tasek Bera (1970–1972). (R. P. Lim and J. I. Furtado)

(Johnson 1967a; Jothy 1968). They are much lower than other waters in Peninsular Malaysia (Ho 1970, Arumugam 1971) and than other tropical waters (Talling and Talling 1965, Carter 1955 and Carter and Beadle 1930). This conductivity indicates an overall poverty of nutrients in this swamp.

Fluctuations in conductivity appeared to be related to the onset of monsoon and dry seasons (Fig. 36). The conductivity was high in September-October and December-January (northeast monsoon), April-May (southeast monsoon) and June-August (dry season). This is probably due to a number of phenomena, particularly the influx of nutrients from the watershed during heavy rains in the monsoon season floods causing reverse flow and nutrient influx from downstream (December-January), to the decomposition and

Table 20 Mean, minimum and maximum conductivity at Stations I-V in Tasek Bera (1970–1972). (R. P. Lim and J. I. Furtado)

Stations	Conductivity (μmhos·cm⁻¹)		
	Mean	Minimum	Maximum
Limnetic, man-modified			
Station I	14.3	10.7	19.6
(Main channel)		(May 1970)	(May 1971)
Sublittoral, man-modified			
Station II	13.9	10.7	17.6
(Pandanus-Utricularia)		(May 1970)	(May 1971)
Station III	14.3	10.5	23.0
(Utricularia)		(May 1970)	(Jan. 1972)
Littoral, man-modified			
Station IV	14.4	10.8	22.5
(Lepironia)		(May 1970)	(Aug. 1971)
Limnetic, swamp-forest			
Station V	14.1	10.8	19.0
(Main-channel)		(May 1970)	(Aug. 1971)
Overall	14.2	10.5	23.0
		(May 1970)	(Jan. 1972)

mineralisation of allochthonous organic matter in the dry season, and to the concentration of nutrients by evaporation in the drier months.

Conductivity is a measure of the total content of electrolytes present, and is taken usually as an index of total dissolved salts. Rhode (1949) observed a direct correlation between conductivity and the different major constituents namely sodium, potassium, calcium, magnesium, sulphate, chlorides and bicarbonates for most of the bicarbonate lakes. Diel and daily changes in conductivity are discussed later.

3.2.5 *Cations*
(R. P. Lim and J. I. Furtado)

3.2.5.1 *Calcium and magnesium.* Disregarding values obtained titrimetrically, the calcium content of Tasek Bera was low at Stations I-V averaging 0.41 ppm with a maximum of 0.93 ppm and a minimum of 0.12 ppm (Fig. 37; Table 21). Under Williams's (1964) modification of Ohle's (1934) classifi-

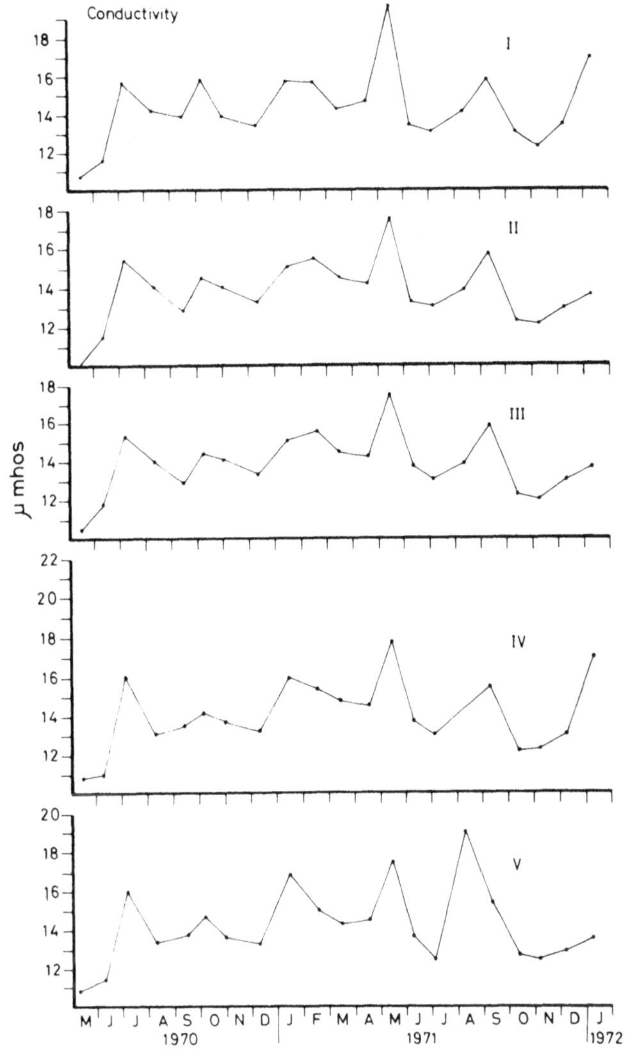

Fig. 36 Conductivity (µmhos) at Stations I-V in Tasek Bera (1970–1972). (R. P. Lim and J. I. Furtado)

cation, the values for calcium at Tasek Bera are extremely poor. This poverty is reflected by the scarcity of 'shell' animals. Only one species of mollusc, *Pila ampullacea* (L.), two species of shrimps and a few species of other crustaceans occur, being limited to the sub-littoral region. The primary producers thus may supply the calcium required by these consumers. The trophic cycle of calcium is being studied (Mizuno and Yoshimi, Section 3.5.3).

Table 21 Mean, minimum and maximum calcium concentrations at Stations I-V in Tasek Bera (1970–1972). (R. P. Lim and J. I. Furtado)

Stations	Calcium (ppm)		
	Mean	Minimum	Maximum
Limnetic, man-modified			
Station I	0.40	0.17	0.88
(Main channel)		(Apr. 1971)	(Dec. 1970)
Sublittoral, man-modified			
Station II	0.43	0.15	0.93
(Pandanus-Utricularia)		(May 1971)	(Dec. 1970)
Station III	0.40	0.17	0.74
(Utricularia)		(Apr. 1971)	(Dec. 1970)
Littoral, man-modified			
Station IV	0.42	0.14	0.88
(Lepironia)		(Apr. 1971)	(Dec. 1970)
Limnetic, swamp-forest			
Station V	0.41	0.12	0.74
(Main channel)		(May 1971)	(Aug. 1971)
Overall	0.41	0.12	0.93
		(May 1971)	(Dec. 1970)

The magnesium content at all stations was also low, averaging 0.28 ppm with a maximum of 0.38 ppm and a minimum of 0.18 ppm (Fig. 38; Table 22). The concentration of magnesium was poorer than calcium in the swamp. Although Johnson (1967) finds magnesium concentrations exceeding calcium in Malayan blackwaters, Tasek Bera shows the obverse pattern which agrees with the values obtained for the Amazon basin (Sioli 1967, 1968). The Tasek Bera pattern could be due to the lack of significant selective adsorption of calcium to organic matter, as demonstrated in humic waters by Ohle (1955), Klinge and Ohle (1964) and Johnson (1967). The paucity of calcium and magnesium in Tasek Bera is similar to that of the Amazon basin (Sioli 1967, 1968; Klinge and Ohle 1964), and is lower than the levels recorded by Johnson (1967) except at Batu Berendam, Malacca (Locality 8) which approaches that of Tasek Bera.

The calcium and magnesium concentrations in Tasek Bera showed somewhat seasonal fluctuations (Figs. 37 and 38). Concentrations increased during the dry seasons, being maximal at the end of the period. They then decreased with the onset of the monsoon but again increased slightly during the mon-

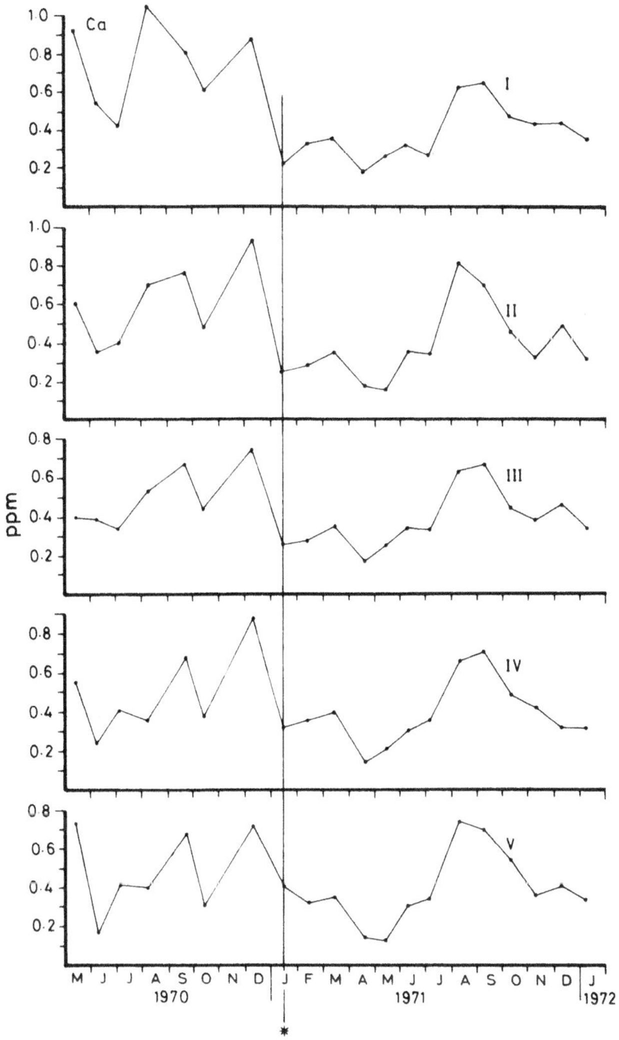

Fig. 37 Calcium concentrations (ppm) at Stations I-V in Tasek Bera (1970–1972). (N.B.: * denotes change in technique from EDTA method to A.A. Spectrophotometer method.) (R. P. Lim and J. I. Furtado)

soon season, and then declined at the end of the monsoon and the onset of the dry season. This suggests a complex interaction of phenomena particularly the release of these minerals by decomposing organic matter in the dry season, their influx in the monsoons from the watershed, their efflux with flood discharge and their use by organisms during population growth.

Table 22 Mean, minimum and maximum magnesium concentrations at Stations I-V in Tasek Bera (1970–1972). (R. P. Lim and J. I. Furtado)

Stations	Magnesium (ppm)		
	Mean	Minimum	Maximum
Limnetic, man-modified			
Station I	0.29	0.18	0.38
(Main channel)		(June 1971)	(Febr. 1971)
Sublittoral, man-modified			
Station II	0.28	0.18	0.37
(Pandanus-Utricularia)		(June 1971)	(Febr. 1971)
Station III	0.28	0.19	0.36
(Utricularia)		(June 1971)	(Febr. 1971)
Littoral, man-modified			
Station IV	0.29	0.19	0.37
(Lepironia)		(June 1971)	(Febr. 1971)
Limnetic, swamp-forest			
Station V	0.29	0.18	0.37
(Main channel)		(June 1971)	(Febr. 1971)
Overall	0.28	0.18	0.38
		(June 1971)	(Febr. 1971)

3.2.5.3 *Sodium and potassium.* The sodium concentration in Tasek Bera averaged 1.10 ppm with a maximum of 4.23 ppm and a minimum of 0.51 ppm (Table 23). The potassium concentration at Tasek Bera averaged at 0.50 ppm with a maximum of 1.23 ppm and a minimum of 0.19 ppm (Table 24). The concentrations of both these monovalent cations were somewhat similar except that the maximum for sodium was significantly higher at Station I probably due to human contamination, since this limnetic region is intensely used by the Semalai aborigines (Tables 23 and 24).

These two monovalent cations fluctuated somewhat seasonally at Stations I-V in Tasek Bera (Figs. 39 and 40). Sodium concentrations were high in June-August (dry season) and October-November (northeast monsoon) being maximal in the dry season; whereas potassium concentrations were high in August-September (dry season) and November-January (northeast monsoon), and extending to April-May (southwest monsoon), being a maximal in the major monsoon. Thus, sodium and potassium showed an inverse relationship with concentrations of potassium increasing and high when those of sodium decrease and were low. This may be attributed to the greater mobility

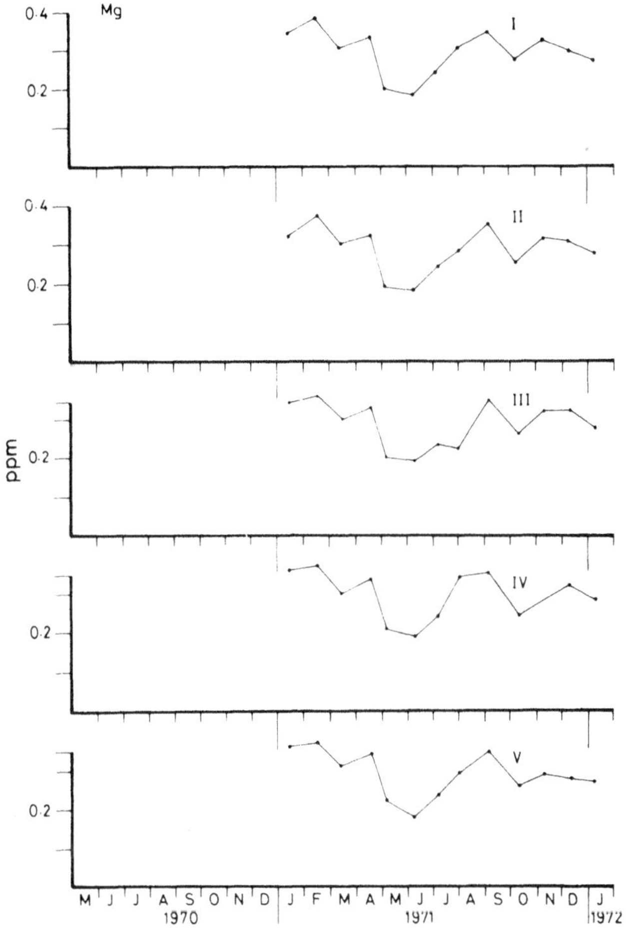

Fig. 38 Magnesium concentrations (ppm) at Stations I-V in Tasek Bera (1970–1972). (N.B.: * denotes change in technique from EDTA method to A.A. Spectrophotometer method.) (R. P. Lim and J. I. Furtado)

of potassium in comparison with sodium, so that potassium was highest in the northeast monsoon due to nutrient influx from the watershed. The peak abundance of sodium in dry season was due probably more to the decomposition of organic matter and mineralisation.

Sodium was normally the dominant cation in the Tasek Bera soft waters, although its absolute concentrations were still very low. This agrees with the observations of Johnson (1967), Ho (1970) and Arumugam (1971) for Malaysian waters, of Kobayashi (1959) for Thai waters, of Klinge and Ohle (1964) and Sioli (1967) for Amazonian waters and of Talling and Talling (1965) for African lakes.

Table 23 Mean, minimum and maximum sodium concentrations at Stations I-V in Tasek Bera (1970–1972). (R. P. Lim and J. I. Furtado)

Stations	Sodium (ppm)		
	Mean	Minimum	Maximum
Limnetic, man-modified			
Station I	1.24	0.52	4.23
(Main channel)		(Jan. 1971)	(Aug. 1970)
Sublittoral, man-modified			
Station II	1.06	0.51	2.21
(*Pandanus-Utricularia*)		(Jan. 1971)	(July 1970)
Station III	1.08	0.52	2.04
(*Utricularia*)		(Jan. 1971)	(July 1970)
Littoral, man-modified			
Station IV	1.09	0.56	2.15
(*Lepironia*)		(June 1971)	(July 1970)
Limnetic, swamp-forest			
Station V	1.04	0.54	2.14
(Main-channel)		(June 1971)	(July 1970)
Overall	1.10	0.51	4.23
		(Jan. 1971)	(Aug. 1970)

Table 24 Mean, minimum and maximum potassium concentrations at Stations I-V in Tasek Bera (1970–1972). (R. P. Lim and J. I. Furtado)

Stations	Potassium (ppm)		
	Mean	Minimum	Maximum
Limnetic, man-modified			
Station I	0.55	0.19	1.22
(Main channel)		(June 1971)	(Jan. 1971)
Sublittoral, man-modified			
Station II	0.51	0.20	1.23
(*Pandanus-Utricularia*)		(June 1971)	(Jan. 1971)
Station III	0.51	0.23	1.18
(*Utricularia*)		(June 1971)	(Jan. 1971)
Littoral, man-modified			
Station IV	0.49	0.20	1.20
(*Lepironia*)		(June 1971)	(Jan. 1971)
Limnetic, swamp-forest			
Station V	0.47	0.20	1.22
(Main-channel)		(June 1971)	(Jan. 1971)
Overall	0.50	0.19	1.23
		(June 1971)	(Jan. 1971)

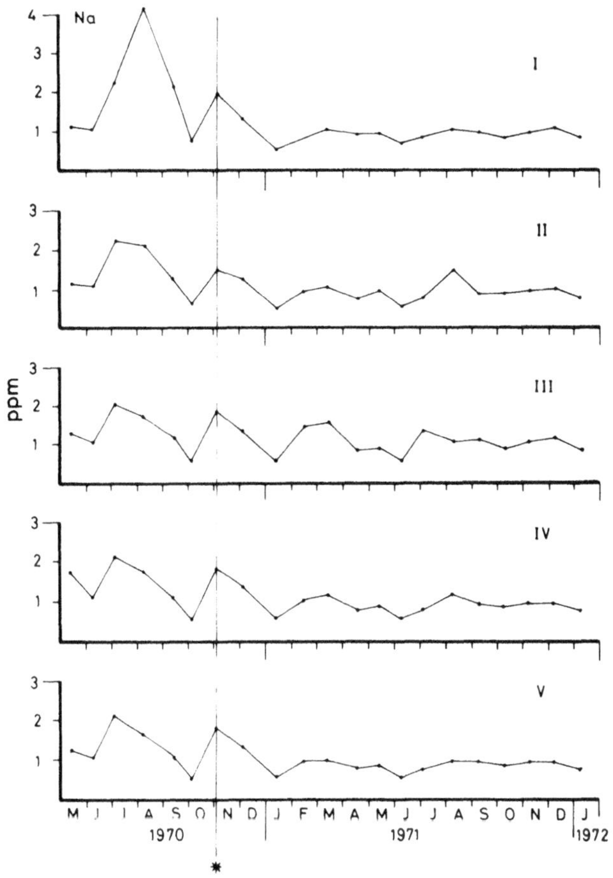

Fig. 39 Sodium concentrations (ppm) at Stations I-V in Tasek Bera (1970–1972). (N.B.: * denotes change in technique from flame photometer to A.A. Spectrophotometer.) (R. P. Lim and J. I. Furtado)

3.2.5.3 *Pearsall's ratio: monovalent/divalent ions.* In analysing 400 different waters, Pearsall (1922) derived a chemical ratio (Na + K)/(Ca + Mg) to classify water bodies chemically, and in relation to their phytoplankton. This ratio is of limited value since it does not take into consideration other chemical, physical and biological characteristics of a water body, which contribute to its trophicity. Pearsall's ratio for Tasek Bera ranged between 1.56 and 3.60 with an average of 2.31 (Table 25). Zafar (1959) adopts Pearsall's ratio as the primary basis for classifying waters into eutrophic, mesotrophic and oligotrophic types. However, this classification is not very useful in the humid tropics (Johnson 1967), especially for complex swamp systems.

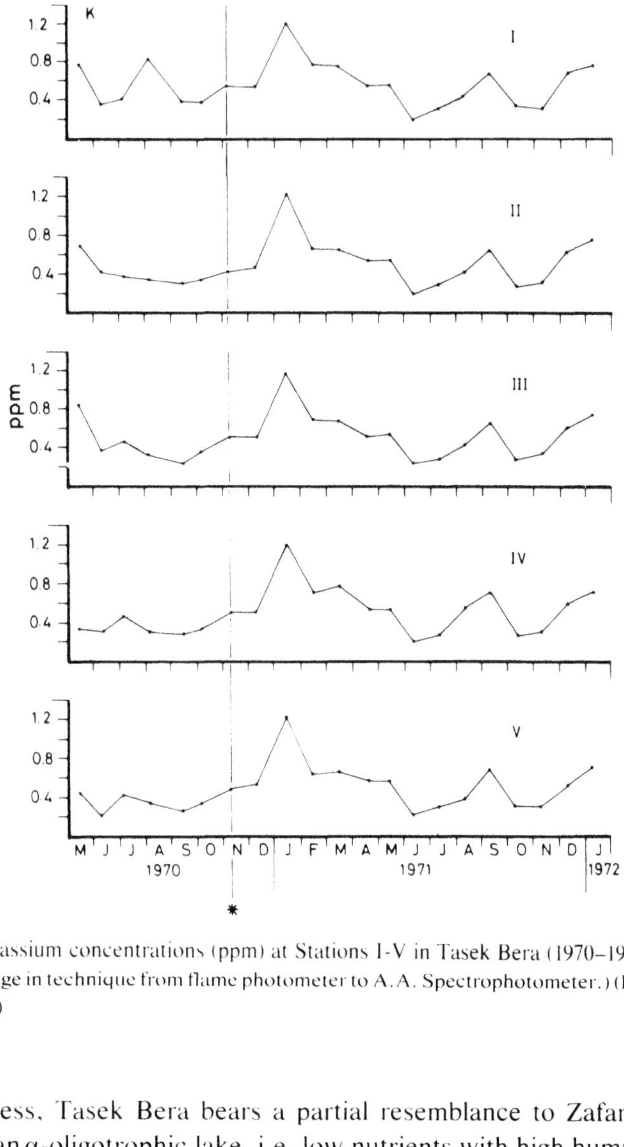

Fig. 40 Potassium concentrations (ppm) at Stations I-V in Tasek Bera (1970–1972). (N.B.: * denotes change in technique from flame photometer to A.A. Spectrophotometer.) (R. P. Lim and J. I. Furtado)

Nevertheless, Tasek Bera bears a partial resemblance to Zafar's classification of an α-oligotrophic lake, i.e. low nutrients with high humus and low production.

3.2.5.4 *Iron and aluminium.* Fluctuations in iron concentrations in Tasek Bera were relatively large with a mean of 0.64 ppm, a maximum of 1.95 ppm and a minimum of 0.18 ppm (Table 26; Fig. 41). These variations were similar at Stations I-V with the forested limnetic (Station V) showing a slightly wider range (Table 26). The mean, maximum and minimum concentrations for

Table 25 Average concentrations of major cations and Pearsall's ratio at Tasek Bera, Stations I-V (1970–1972). (R. P. Lim and J. I. Furtado)

Date	Ca	Mg	Na	K	Pearsall's ratio
11 Jan. 1971	0.29	0.34	0.54	1.21	2.78
14 Febr. 1971	0.31	0.34	0.91	0.69	2.46
12 Mar. 1971	0.36	0.30	1.05	0.70	2.65
17 Apr. 1971	0.16	0.33	0.82	0.54	2.78
12 May 1971	0.20	0.20	0.89	0.55	3.60
7 June 1971	0.32	0.18	0.58	0.20	1.56
4 July 1971	0.32	0.24	0.81	0.29	2.20
9 Aug. 1971	0.69	0.29	1.04	0.44	1.61
8 Sept. 1971	0.67	0.35	0.96	0.68	1.61
6 Oct. 1971	0.47	0.26	0.86	0.29	1.58
7 Nov. 1971	0.40	0.30	0.96	0.31	1.81
7 Dec. 1971	0.42	0.30	1.02	0.60	2.75
8 Jan. 1972	0.32	0.27	0.78	0.74	2.58
Average	0.38	0.29	0.86	0.56	2.31

Table 26 Mean, minimum and maximum iron concentrations at Stations I-V in Tasek Bera (1970–1972). (R. P. Lim and J. I. Furtado)

Stations	Iron (ppm)		
	Mean	Minimum	Maximum
Limnetic, man-modified			
Station I (Main channel)	0.63	0.15 (Oct. 1970)	1.45 (Febr. 1971)
Sublittoral, man-modified			
Station II (*Pandanus-Utricularia*)	0.64	0.12 (Oct. 1970)	1.55 (Febr. 1971)
Station III (*Utricularia*)	0.64	0.103 (Oct. 1970)	1.61 (Febr. 1971)
Littoral, man-modified			
Station IV (*Lepironia*)	0.62	0.19 (Dec. 1971)	1.56 (Febr. 1971)
Limnetic, swamp-forest			
Station V (Main channel)	0.68	0.15 (Dec. 1970)	1.95 (Febr. 1971)
Overall	0.64	0.103 (Oct. 1970)	1.95 (Febr. 1971)

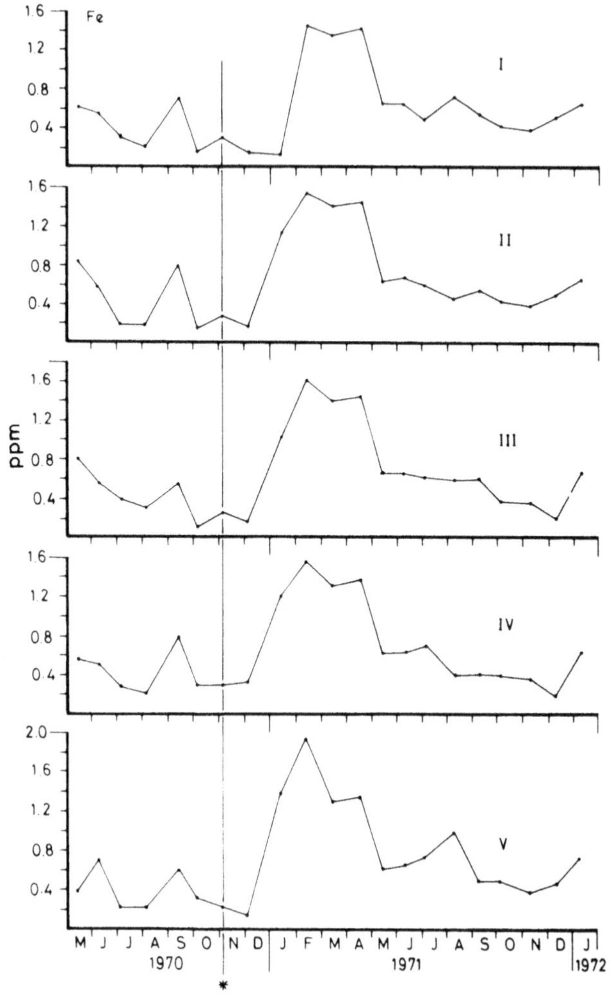

Fig. 41 Total iron concentration (ppm) at Stations I-V in Tasek Bera (1970–1972). (N.B.: * denotes change in technique from o-phenanthroline to A.A. Spectrophotometer method.) (R. P. Lim and J. I. Furtado)

aluminium at Stations I-V were 0.197, 0.370 ppm and 0.045 ppm (Table 27). These fluctuations were not as large as in iron.

Variations in iron concentrations were principally governed by rainfall with the highest values occurring in the dry months following heavy rains (January-April and July-August) (Table 26; Fig. 41). This was probably due to heavy northeast monsoonal rains enhancing the leaching of colloidal and precipitated iron from the soil and the leaf litter on the forest floor. The

Table 27 Mean, minimum and maximum aluminium concentrations at Stations I-V in Tasek Bera (1970–1972). (R. P. Lim and J. I. Furtado)

Stations	Aluminium (ppm)		
	Mean	Minimum	Maximum
Limnetic, man-modified			
Station I	0.189	0.045	0.357
(Main channel)		(Dec. 1970)	(Oct. 1971)
Sublittoral, man-modified			
Station II	0.190	0.050	0.370
(Pandanus-Utricularia)		(Febr. 1971)	(Oct. 1971)
Station III	0.195	0.051	0.370
(Utricularia)		(Dec. 1970)	(Oct. 1971)
Littoral, man-modified			
Station IV	0.206	0.051	0.370
(Lepironia)		(Dec. 1970)	(Oct. 1971)
Limnetic, swamp-forest			
Station V	0.206	0.058	0.370
(Main channel)		(Dec. 1970)	(Oct. 1971)
Overall	0.197	0.045	0.370
		(Dec. 1970)	(Oct. 1971)

converse happens in the dry season when the mobility of iron and aluminium is reduced (Richards 1957, Sioli 1968) tending to remain bound as sesquioxides in the lateritic soil. High iron values in the February to April dry season were due also to decomposition exceeding all other processes and resulting in partial or total anaerobic conditions. In these conditions, iron would be reduced to the soluble ferrous state, and this has been observed in the occurrence of such heavily laden water especially in the littoral with a film of iron bacteria present near the landing point. Iron could persist in this ferrous state if humic acids were capable of peptising and holding it in solution under conditions where it would normally precipitate (Shapiro 1966). This could account for the greater values obtained for iron in comparison to aluminium. The significant correlation ($r = 0.658$; $p < 0.05$) between pH and iron content further supported this.

Aluminium fluctuations were somewhat asynchronous with those of iron (Fig. 42). An increase in aluminium occurred in the monsoon seasons especially the initial months (April-May and September-January) with a maximum in September-October of 0.370 ppm. This was probably due to leaching from the watershed and influx into the swamp in the monsoons. The poor alumi-

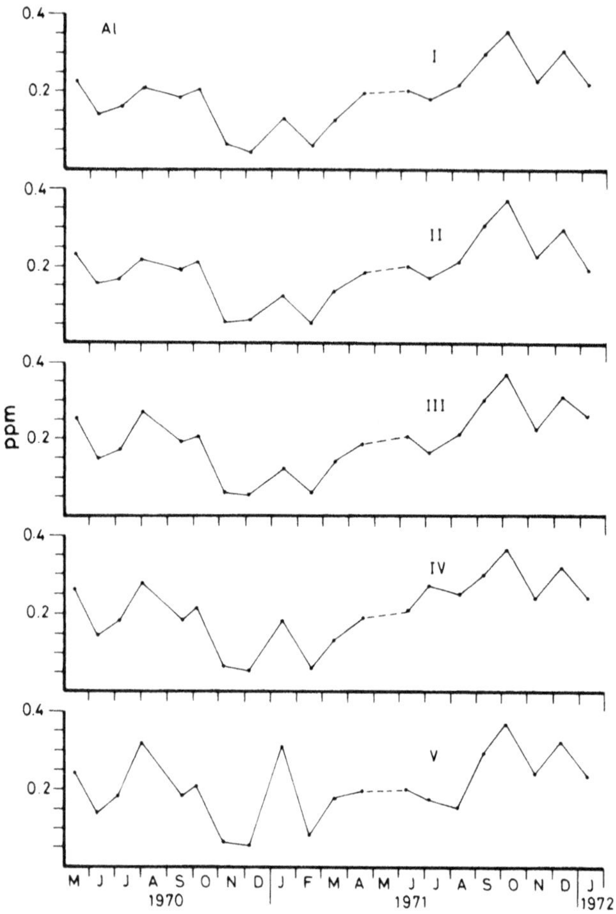

Fig. 42 Aluminium concentration (ppm) at Stations I-V in Tasek Bera (1970–1972). (R. P. Lim and J. I. Furtado)

nium content of the latosol soil of the watershed together with the large volume of drainage water in the monsoon season probably caused a dilution to follow the initial enrichment. Although the primary source of iron and aluminium is the lateritic soil, the asynchrony of their concentrations in Tasek Bera suggests that iron originates from a secondary source such as the forest litter, whereas aluminium originates directly from the parent soil.

3.2.5.5 *Silica.* The mean, maximum and minimum silica content at Tasek Bera was 3.15, 7.80 ppm and 1.10 ppm respectively (Table 28). Silica determined and expressed as silicon dioxide was the highest solute at Tasek Bera, partly because dissolved silica was dependent upon its mobility in the sur-

Table 28 Mean, minimum and maximum silica concentrations at Stations I-V in Tasek Bera (1970–1972). (R. P. Lim and J. I. Furtado)

Stations	Silica (ppm)		
	Mean	Minimum	Maximum
Limnetic, man-modified			
Station I	3.21	2.10	4.64
(Main channel)		(Oct. 1970)	(Febr. 1971)
Sublittoral, man-modified			
Station II	3.12	1.81	4.60
(Pandanus-Utricularia)		(Nov. 1971)	(July 1971)
Station III	3.22	1.10	7.80
(Utricularia)		(Jan. 1972)	(Aug. 1971)
Littoral, man-modified			
Station IV	3.03	1.50	5.21
(Lepironia)		(Nov. 1971)	(Sept. 1971)
Limnetic, swamp-forest			
Station V	3.17	1.81	5.53
(Main-channel)		(Nov. 1971)	(July 1971)
Overall	3.15	1.10	7.80
		(Jan. 1972)	(Aug. 1971)

rounding soils (Kobayashi 1960, Bishop 1971). This silica value was similar at Stations I-V with a wider range at Station III, and resembles the limnetic waters in Peninsular Malaysia (Arumugam 1971) and Ugandan swamps (Carter 1955), but is higher than the values for the 'blackwaters' of the Amazon basin (Sioli 1968).

The silica concentration fluctuated seasonally (Fig. 43). Silica concentrations rose slightly during the monsoons due to leaching from watershed soil by the increased runoff, and were maximal in the dry seasons (January-February and July-August) due to mineralisation by decomposing diatoms and plants.

3.2.6 *Anions*
(R. P. Lim and J. I. Furtado)

3.2.6.1 *Chlorides and sulphates.* The mean, maximum and minimum concentrations of chlorides and sulphates were 1.93, 4.32 ppm and 0.57 ppm

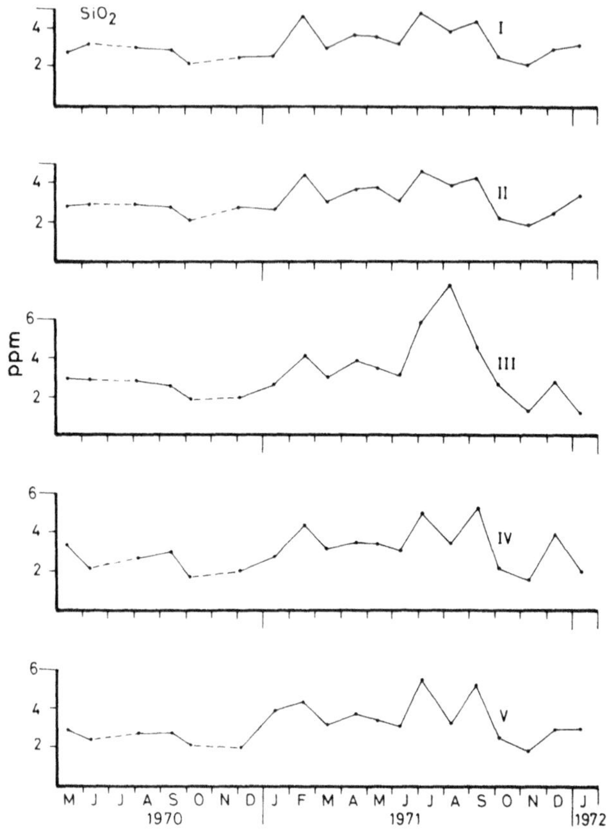

Fig. 43 Silica concentration (SiO_2 mg/l) at Stations I-V in Tasek Bera (1970–1972). (R. P. Lim and J. I. Furtado)

(Table 29), and 319, 5.59 ppm and 0.96 ppm (Table 30), respectively. The chloride and sulphate values were similar at all stations and rather low compared to most 'blackwaters' of southern Peninsular Malaysia (Johnson 1967) except for localities 6 (a ditch, mile 17 Jurong Road, Singapore), 7 (S. Gembol, Sedili Road, Johore) and 8 (5/4 Bachang-Batu Berendam Road, Malacca) for chlorides. However, these values are similar to those of the Amazon basin (Sioli 1968). The sulphate content was generally slightly higher than chloride, and conforms with the observations of Johnson (1967) and Talling and Talling (1965). The low value for sulphate could be due to the small amounts commonly present in soils (Beauchamp 1953), although there is a need to determine the sulphate content of the soils of the watershed of Tasek Bera.

Table 29 Mean, minimum and maximum chloride concentration at Stations I-V in Tasek Bera (1970–1971). (R. P. Lim and J. I. Furtado)

Stations	Chloride (ppm)		
	Mean	Minimum	Maximum
Limnetic, man-modified			
Station I	0.02	0.57	4.32
(Main channel)		(Nov. 1970)	(Dec. 1970)
Sublittoral, man-modified			
Station II	1.87	0.95	3.99
(Pandanus-Utricularia)		(Nov. 1970)	(Apr. 1971)
Station III	1.78	0.64	3.92
(Utricularia)		(Sept. 1970)	(Apr. 1971)
Littoral, man-modified			
Station IV	1.88	0.71	3.94
(Lepironia)		(Sept. 1970)	(Mar. 1971)
Limnetic, swamp-forest			
Station V	1.97	1.08	4.12
(Main channel)		(May 1971)	(Apr. 1971)
Overall	1.93	0.57	4.32
		(Nov. 1970)	(Dec. 1970)

Table 30 Mean, minimum and maximum sulphate concentration at Stations I-V in Tasek Bera (1970–1972). (R. P. Lim and J. I. Furtado)

Stations	Sulphate (ppm)		
	Mean	Minimum	Maximum
Limnetic, man-modified			
Station I	3.10	1.87	4.48
(Main channel)		(June 1971)	(Sept. 1970)
Sublittoral, man-modified			
Station II	3.05	1.79	4.87
(Pandanus-Utricularia)		(June 1971)	(Oct. 1970)
Station III	3.05	1.92	4.80
(Utricularia)		(July 1971)	(May, Oct. 1970)
Littoral, man-modified			
Station IV	3.33	0.96	5.59
(Lepironia)		(June 1971)	(Oct. 1970)
Limnetic, swamp-forest			
Station V	3.41	1.20	5.28
(Main channel)		(July 1971)	(Oct. 1971)
Overall	3.19	0.96	5.59
		(June 1971)	(Oct. 1970)

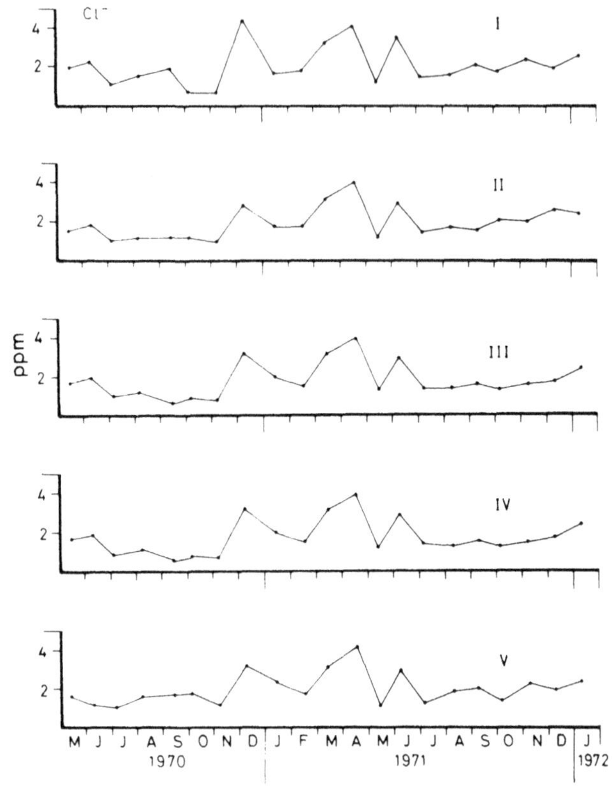

Fig. 44 Chloride concentrations (ppm) at Stations I-V in Tasek Bera (1970–1972). (R. P. Lim and J. I. Furtado)

Chlorides and sulphates fluctuated somewhat seasonally (Figs. 44 and 45). Chlorides were abundant during the monsoons (November-January and April-May) and the beginning of the dry season (March and June), being maximal in the intermonsoonal period (March-April). This suggests an input of chlorides from the watershed due to chloride mineralisation during organic decomposition in the dry season. Sulphates were abundant in the monsoons (September-October and May) and the dry seasons (July-August and February-March) being maximal at the beginning (September-October) of the northeast monsoon (Fig. 45). This again suggests a sulphate input from the watershed during the monsoon as well as the release of sulphur during decomposition, probably in the sulphide state under anaerobic conditions as evidenced by negative correlation with dissolved oxygen ($r = 0.67$; $p < 0.05$).

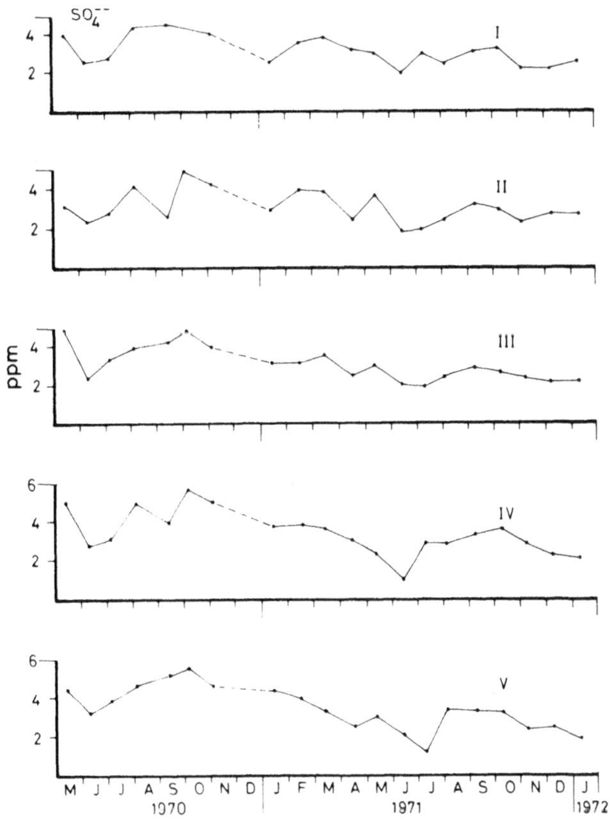

Fig. 45 Sulphate concentrations (ppm) at Stations I-V in Tasek Bera (1970–1972). (R. P. Lim and J. I. Furtado)

3.2.6.2 *Phosphates.* The mean, minimum and maximum phosphate-phosphorus concentrations in Tasek Bera are 0.0240, 0.000 and 0.1050 ppm respectively (Table 31). It is fairly similar at all stations with the greatest range in the littoral (Station IV), and is low being similar to tropical Malaysian waters (Ho 1970; Bishop 1971; Kobayashi 1960) and Amazon blackwaters (Sioli 1968; Klinge and Ohle 1964). This poverty is related to the poor availability of phosphorus from the ultisol soil (Bishop 1971). Although the phosphate content lies within the range of some tropical lakes (Ho 1970; Talling 1966) it does not support as much plankton as would be expected, suggesting that other factors are limiting production. Phosphorus poor soils are known to support the highest productivity of any forest (Richards 1957) because of an efficient closed recycling system. It is probable that a similar system operates in Tasek Bera, particularly with the swamp-forest.

Table 31 Mean, minimum and maximum phosphate-phosphorus concentration at Stations I-V in Tasek Bera (1970–1972). (R. P. Lim and J. I. Furtado)

Stations	Phosphate-phosphorus (ppm)		
	Mean	Minimum	Maximum
Limnetic, man-modified			
Station I (Main channel)	0.0213	0.0000 (Sept. 1970)	0.0650 (May 1971)
Sublittoral, man-modified			
Station II (*Pandanus-Utricularia*)	0.0212	0.0000 (Oct.-Nov. 1970)	0.0680 (Sept. 1971)
Station III (*Utricularia*)	0.0327	0.0000 (May, Sept.-Nov. 1970)	0.0770 (July. 1970)
Littoral, man-modified			
Station IV (*Lepironia*)	0.0217	0.0000 (May, Sept.-Nov. 1970)	0.1050 (Sept. 1971)
Limnetic, swamp-forest			
Station V (Main channel)	0.0332	0.0000 (May, Oct.-Nov. 1970)	0.0790 (Sept. 1971)
Overall	0.0240	0.0000 (May, Sept.-Nov. 1970)	0.1050 (Sept. 1971)

Phosphate-phosphorus concentrations showed a seasonal fluctuation being abundant in the monsoon seasons (April-May and September-December) as well as the dry seasons (February-March and June-August) (Fig. 46). Phosphate concentrations were maximal either at the onset of rains (May and September) or in the dry months (July). This suggests that phosphate concentrations are dependent on nutrient enrichment from the watershed in the monsoons and on mineralisation of decomposing organic matter in the dry seasons. The undetectable concentrations of PO_4-P in May, and between September and November 1970, indicate that phosphate consumption overrides replenishment (Fig. 46). The exceptional maximum value obtained in the wet month of September 1970 was due to remineralisation after the dry season (July-August) and burning of the littoral *Lepironia* and *Pandanus* by the Semalai aborigines particularly at Station IV. Temporal PO_4-P fluctuations were relatively stable reflecting either a poor input or a rapid utilisation

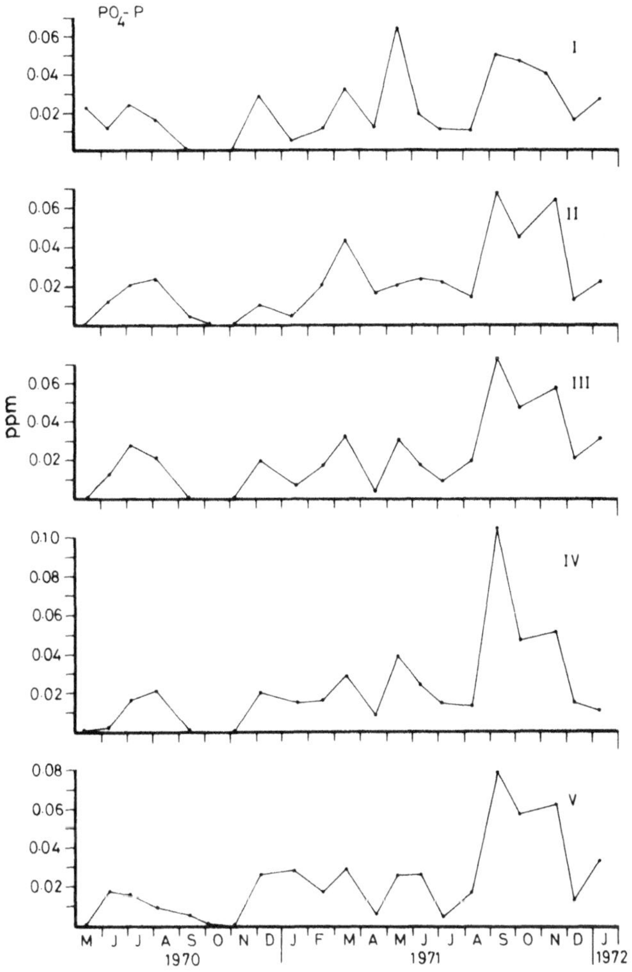

Fig. 46 Phosphate-phosphorus concentrations (ppm) at Stations I-V in Tasek Bera (1970–1972). (R. P. Lim and J. I. Furtado)

by growing cells (Sreenivasan 1964, quoting Benoit).

The mean, minimum and maximum total phosphate concentrations at Tasek Bera are 0.345, 0.033 and 0.781 ppm respectively (Table 32). Total phosphate was relatively similar at all the stations. The forms of 'reactive' and 'unreactive' phosphates indicate potential supply of phosphorus (Strickland and Parsons, quoted by Olsen 1966), irrespective of biological utilisation. The proportion of 'reactive' to 'unreactive' phosphate at Stations I-V was about 1:21, ranging from 1:7 to 1:180 (Table 33). This is expected for humus rich waters. Plankton studies suggest that the 'unreactive' phosphate contribution is mainly from dead tripton.

Table 32 Mean, minimum and maximum total phosphate concentrations at Stations I-V in Tasek Bera (1970–1972). (R. P. Lim and J. I. Furtado)

Stations	Total phosphate (ppm)		
	Mean	Minimum	Maximum
Limnetic, man-modified			
Station I	0.258	0.146	0.437
(Main channel)		(Sept. 1970)	(Oct. 1970)
Sublittoral, man-modified			
Station II	0.343	0.098	0.781
(Pandanus-Utricularia)		(May 1970)	(Aug. 1970)
Station III	0.252	0.033	0.592
(Utricularia)		(Oct. 1970)	(July 1970)
Littoral, man-modified			
Station IV	0.370	0.056	0.739
(Lepironia)		(June 1970)	(May 1970)
Limnetic, swamp-forest			
Station V	0.326	0.107	0.764
(Main-channel)		(Oct. 1970)	(May 1970)
Overall	0.342	0.033	0.764
		(Oct. 1970)	(Aug. 1970)

Table 33 Mean, minimum and maximum total 'reactive' and 'unreactive' Stations I-V in phosphate-phosphorus (ppm) at Tasek Bera (1970). (R. P. Lim and J. I. Furtado)

	Phosphate-phosphorus (ppm)		
Total	0.342	0.194	0.462
		(Sept. 1970)	(Oct. 1970)
'Reactive'	0.017	0.000	0.070
		(Oct. 1970)	(July 1971)
'Unreactive'	0.351	0.185	0.462
		(June 1970)	(Oct. 1970)

3.2.6.3 *Nitrogen.* The mean, minimum and maximum nitrate-nitrogen concentrations at Tasek Bera are 0.107, 0.010 and 0.290 ppm respectively (Table 34). It was fairly similar at all stations. This high nitrate-nitrogen value at Tasek Bera differs from the 'blackwaters' of the Amazon (Klinge and Ohle 1964; Sioli 1967, 1968) and the African lakes (Talling and Talling 1965), but is similar to the Ugandan swamps (Carter 1955).

Table 34 Mean, minimum and maximum nitrate-nitrogen concentrations at Stations I-V in Tasek Bera (1970–1972). (R. P. Lim and J. I. Furtado)

Stations	Nitrate-nitrogen (ppm)		
	Mean	Minimum	Maximum
Limnetic, man-modified			
Station I	0.111	0.019	0.290
(Main channel)		(Oct. 1971)	(Febr. 1971)
Sublittoral, man-modified			
Station II	0.108	0.014	0.260
(*Pandanus-Utricularia*)		(Nov. 1970)	(Jan. 1971)
Station III	0.101	0.010	0.240
(*Utricularia*)		(Sept. 1970)	(Febr. 1971)
Littoral, man-modified			
Station IV	0.104	0.010	0.250
(*Lepironia*)		(Oct. 1971)	(Febr. 1971)
Limnetic, swamp-forest			
Station V	0.109	0.020	0.240
(Main channel)		(Nov. 1970)	(Febr. 1971)
Overall	0.107	0.010	0.290

The nitrate-nitrogen showed a seasonal fluctuation with greater concentrations in the dry seasons (February-March and June-September) with a low concentration during the September-November northeast monsoon (Fig. 47). This suggests a dilution due to initial monsoonal rainfall, and a nitrate enrichment due to the decomposition and remineralisation of allochthonous organic materials. The exceptionally high nitrate enrichment in February 1971 was probably due to organic input during the unusual floods in January 1971, and its subsequent decomposition and remineralisation. The main source of nitrates, thus, appears to be biological. The abundance of soil bacteria and fungal spores, washed into the swamp from the watershed, is probably related to this nitrate enrichment (see later). The abundance of nitrate-nitrogen was not affected by the burning of littoral vegetation, as in August 1971.

The mean, minimum and maximum nitrite-nitrogen concentrations at Tasek Bera are 0.0074, 0.000 and 0.059 ppm respectively (Table 35). This low value was similar at all stations, and indicates the relative instability of these compounds. It was most abundant in the northeast monsoon probably because of influx from the watershed.

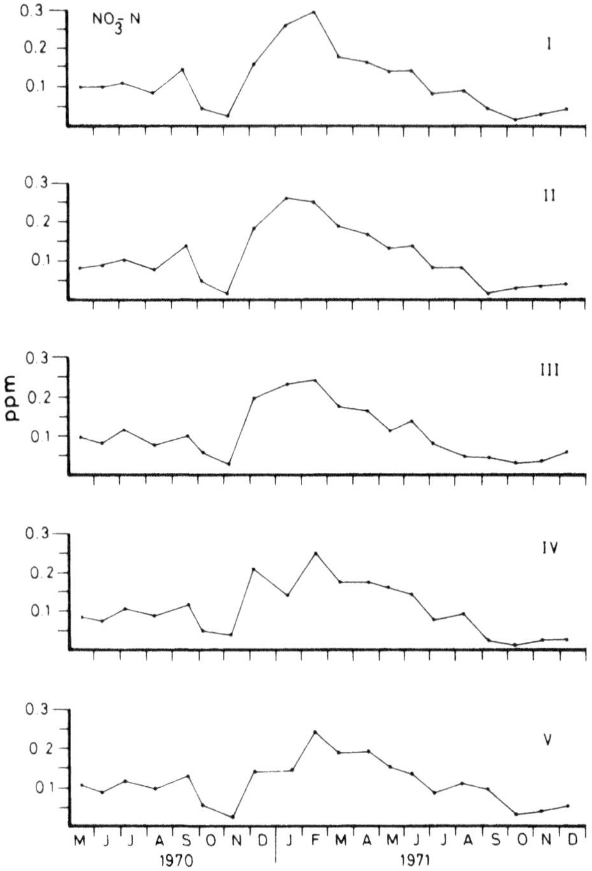

Fig. 47 Nitrate-nitrogen concentrations (ppm) at Stations I-V in Tasek Bera (1970–1972). (R. P. Lim and J. I. Furtado)

The mean, minimum and maximum ammonium-nitrogen concentrations at Tasek Bera are 0.304, 0.000 and 0.767 ppm, respectively (Table 36). It was more abundant than nitrate-nitrogen, and was fairly similar at all stations. It showed a seasonal fluctuation being abundant during the dry periods (February-March and July-September) and low during the monsoons except for May (southwest monsoon) (Fig. 48). This suggests that reduction processes predominating during the dry seasons enhance the oxidation of nitrogen from the nitrate to ammonium state, and enrichment from the watershed during the southwest monsoon.

The mean, minimum and maximum organic nitrogen concentrations in Tasek Bera are 0.501, 0.056 and 1.526 ppm respectively (Table 37). These

Table 35 Mean, minimum and maximum nitrite-nitrogen concentrations at Stations I-V in Tasek Bera (1970–1972). (R. P. Lim and J. I. Furtado)

Stations	Nitrite-nitrogen (ppm)		
	Mean	Minimum	Maximum
Limnetic, man-modified			
Station I (Main channel)	0.0076	0.000 (Febr. 1971, Apr.-Dec. 1971, Jan. 1972)	0.059 (Dec. 1970)
Sublittoral, man-modified			
Station II (*Pandanus-Utricularia*)	0.0078	0.000 (Febr. 1971, Apr.-July 1971, Sept.-Dec. 1971, Jan. 1972)	0.055 (Dec. 1970)
Station III (*Utricularia*)	0.0068	0.000 (Febr. 1971, Apr.-Dec. 1971, Jan. 1972)	0.039 (Dec. 1970)
Littoral, man-modified			
Station IV (*Lepironia*)	0.0082	0.000 (Febr. 1971, Apr.-July 1971, Sept.-Dec. 1971, Jan. 1972)	0.038 (Dec. 1970)
Limnetic, swamp-forest			
Station V (Main channel)	0.0062	0.000 (Febr. 1971, Apr.-Dec. 1971, Jan. 1972)	0.039 (Dec. 1970)
Overall	0.0074	0.000 (Febr. 1971-Jan. 1972)	0.059 (Dec. 1970)

concentrations were slightly greater than ammonium-nitrogen, and were fairly similar at all stations. Organic nitrogen showed a seasonal fluctuation, being abundant during the dry season due to the decomposition of allochthonous organic materials which had entered the swamp from the watershed during the monsoonal rains (Fig. 49). The potential of nitrogenous sources are thus biological and rich.

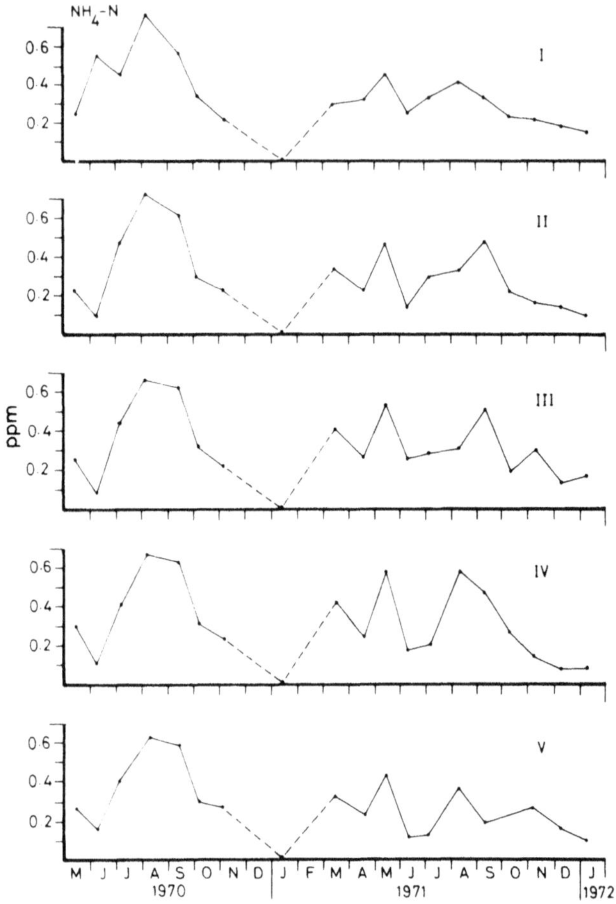

Fig. 48 Ammonium-nitrogen concentrations (ppm) at Stations I-V in Tasek Bera (1970–1972). (R. P. Lim and J. I. Furtado)

The mean, mimimum and maximum total nitrogen concentrations at Stations I-V in Tasek Bera are 0.986, 0.237 at Station III (June 1970) and 2.380 ppm at Station I (August 1970), respectively (Table 38). As is evident from the preceding discussions and from a summary of nitrogen compounds (Table 39), organic and ammonium-nitrogen are the dominant forms present at Tasek Bera. This conforms to studies in most tropical areas, which indicate that nitrogen sources are biological rather than mineral. Biological nitrogen sources are in the form of (1) leachates directly from aquatic macrophytes (cf. Nye 1961), (2) leachates from allochthonous and autotochthonus organic material decomposing in the water (e.g. Hynes and Kaushik 1969), and (3) nitrogen-fixing bacteria and algae both of terrestrial and aquatic origin.

Table 36 Mean, minimum and maximum ammonium-nitrogen concentrations at Stations I-V in Tasek Bera (1970–1972). (R. P. Lim and J. I. Furtado)

Stations	Ammonium-nitrogen (ppm)		
	Mean	Minimum	Maximum
Limnetic, man-modified			
Station I	0.332	0.000	0.767
(Main channel)		(Jan. 1971)	(Aug. 1970)
Sublittoral, man-modified			
Station II	0.291	0.000	0.724
(Pandanus-Utricularia)		(Jan. 1971)	(Aug. 1970)
Station III	0.313	0.000	0.660
(Utricularia)		(Jan. 1971)	(Aug. 1970)
Littoral, man-modified			
Station IV	0.308	0.000	0.662
(Lepironia)		(Jan. 1971)	(Aug. 1970)
Limnetic, swamp-forest			
Station V	0.276	0.000	0.637
(Main channel)		(Jan. 1971)	(Aug. 1970)
Overall	0.304	0.000	0.767
		(Jan. 1971)	(Aug. 1970)

Table 37 Mean, minimum and maximum organic nitrogen concentrations at Stations I-V in Tasek Bera (1970). (R. P. Lim and J. I. Furtado)

Stations	Organic nitrogen (ppm)		
	Mean	Minimum	Maximum
Limnetic, man-modified			
Station I	0.577	0.062	1.526
(Main channel)		(July 1970)	(Aug. 1970)
Sublittoral, man-modified			
Station II	0.505	0.071	1.097
(Pandanus-Utricularia)		(June 1970)	(Aug. 1970)
Station III	0.433	0.056	0.903
(Utricularia)		(June 1970)	(Aug. 1970)
Littoral, man-modified			
Station IV	0.538	0.098	1.176
(Lepironia)		(June 1970)	(Aug. 1970)
Limnetic, swamp-forest			
Station V	0.454	0.243	0.858
(Main channel)		(May 1970)	(Sept. 1970)
Overall	0.501	0.056	1.526
		(June 1970)	(Aug. 1970)

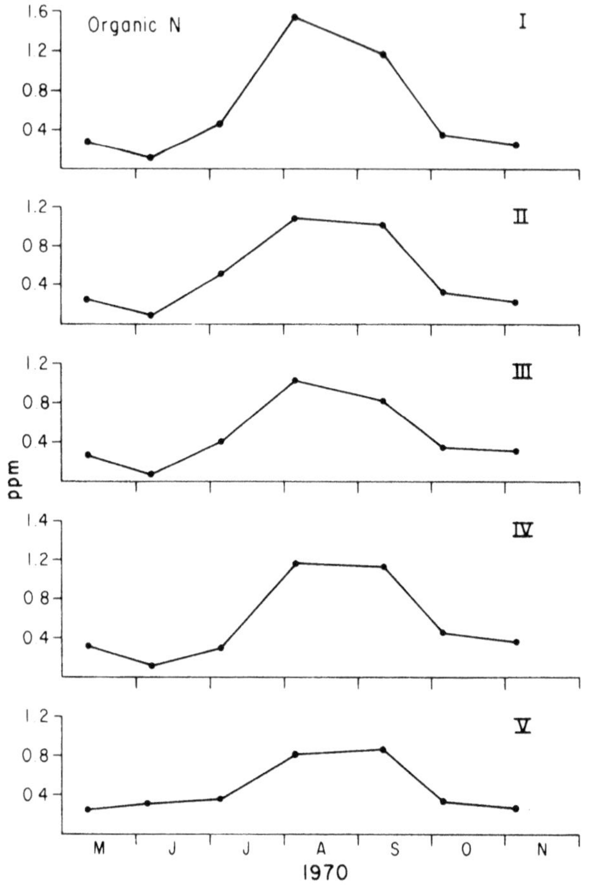

Fig. 49 Organic nitrogen concentrations (ppm) at Stations I-V in Tasek Bera (1970–1972). (R. P. Lim and J. I. Furtado)

3.3 Diel behaviour (R. P. Lim and J. I. Furtado)

Diel studies of physico-chemical conditions of aquatic environments in the tropics are several (Vaas and Sachlan 1953, Ganapati 1955, Talling 1957, George 1961, Rao and Rao 1962, Dunn 1967, Verma 1967 and Khan et al. 1970). However, these are limited mainly to eutrophic environments. A survey of the diel behaviour of the waters at Tasek Bera was made at Stations I and IV on 8-9 March 1971. The major physical and chemical factors were measured, namely temperature, alkalinity, pH, conductivity and oxygen (Fig. 50).

Table 38 Mean, minimum and maximum total nitrogen concentrations at Stations I-V in Tasek Bera (1970). (R. P. Lim and J. I. Furtado)

Stations	Total nitrogen (ppm)		
	Mean	Minimum	Maximum
Limnetic, man-modified			
Station I	1.121	0.500	2.380
(Main channel)		(Nov. 1970)	(Aug. 1970)
Sublittoral, man-modified			
Station II	0.972	0.270	1.903
(Pandanus-Utricularia)		(June 1970)	(Aug. 1970)
Station III	0.895	0.237	1.645
(Utricularia)		(June 1970)	(Aug. 1970)
Littoral, man-modified			
Station IV	1.015	0.299	1.934
(Lepironia)		(June 1970)	(Aug. 1970)
Limnetic, swamp-forest			
Station V	0.928	0.554	1.578
(Main-channel)		(June 1970)	(Sept. 1970)
Overall	0.986	0.237	2.380
		(June 1970)	(Aug. 1970)

Table 39 Mean, minimum and maximum concentrations of various nitrogen compounds at Stations I-V in Tasek Bera (1970). (R. P. Lim and J. I. Furtado)

	Nitrogen compounds (ppm)		
	Mean	Minimum	Maximum
Nitrate-nitrogen	0.081	0.023	0.125
		(Nov. 1970)	(Sept. 1970)
Nitrite-nitrogen	0.013	0.003	0.028
		(Sept. and Nov. 1970)	(July 1970)
Ammonium-nitrogen	0.390	0.193	0.692
		(June 1970)	(Aug. 1970)
Organic nitrogen	0.501	0.116	1.102
		(June 1970)	(Aug. 1970)
Total nitrogen	0.985	0.418	1.881
		(June 1970)	(Aug. 1970)

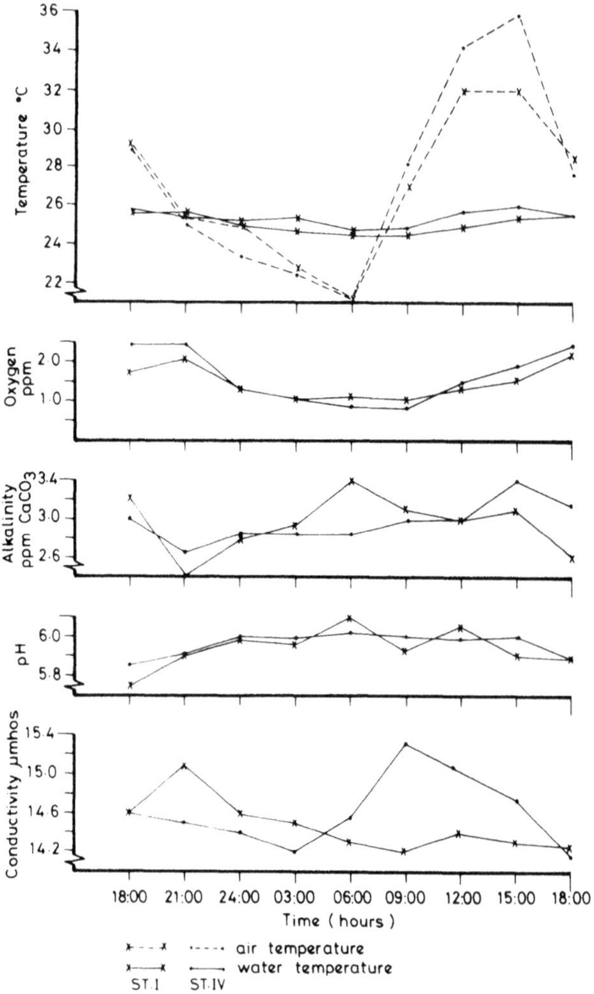

Fig. 50 Diel behaviour of the major physico-chemical parameters (temperature, dissolved oxygen, alkalinity, pH and conductivity) at Stations I and IV in Tasek Bera (8–9 March 1971). (R. P. Lim and J. I. Furtado)

Diel changes in water temperature were slight, when compared with air temperature (Fig. 50). A slow cooling of the water in the late afternoon (1800 hours) and throughout the night was followed by a gradual increase in temperature as the intensity of radiation increased. At Station IV, water temperature closely corresponded to increasing radiation and reached a maximum at 1500 hours. However, the slight temperature rise at 0300 hours is difficult to account, although relatively higher temperatures at greater depths could have increased the average values, thus creating an artefact. Isotherms exhibit a daily turnover at 2100-0600 hours along the time-depth profile (Fig. 51). Slight stratification occured at Station I from 1500 to 1800 hours, bur broke down within a few hours. Diel temperature observations further indicate (1) very little change in water temperature at all depths, (2) a maximum temperature at the surface at 1500 hours, and (3) a stratification a 1500 hours. There was superficial heating of the water body during the day followed by a loss of heat and complete mixing during the night. As such, the swamp is polymictic (Hutchinson and Loffler 1956, as cited by Hutchinson 1957) and conforms the other shallow water bodies. There was heat dissipation from the water to the bottom sediments of the order of 1°C, indicating that most of the cooling of the water at night is at the surface (Carter and Beadle 1930).

Changes in alkalinity over the 24-hour period showed an increase during the early morning hours and a gradual decrease in the late afeternoon (Fig. 50). The change were due to the increased carbon dioxide input by respiring

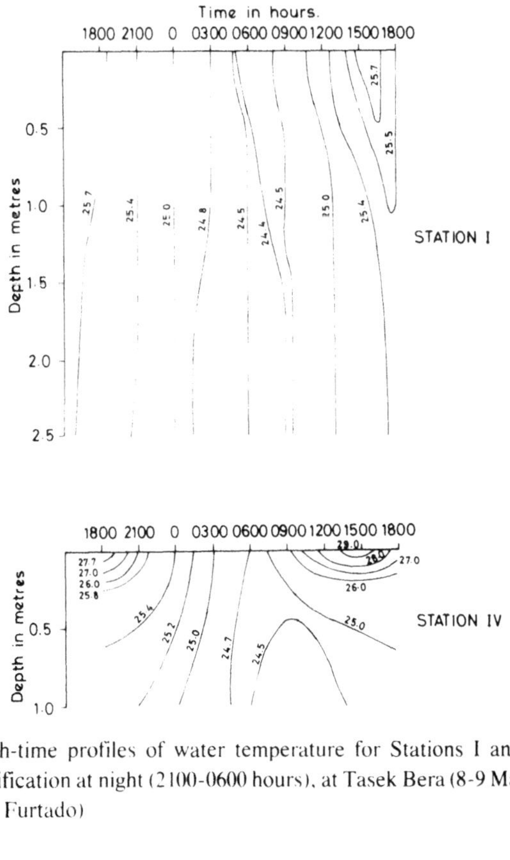

Fig. 51 Diel depth-time profiles of water temperature for Stations I and IV, showing the breakdown of stratification at night (2100-0600 hours), at Tasek Bera (8-9 March 1971).
(R. P. Lim and J. I. Furtado)

organisms in the water during the night and decrease during the day due to photosynthetic uptake. These changes were greater in the littoral (Station IV) than in the open water (Station I) regions, however, indicating a richer fauna and flora in the former.

pH was almost constant throughout the 24 hours with only very slight increases during the night. The lack of fluctuation in pH is due to the fact that humic acids contribute mainly to pH measurements, thus masking any changes that would be contributed by biological processes.

Conductivity gradually decreased during the day (Fig. 50), probably due to uptake of nutrients. However, it increased at night (at 0900 hours in Station IV, at 2100 hours in Station I) probably because of a turnover.

Dissolved oxygen showed no stratification at all at Station I, but a marked surface stratification at Station IV over the 24 hours (Fig. 52). The decrease in oxygen concentration during the night is less marked in the open water, because community respiration was greater in the littoral water (Station IV) where oxygen decreased from 2.83 ppm at 1800 hours to 0.82 ppm at 0600 hours resulting in a transformation from the clinograde to the orthograde curve.

3.4 Daily behaviour
(R. P. Lim and J. I. Furtado)

The daily behaviour of major physico-chemical parameters at Tasek Bera

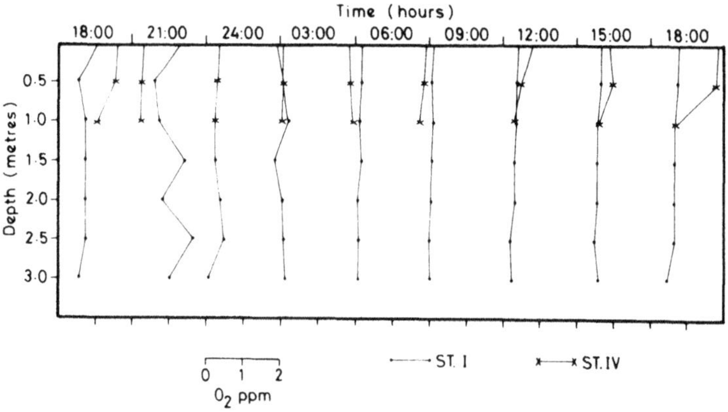

Fig. 52 Diel depth-time profiles of dissolved oxygen for Stations I and IV, showing the breakdown of stratification at night (2100-0900 hours), at Tasek Bera (7-8 March 1971). (R. P. Lim and J. I. Furtado)

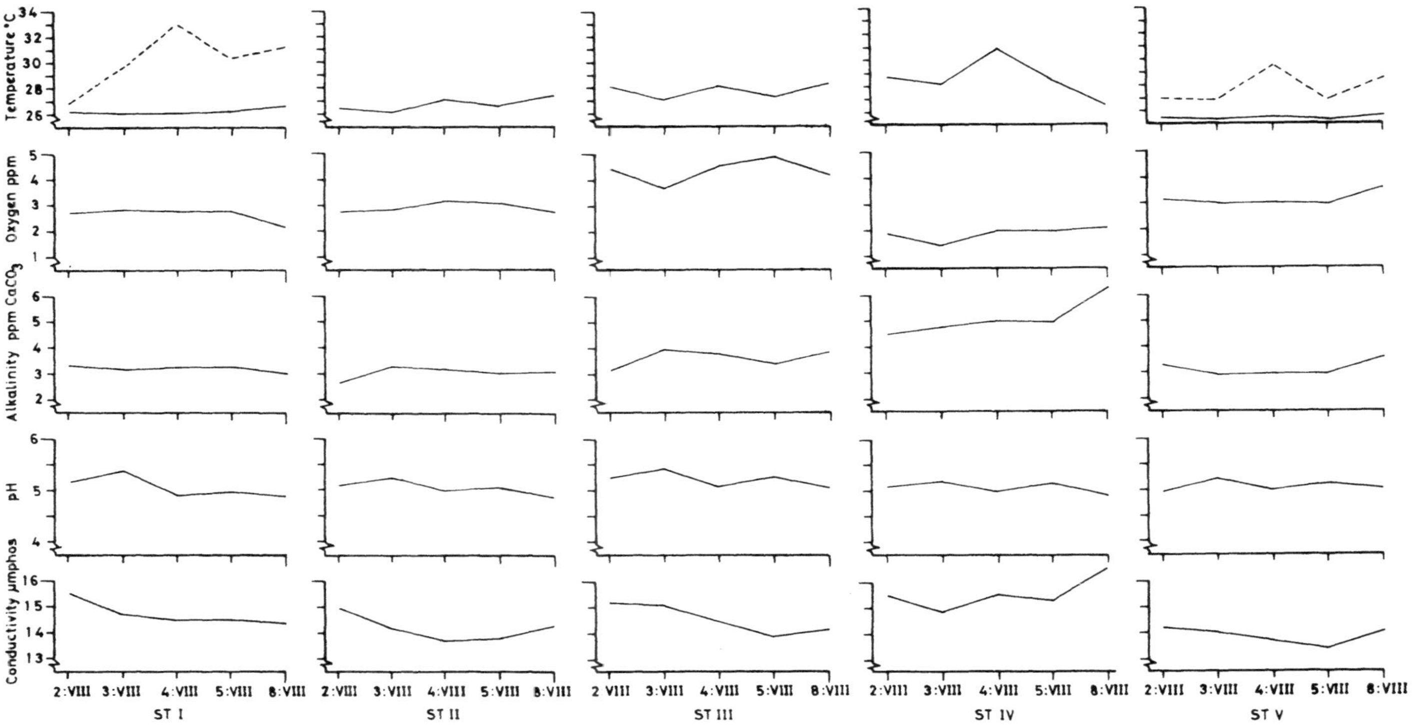

Fig. 53 Daily variations in major physico-chemical parameters (water temperature, air temperature, dissolved oxygen alkalinity, pH and conductivity) at all stations (I-V) in Tasek Bera (2-3 August 1971). - - - - - air tempeature, ——— water temperature. (R. P. Lim and J. I. Furtado).

from 2-8 August 1971 showed very slight fluctuations (Fig. 53). However, the littoral region (Stations III and IV) showed greater variability than the open water region, probably due to greater complexity of this region. The relative stability of the various parameters from day to day showed that they did not contribute to seasonal fluctuations.

3.5 Geochemical Aspects of the Aquatic Environment (O. Sato)

3.5.1 *Dynamics of oxygen and carbon dioxide in relation to decomposition of organic matter*

The formation of peat in the swamp especially in the vegetated littoral regions indicates that the rate of production of organic matter exceeds that of decomposition. The process of peat formation is closely related to the rate of mineralisation of organic matter which is controlled by the physical and chemical properties of the water. In the swamp, low dissolved oxygen, low pH, high iron and other variables could be responsible for reducing the rate of organic decomposition. The rate of oxygen abstraction from, and carbon dioxide evolution to the water at various locations in the swamp were assessed in relation to estimating the rate of organic decomposition. First-order models were adopted to assess the rates of mass transfer in the following processes:

1) abstraction of oxygen from the atmosphere into the water,
2) consumption of oxygen in the water,
3) consumption of oxygen in the mud, and
4) loss of carbon dioxide from the water to the atmosphere (Fig. 54).

Absorption of oxygen across the air-water interface was determined using the floating dome method. Plastic domes (bottom area 133 cm^2) containing 200 ml of air were placed on the water surface within the *Lepironia* reed stand, within the swamp forest and on the stream channel in the swamp-forest. 20 ml of air was extracted with a syringe at regular intervals. The concentration of oxygen was then measured by the gas volumetric method using a 10 ml burette and a strong alkaline solution of pyrogallol as the oxygen absorbant.

The rate of oxygen consumption was measured by the dark bottle technique. 100 ml bottles were filled with swamp water, covered with aluminium foil and incubated for specific time intervals. Dissolved oxygen was measured using the Winkler method and the decrease in oxygen concentration was

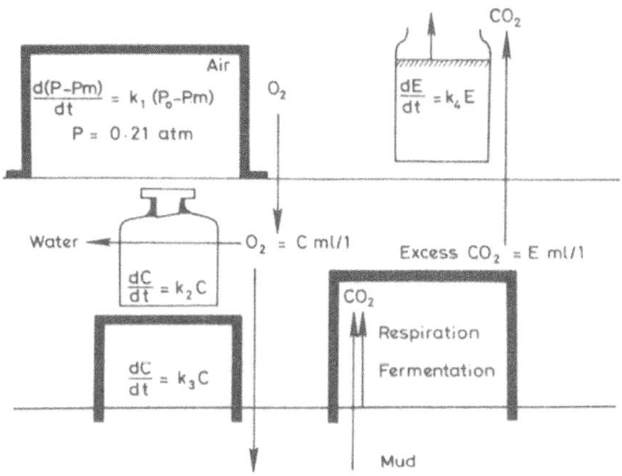

Fig. 54 Models of mass transfer in a water column. (O. Sato)

P : Partial pressure of O_2 in the dome
P_m : Partial pressure of O_2 in equilibrium with dissolved O_2
C : Dissolved O_2 ml/l
E : Excess CO_2 ml/l
K_1 : O_2 absorption coefficient
K_2 : O_2 consumption coefficient in water
K_3 : O_2 consumption coefficient of mud
K_4 : CO_2 evolution coefficient from water into air.

calculated. Determination of oxygen consumption in the swamp mud was carried out with bottomless plastic boxes (volume 11.6 cm^3, bottom area 642 cm^2). The boxes were placed with their open ends down on the mud surface. The change in oxygen concentration in the box was observed at regular intervals.

The rate of carbon dioxide production in the water was calculated from oxygen consumption rates mentioned above by multiplying with a respiratory quotient of 0.85. The increase in total carbon dioxide, which estimates the evolution of carbon dioxide from the mud, was measured concurrently with measurements of oxygen decrease in the plastic boxes. Total carbon dioxide was determined by the modified Koyama method (1953).

The rate of release of gaseous carbon dioxide from the water to the atmosphere was estimated through laboratory experiments. Swamp water collected before dawn (250 ml in 500 ml capacity bottles) was allowed to stand uncapped in the laboratory at ambient temperature. The decrease in total carbon dioxide was measured at regular time intervals.

Soil respiration was measured using the absorption method. An absorbent holder made of plastic sponge was used for measuring the respiration rate of the swamp-forest soil and the swamp mud during the dry season (Kirita 1971). Respiration of the mud was measured in the field and under laboratory conditions. In the laboratory, freshly collected mud with plant roots removed were placed in plastic basins to a depth of 5 cm. The absorbent holders were placed on the mud and allowed to stand for five days. Respiration was calculated following the procedures of Kirita (1971).

Diurnal variation of atmospheric carbon dioxide just above the swamp water and forest soil were measured. Air samples 15 cm above the water surface and soil surface were collected in plastic balloons using a foot pump. Carbon dioxide concentrations of the samples were measured using an infrared absorption gas analyser (Hitachi-Horiba, Type ASSA A-21).

Results and discussion

No equation has yet been developed for predicting oxygen absorption rates of flowing water with sufficient accuracy under laboratory conditions. Thus absorption coefficients had to be determined *in situ* using plastic domes which were not affected by surface movements. Wilson and Macleod (1974) note that the transport of oxygen is associated with the process on the liquid side of the air-water interface, and hence winds or other air movements do not affect the rate of river reoxygenation except by their mechanical influence which is normally very limited. The calm water conditions at Tasek Bera provide suitable conditions for the use of domes for measuring fairly accurately the oxygen absorption coefficient of flowing waters.

It is assumed that the oxygen absorption rate of water obeys a first-order process as described by the following equation:

$$\frac{d\,(P - P_m)}{dt} = k_l\,(P_o - P_m) \cdot \qquad (1)$$

Integrating

$$\log_e \frac{P - P_m}{P_o - P_m} = -\,kt \qquad (2)$$

where P is the partial pressure of oxygen in the dome at time t, P_o is the partial pressure of oxygen in the atmosphere, P_m is the partial pressure of dissolved oxygen, and k is the absorption coefficient.

There is a linear relationship between the logarithm of $(P - P_m/P_o - P_m)$ with time (Fig. 55) indicating that the foregoing assumption is reasonable for Tasek Bera. Absorption rates of oxygen were calculated from the absorption coefficient determined from the slope of the curve and experimental variables like the cross-section and volume of the dome. The absorption coefficient varies

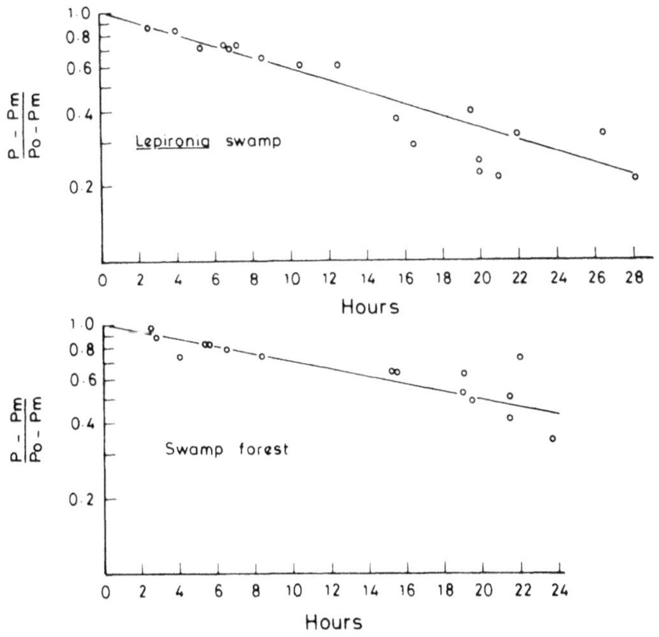

Fig. 55 Absorption of oxygen across the air-water interface in the *Lepironia* swamp and the swamp-forest approximated as a first-order reaction. (O. Sato)

with the physical conditions of the stream such as its depth and flow rate. The coefficient obtained through these experiments can therefore be used only for estimations of oxygen absorption under similar conditions. Oxygen absorption rates were calculated from the coefficient and partial pressure of dissolved oxygen for a particular observation. The oxygen absorption rate is highest at Station V where it is deep, the flow rate is high and the dissolved oxygen content low. The water in the swamp-forest has similar oxygen absorption rates as the *Lepironia* reed stand although their physical conditions are different. The shallowness of the forest waters and its slow flow rate limits oxygen absorption, but the extremely low oxygen content increases the rate of absorption resulting in a comparatively high rate of 1200 $ml \cdot m^{-2} \cdot day^{-1}$. On the other hand, in the *Lepironia* reed swamp with its higher oxygen concentration, the more rapid flow rate results in a similar absorption rate of 1300 $ml \cdot m^{-2} \cdot day^{-1}$.

An inverse linear relationship between the logarithm of oxygen concentration in the experimental bottle and time (Fig. 56) approximates a first-order process. The oxygen consumption is 0.0026 c $ml \cdot l^{-1} \cdot h^{-1}$ where c is the dissolved oxygen concentration expressed as $ml \cdot l^{-1}$.

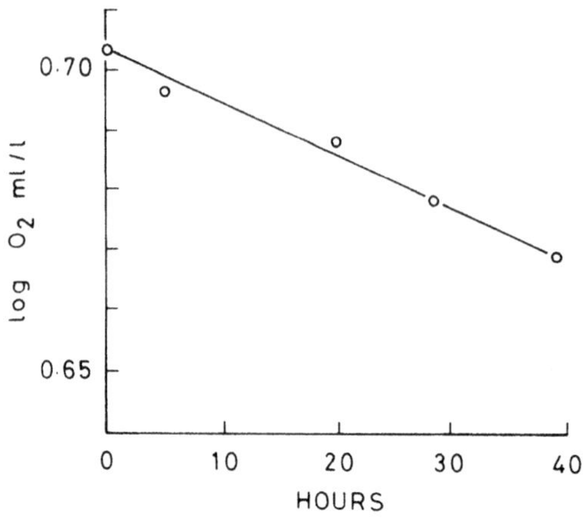

Fig. 56 Consumption of oxygen in water as a first-order process (O. Sato)

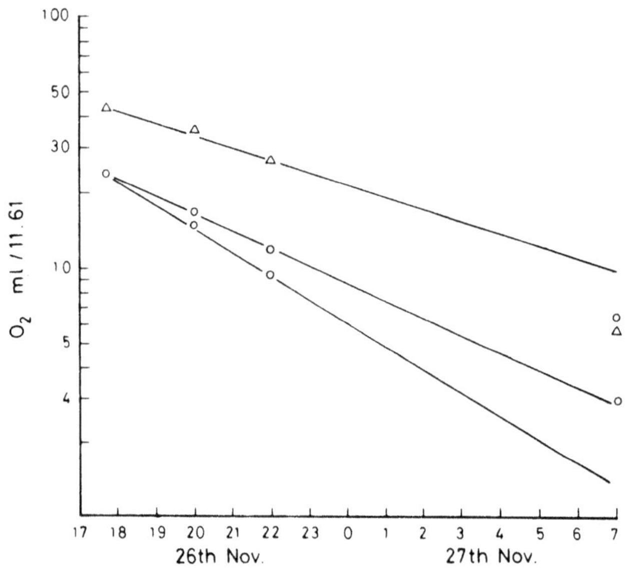

Fig. 57 Oxygen consumption by swamp mud plotted to the first-order reaction model. (O. Sato)

Figure 57 shows a negative linear relationship between the logarithm of oxygen concentration in the mud and time. The consumption rate of oxygen in the mud was calculated using the mean of three experiments. An attempt was made to compare the observed oxygen content with the value obtained through calculations using a first-order model. This was aimed at determining the validity of the first-order model used in estimating the oxygen consumption rate in the mud. Water containing various concentrations of dissolved oxygen were placed in contact with the mud for 12 hours after which the concentration of dissolved oxygen was determined. The experiment shows that the predicted values coincide fairly well with the observed values (Table 40) and that consumption rates in mud approximates a first-order process. The estimated rates of oxygen consumption are 100 ml $O_2 \cdot m^{-2} \cdot day^{-1}$ and 1100 ml $O_2 \cdot m^{-2} \cdot day^{-1}$ for the water and mud, respectively.

Carbon dioxide in the mud is produced mainly through aerobic decomposition of organic matter, fermentation, and respiration by *Lepironia* roots. The rate of carbon dioxide production through aerobic decomposition is 1360 ml $CO_2 \cdot m^{-2} \cdot day^{-1}$ while the rate of carbon dioxide production through fermentation processes and root respiration is 385 ml $CO_2 \cdot m^{-2} \cdot day^{-1}$ (Table 41). From these processes it is estimated that the rate of carbon dioxide released into the water from the mud is 1745 ml $CO_2 \cdot m^{-2} \cdot day^{-1}$. (Table 41).

Since the swamp water is supersaturated with carbon dioxide when compared with its partial pressure in the air, carbon dioxide will be released into the atmosphere. The diffusion rate of carbon dioxide across the air-water interface is, however, dependent on the excess carbon dioxide concentration

Table 40 Oxygen consumption rate of swamp mud. (O. Sato)

Plot	Initial ml $O_2 \cdot l^{-1}$	After 12 hours	
		Observed ml $O_2 \cdot l^{-1}$	Predicted ml $O_2 \cdot l^{-1}$
1	4.34	0.78	0.72
2	5.24	1.17	1.17
3	2.19	0.54	0.36
4	2.36	0.18	0.38

Table 41 Carbon dioxide production and oxygen consumption on bottom surface. (O. Sato)

Plot	Production of CO_2 ml $CO_2 \cdot m^{-2} \cdot day^{-1}$	Consumption of O_2 ml $O_2 \cdot m^{-2} \cdot day^{-1}$	Aerobic decomposition ml $CO_2 \cdot m^{-2} \cdot day^{-1}$	Fermentation and others ml $CO_2 \cdot m^{-2} \cdot day^{-1}$
1	1400	1350	1150	250
2	2090	1850	1570	520
Average	1745	1600	1360	385

(i.e. the difference between the actual concentration of carbon dioxide in the water in relation to the concentration at saturation in the water). A negative linear relationship between the logarithm of excess carbon dioxide and time was observed which indicates that a first-order process occurs (Fig. 58). The diffusion rate of carbon dioxide to the atmosphere is 4.5 E ml $CO_2 \cdot m^{-2} \cdot day^{-1}$. However, this rate is an underestimate since it was determined under laboratory conditions where there was no movement of the water, a situation which does not occur under field conditions where mixing due to currents and wind occurs. The calculated value based on the concentration of carbon dioxide in the air and the diffusion coefficient is 600 ml $CO_2 \cdot m^{-2} \cdot day^{-1}$.

The diurnal fluctuations of carbon dioxide concentrations in the air just above the water surface within the *Lepironia* reed swamp and above the soil surface in the swamp-forest follows the same pattern (Fig. 59). In general the carbon dioxide content decreases during the day and increases over the night (Fig. 59). This is due to absorption by plants during photosynthesis. In comparison the atmospheric carbon dioxide content at the swamp is higher than at Kuala Pilah. The higher concentration at the swamp further substantiates the hypothesis that the swamp environment produces more carbon dioxide than other forest types. The relatively high carbon dioxide in the atmosphere results in a reduced release of carbon dioxide from the water to the atmosphere. This could explain the relatively high concentration of carbon dioxide in the swamp water. The rate of oxygen production and carbon dioxide accumulation in a water column 80 cm long by 1 m^2 in cross-section is estimated at 300 ml $O_2 \cdot day^{-1}$ and 1000 ml $CO_2 \cdot day^{-1}$ respectively.

From the above data a summarised budget of oxygen and carbon dioxide in the water within the *Lepironia* reed swamp can be developed (Fig. 60), on the

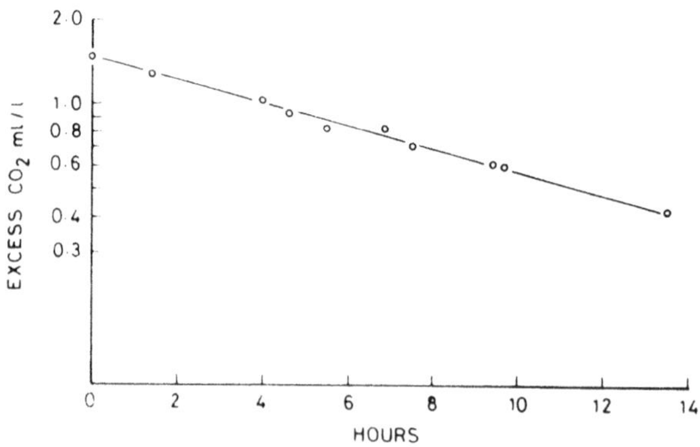

Fig. 58 Diffusion of CO_2 across the air-water interface as a first-order process. (O. Sato)

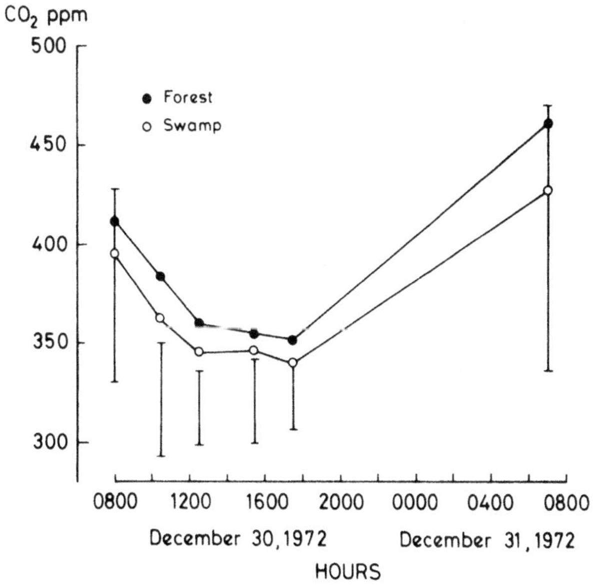

Fig. 59 Diurnal fluctuation of atmospheric CO_2 above the swamp water (o), in the forest (•) and Kuala Pilah (bars show the ranges of CO_2 concentrations observed by Koyama between March and September 1972). (O. Sato)

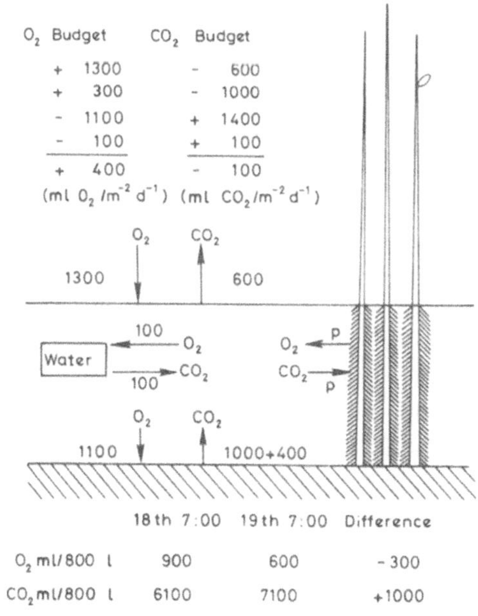

Fig. 60 Summarised oxygen and carbon dioxide budget in the water within the *Lepironia* reed swamp. (O. Sato)

assumption that oxygen and carbon dioxide of the inflow is the same as that of the outflow. The flow diagram indicates:

1) that oxygen available for respiration from dissolved atmospheric oxygen and dissolved oxygen from photosynthetic activities is 1600 ml $O_2 \cdot m^{-2} \cdot day^{-1}$, of which 1200 ml $O_2 \cdot m^{-2} \cdot day^{-1}$ is consumed through respiratory activities in the water and mud;
2) that carbon dioxide produced through respiratory and fermentation activities in the water and in the mud is 1500 ml $CO_2 \cdot m^{-2} \cdot day^{-1}$;
3) and that a portion of the carbon dioxide (600 ml $CO_2 \cdot m^{-2} \cdot day^{-1}$) in the water is released to the atmosphere as gaseous carbon dioxide while the remaining 100 ml $CO_2 \cdot m^{-2} \cdot day^{-1}$ is retained in the water.

The difference between the estimated total carbon dioxide and the two carbon dioxide compartments is 100 ml $CO_2 \cdot m^{-2} \cdot day^{-1}$. The excess oxygen supply for respiratory activities and the shortage of carbon dioxide production suggests that respiration by algae, aquatic macrophytes and microbes on dead organic matter exceed production of oxygen through photosynthesis in the water column on that day.

The mean respiration rate of non-water logged soils in Tasek Bera is estimated at 538 mg $CO_2 \cdot m^{-2} \cdot h^{-1}$ which is somewhat lower than the rate (624 mg $CO_2 \cdot m^{-2} \cdot h^{-1}$) observed in tropical rain forests in Thailand (Yoda and Kira 1969). The mean respiration rate of water-logged soils is 145 mg $CO_2 \cdot m^{-2} \cdot h^{-1}$ (range: 83-200 mg $CO_2 \cdot m^{-2} \cdot h^{-1}$), which corresponds to the mean rate in the muds of the *Lepironia* reed swamp (mean: 143 mg $CO_2 \cdot m^{-2} \cdot h^{-1}$). This similarity in rates is due mainly to the restriction of oxygen supply through the water overlying the soil and mud.

Laboratory experiments indicate that mud does not contain gaseous oxygen and that the supply must be derived from diffusion processes at the mud-water interface. The estimated mean respiration rate from these experiments is 101 mg $CO_2 \cdot m^{-2} \cdot h^{-1}$ (range: 72-129 mg $CO_2 \cdot m^{-2} \cdot h^{-1}$) which is less than the rates estimated from *in situ* experiments. This is mainly due to the exclusion of root respiration and fermentation processes from the experiment.

In situ measurements of respiration of the mud when submerged (1-3 cm) and when dry show that lowering of ground water level causes a remarkable increase in respiration rate (Table 42). This respiration rate increases with distance from the shore further substantiating the fact that soil respiration increases with decreasing ground-water level (Fig. 61).

Two factors are known to affect soil respiration rates: these are temperature and soil humidity. Since temperature in the swamp is consistently higher than 20°C, water content might be the regulatory factor in soil respiration. In water-logged soils oxygen supply to the microbes inhabiting the soil is low

Table 42 Respiration of soils and muds. (O. Sato)

A.	Soils in the swamp-forest		Muds in the *Lepironia* swamp	
	Normal	Water-logged	In situ	At laboratory
Respiration ($mg\ CO_2 \cdot m^{-2} \cdot h^{-1}$)	538	144	143	101
Number of observations	18	9	6	4

B. Changes of respiration accompanied with changes in water condition ($mg\ CO_2 \cdot m^{-2} \cdot h^{-1}$).

(1) Plot	Normal soil	Normal soil	(2) Plot	Water-logged soil	Normal soil
	8-9 Sept.	24-25 Sept.		8-9 Sept.	24-25 Sept.
1	937	321	9	128	259
3	491	480	10	135	888
4	347	281	12	83	463
6	210	229	13	199	354
7	340	651			
8	335	654			
Mean	443	436	Mean	136	463

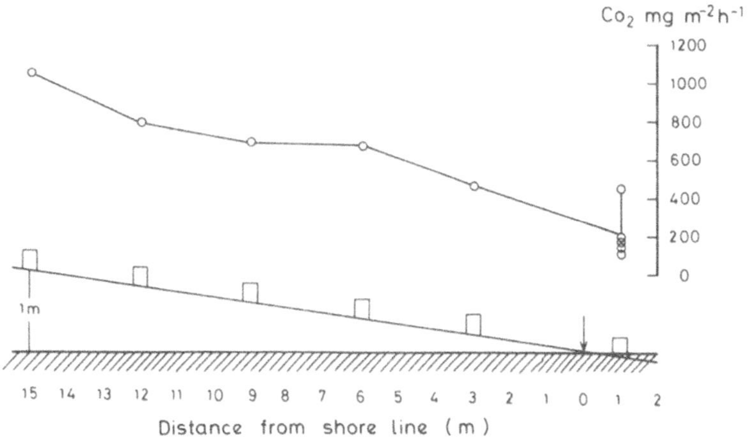

Fig. 61 Relation of soil respiration to distance from the shore line or ground water level. (O. Sato)

because of the low diffusion coefficient of oxygen in water. This factor coupled with the fact that higher temperatures reduce oxygen solubility, could result in anaerobic conditions observed in the soils of Tasek Bera, thus partly accounting for the slow decomposition rate of organic matter despite the high soil temperature.

Figure 62 shows the seasonal variation in water level, temperature of the mud and oxygen consumption rates in the mud. Low oxygen consumption rates in January and February reflect low oxygen concentrations in the water above the mud. The sustained low level of oxygen during this period is due to heavy rains which flooded the whole swamp (see earlier), which washed away a great proportion of non-rooted autotrophs and resulted in reduced photosynthesis. Integrating the area under the curve of oxygen consumption rate the annual decomposition rate in mud is estimated at 1.1 kg $O_2 \cdot m^{-2} \cdot h^{-1}$ (Fig. 62).

From the above-mentioned experiments the relationship between oxygen concentration and oxygen consumption rate in a compartmental process,

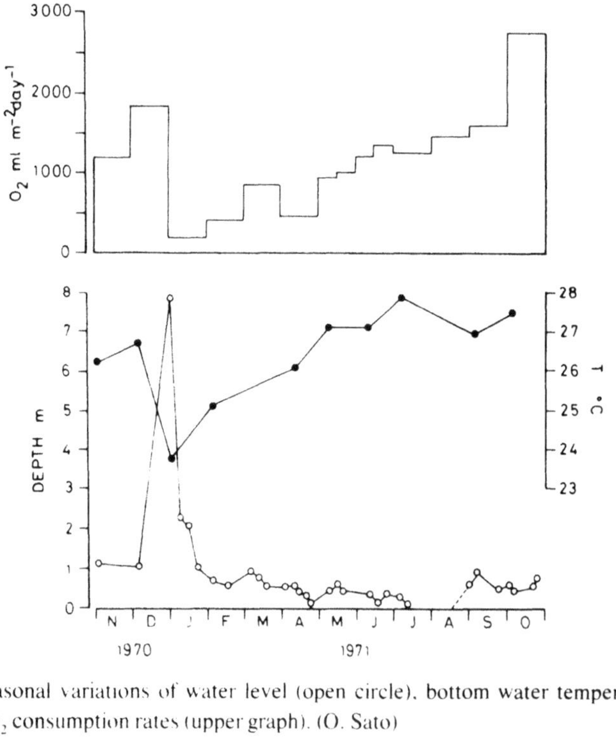

Fig. 62 Seasonal variations of water level (open circle), bottom water temperature (closed circle) and O_2 consumption rates (upper graph). (O. Sato)

such as respiration of organisms in water and respiration in mud, can be expressed as:

$$\frac{dC}{dt} = k_i C \tag{3}$$

where C is the oxygen concentration in water and k_i the deoxygenation rate coefficient of the unit system i.

Respiration of all compartmental processes can be expressed as follows:

$$\frac{dC}{dt} = -k_1 C - k_2 C \ldots - k_i C = -KC \tag{4}$$

or $\quad K = k_1 + k_2 \ldots + k_i$

Oxygen absorption rate is expressed as:

$$\frac{dC}{dt} = A\,(C_s - C) \tag{5}$$

where A is the oxygenation rate coefficient, C_s is the degree of oxygen saturation and C is the oxygen concentration in the water. Under natural conditions, oxygen is simultaneously absorbed into the water and consumed by the organisms. This process can be expressed as:

$$\frac{dC}{dt} = A\,(C_s - C) - lc \tag{6}$$

If we assume the water temperature and the degree of oxygen saturation to be constant, integration of equation (6) gives:

$$-\frac{1}{A + K} \ln AC_s - (A + K)\,C = t + \text{constant.} \tag{7}$$

If at t = 0, the swamp water does not contain any free oxygen, then C = 0 and

$$\text{constant} = -\frac{1}{A + K} \ln AC_s \tag{8}$$

The oxygen concentration at any time t is

$$C = \frac{AC_s}{A + K}\; 1 - \exp. - (A + K)\,t \tag{9}$$

If at t = 0, the swamp water is saturated with oxygen, then $C = C_s$, and

$$\text{constant} = -\frac{1}{A + K} \ln(-KC_s) \tag{10}$$

Thus, the oxygen concentration at any time t:

$$C = \frac{AC_s}{A + K}\ 1 + \frac{K}{A} \exp. -(A + K)\,t \tag{11}$$

In cases where the initial oxygen saturation is 0% and increases with time, or where the initial oxygen saturation is 100% and decreases with time, the exponential term becomes zero at infinite time and the oxygen concentration reaches constant values of $AC_s/(A + K)$ (Fig. 63). The changes in oxygen concentration are calculated according to equation (1) under the following conditions: A = 0.033 and K = 0.386 calculated on the assumption that water temperature and oxygen concentration are constant, although this is strictly not acceptable under natural conditions. However these assumptions are acceptable for approximate measurements in the swamp-forest. The mean oxygen concentration on 6 November 1972 in the swamp-forest was 0.52 $ml \cdot l^{-1}$ (range: 0.42–0.76 $ml \cdot l^{-1}$), the mean temperature was 27.0°C (range: 26.5–28.0°C) and the depth of the water was 30 cm. Using the assumption that the oxygenation rate is nearly constant at any given time, the oxygenation rate coefficient (A) and the deoxygenation rate coefficient (K) (i.e. respira-

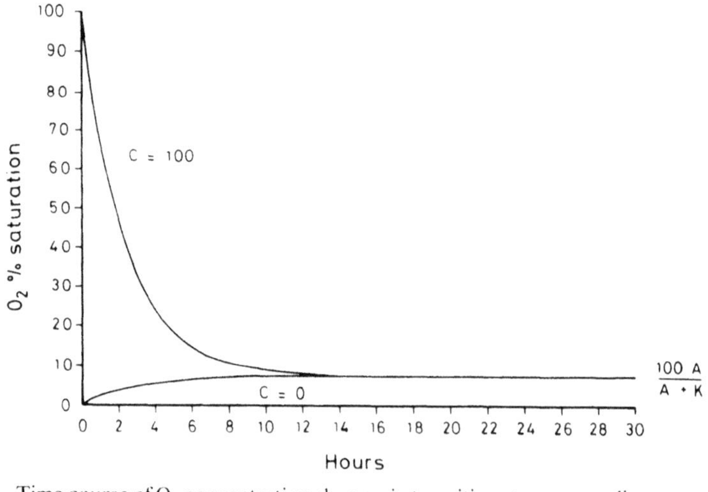

Fig. 63 Time course of O_2 concentration changes in transition stages according to equation (9). (O. Sato)

tion rate coefficient of the aquatic system) calculated from the oxygen absorption rate (1200 ml · m^{-2} · day^{-1}), oxygen concentration, mean temperature and the volume of water oxygenated (300 l) are 0.033 and 0.386 using equations 4 and 5 respectively. Equation 11 assumes that the oxygen concentration of the swamp-forest should never be zero even if the deoxygenation rate is very high.

The calculated rate coefficients indicate that oxygen consumption is approximately 10 times faster than oxygen absorption by water when dissolved oxygen saturation is at 50%. In this case, the oxygen concentration can be approximately assessed at one-tenth the saturated oxygen concentration which was calculated with constants derived from $AC_s/(A + K)$. Even if strong wind oxygenates the swamp water, the oxygen concentration of the water in the swamp-forest is maintained at a constant value of 0.4 ml · l^{-1} through the regulatory mechanism of the above dynamic system.

3.5.2 *Studies on the gases dissolved in water and in the bottom sediments* (T. Koyama and O. Sato)

This section attempts to define the mechanism of chemical changes in the swamp, focussing on gaseous metabolism in the water and sediments. The gases that are considered are oxygen, carbon dioxide, nitrogen and methane.

Total dissolved gases and some selected gases were initially measured to determine the status of the gases in the water throughout the day. Total dissolved gases in the water were measured *in situ* using the carbon dioxide technique (Sugawara 1939) modified by Oana (1957) and Koyama and Tomino (1968). Gases in the sediment were measured following Koyama's method (1953). Nitrogen, argon and methane were measured by gas chromatography using oxygen as the carrier.

A series of experiments were carried out at the principal sites of gas production (i.e. selected plants in the water and the sediments) in order to elucidate their effects on the chemistry of the water.

In the first set of experiments the gaseous metabolism of blue-green algae attached to *Nitella* sp. was determined using a modified light and dark bottle technique. One litre bottles were used and the light bottles contained different concentrations of the algae while the dark bottles were without algae. After incubation *in situ* for 8 hours (0730–1520 hours), oxygen, nitrogen and argon concentrations were determined.

Production of gases from sediments was measured in the laboratory as well as *in situ*. For laboratory experiments core samples 30 cm deep were collected and divided according to depth into 4-5 pieces and transported to the laboratory. Subsamples were then added to soil injectors and incubated at 26°C for 12 to 185 days. Gases produced were analysed using Koyama's method (1953).

In situ experiments were carried out using the apparatus shown in Fig. 64. The tunnel was placed on the surface of the bottom mud and the gas outlet (B) was connected to a prefilled bottle (C). Two sets of the apparatus were placed in the *Lepironia* reed swamp. After a few weeks the volume of the gas accumulated in the bottles (C) were measured.

Results and discussion

Oxygen concentration in the water was low (Table 43 and see earlier). Denitrified nitrogen was present in most of the surface water samples indicating that denitrification of organic matter occurs and is lost to the atmosphere (Table 43). The ratio of fixed carbon to fixed nitrogen ranged from 6.5 to 11.1 which was reasonable (Table 44).

The rate of carbon dioxide formation from the sediment cores was higher in the first 12 days of incubation when compared with the rate for the period between 12 to 188 days (Table 45). This may be due to the oxidation of the cores during its preparation for incubation and that the rate of carbon dioxide production between 12 to 188 days may reflect more realistically carbon dioxide production under natural conditions (average of 1.26 ml · day^{-1} · 100 g^{-1} dry sediments).

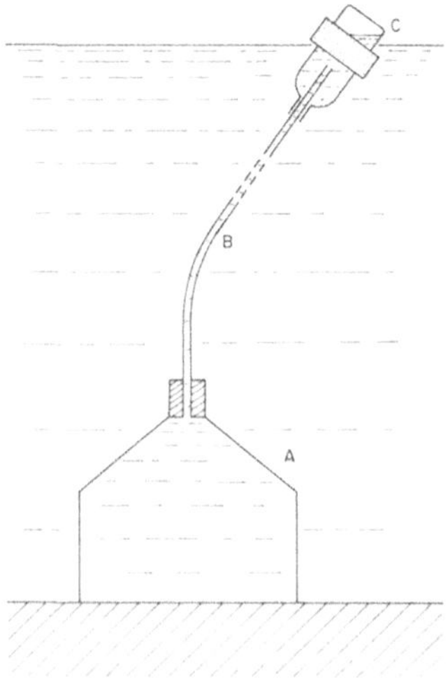

Fig. 64 Apparatus for collecting gas bubbles evolved from bottom sediments. (O. Sato)

Table 43 Diurnal variation of gases dissolved in Tasek Bera (12-13 Aug. 1973). (T. Koyama)

Station	and time	Depth m	Water temp. °C	O_2 ml/l	O_2 %	$N_2 + Ar$ ml/l	Ar ml/l	N_2 ml/l	$N_{2(Ar)}$ ml/l	Denitrified N_2 ml/l	N_2/Ar
St.	**12 Aug.**										
I	0750	0	25.0	0.78	13.5	12.40	0.289	12.11	11.85	+0.26	42.0
		1	25.5	0.75	13.1	12.57	0.308	12.64	12.27	+0.37	41.0
		3	25.5	0.82	14.3	12.95					
V	0810	1	25.3	0.61	10.6	12.57	0.302	12.27	12.07	+0.20	40.7
XI	0835	0.5	25.7	0.92	16.1	12.37	0.296	12.07	11.85	+0.22	40.8
XII	0825	0.5	25.7	0.92	16.1	12.50	0.299	12.20	11.96	+0.24	40.8
I	1200	0	27.1	1.40	25.2	12.57	0.293	12.28	11.76	+0.52	41.9
		1	26.7	1.26	22.5	12.57	0.300	12.27	12.00	+0.27	40.9
		3	26.7	1.12	20.0	12.68	0.296	12.34	+0.50	41.7	
V	1220	1	25.7	0.88	15.4	12.77	0.310	12.46	12.36	+0.10	40.2
XI	1245	0.5	28.6	0.95	17.6	12.64	0.296	12.34	11.84	+0.50	41.7
XII	1235	0.5	27.0	1.97	35.4	12.70	0.311	12.39	12.37	+0.02	39.8
Xi	1520	0.5	28.8	1.94	36.1	12.51					
XII	1500	0.5	27.0	1.97	35.4	12.12					
I	1800	0	28.0	2.18	39.9	12.51	0.302	12.21	12.07	+0.14	40.5
		1	27.2	1.83	33.0	12.46	0.315	12.14	12.54	−0.40	38.5
		3	26.8	1.84	33.0	12.54	0.312	12.23	12.52	−0.19	39.2
V	1820	1	26.1	1.46	25.8	12.43					
XI	1855	0.5	27.7	3.24	59.0	12.23	0.285	11.94	11.48	+0.46	41.8
XII	1840	0.5	26.6	2.00	35.7	12.09					
	13 Aug.										
I	0730	0	25.5	0.85	14.8	12.37					
		1	25.4	0.85	14.8	12.33					
		3	25.6	0.82	14.4	12.33					
V	0805	1	25.4	0.88	15.3	12.53	0.293	12.24	11.75	+0.49	41.8
XI	0830	0.5	25.6	0.78	13.7	12.71					
XII	0820	0.5	25.6	0.82	14.4	12.60	0.311	12.29	12.39	−0.10	39.5

Table 44 Experiments to determine the gaseous metabolism of blue-green algae attached to the surface of *Nitella* sp. from Tasek Bera (14 Aug. 1973) (Water temp.: 30.0°C). (T. Koyama)

Sample no.	Time	Relative amount of blue-green algae	O_2 $ml \cdot l^{-1}$	O_2 %	N_2+Ar $ml \cdot l^{-1}$	Form. O_2 $ml \cdot l^{-1}$	Fix. CO_2-C $mg \cdot l^{-1}$	Fix. N_2 $mg \cdot l^{-1}$	Fix. C/Fix. N
1	**14 Aug.**								
Initial state	0730	none	0.75	14.3	12.40				
2	1520	+	3.24	61.6	12.33	1.67	0.89	0.14	6.5
3	1520	++	8.43	160.3	12.09	6.86	3.67	0.44	8.4
4	1520	+++	9.58	182.1	12.13	8.01	4.29	0.39	11.1
5	Dark bottle	none	1.63	31.0	12.40				
6	Dark bottle	none	1.50	28.5	12.47				

Table 45 Formation of CO_2 in the core samples incubated at 26°C for different periods. (T. Koyama)

Station	Core depth (cm)	Incubation period (days) 0	12	188		
		Total CO_2			Form. CO_2*	Form. rate of CO_2
		$ml \cdot 100\ g^{-1}$ Dry sed.	$ml \cdot 100\ g^{-1}$ Dry sed.	$ml \cdot 100\ g^{-1}$ Dry sed.	$ml \cdot 100\ g^{-1}$	$ml \cdot day^{-1} \cdot 100\ g^{-1}$
I	0–10	37	75	98	23	0.1
	10–15	305	153	346	41	0.2
	15–20	296	568	1310	742	4.2
	20–25	187	170	156	– 14	–0.1
	25–30	66	120	239	119	0.7 (average 1.0)
XI	0–10	120	375	308	– 67	–0.4
	10–15	180	347	388	41	0.2
	15–20	90	174	574	400	2.3
	20–25	118	230	1160	930	5.3 (average 1.8)
XII	0–10	50	251	340	89	0.5
	10–15	32	106	159	53	0.3
	15–20	62	138	360	222	1.3
	20–25	48	122	259	137	0.8
	25–30	40	111	479	368	2.1 (average 1.0)

* Form. CO_2 was calculated from the total CO_2 formed during the incubation period from 12 days to 188 days.

The N_2/Ar ratio of gases produced from the core samples averaged 59.6 (Table 46) which is similar to the ratio obtained for the sediment gas collected in the surface waters (Table 48). However, these ratios are higher (39.5) than that of gases dissolved in water in equilibrium with air at 26°C. The higher values indicate that denitrification processes are taking place in the core samples. The amount of nitrogen production (Formed N_2) was obtained from the difference between the observed nitrogen content and 25.0 $ml \cdot 100\ g^{-1}$ which was the initial average nitrogen content of the samples. The rate of nitrogen production was 0.14 $ml \cdot day^{-1} \cdot 100\ g^{-1}$ dry wt of sediment. This rate was calculated by dividing the Form. N_2 by 173 days, the duration of the experiment (12-188 days).

The average rate of methane production from the sediment cores was 0.20 $ml \cdot day^{-1} \cdot 100\ g^{-1}$ dry wt of sediment (Table 46). This production rate is low when compared with that of Lake Suwa, Japan and sediments in ricefields which had rates of 5 $ml \cdot day^{-1} \cdot 100\ g^{-1}$ dry wt and 1 $ml \cdot day^{-1} \cdot 100\ g^{-1}$ dry wt respectively (Koyama 1964). This suggests that the sediments in Tasek Bera are poor in decomposable organic matter such as proteinaceous compounds when compared with other aquatic systems.

If it is assumed that there is no difference in the rate of gas production up to a depth of 1 m with a reduction in the production rate below 1 m, then the average rates of production of CO_2-C, CH_4-C and N_2-N in the swamp was 543 $mg \cdot m^{-2} \cdot day^{-1}$, 90 $mg \cdot m^{-2} \cdot day^{-1}$ and 137 $mg \cdot m^{-2} \cdot day^{-1}$, respectively (Table 47). The rate of carbon dioxide production is similar to those estimated from *in situ* experiments (see below).

Total C/N ratios of gases formed in the swamp range from 3.7 to 4.2 which are lower than the ratios obtained for general phytoplankton (5-6). This indicates that the mineralisation rate of nitrogen in organic matter is higher in relation to carbon mineralisation. However, this conforms to the rates generally found in lake sediments (Koyama 1954, 1964; Koyama and Tomino 1967, 1968).

The average rate of evolution of total gases from the sediments that was released into the atmosphere was 31.2 $ml \cdot m^{-2} \cdot day^{-1}$, while the evolution rates of gaseous CH_4, N_2 and Ar were 14.1, 16.8, and 0.3 $ml \cdot m^{-2} \cdot day^{-1}$, respectively (Table 48). This is equivalent to 8 and 21 $mg \cdot m^{-2} \cdot day^{-1}$ for CH_4-C and N_2-N, respectively that is released to the atmosphere.

This forms one-twentieth of the average production rates of CH_4-C and N_2-N from the sediment (Table 47). This great difference in rates may be due to the following reasons:

1) the evolution rates as gas bubbles might be an under-estimate due to the possibility that a large amount of gas bubbles evolved (more than 500

Table 46 Formation of Ar. N_2 and CH_{4f} in the core samples incubated at 36°C for 185 days. (T. Koyama)

Station	Core depth (cm)	$Ar+N_2+CH_4$ ml · 100 g^{-1}	Ar ml · 100 g^{-1}	N_2 ml · 100 g^{-1}	Form[a] N_2 ml · 100 g^{-1}	CH_4 ml · 100 g^{-1}	Form. rate[b] of N_2 ml · day^{-1} · 100 g^{-1}	Form. rate[c] of CH_4 ml · day^{-1} · 100 g^{-1}
I	0–10	60	0.7	42	18	17	0.10	0.10
	10–15	43	0.5	31	6	12	0.04	0.07
	15–20	73	0.9	52	27	20	0.16	0.12
	20–25	26	0.3	18	– 7	7	–0.04	0.04
	25–30	29	0.3	21	– 4	8	–0.02	0.05
						Average	0.05	0.08
	Composition of the mixed gas (Nos. 1-5): Ar: 1.18% : N_2: 70.9% : CH_4: 28.0% : N_2/Ar: 60.1							
XI	0–10	106	1.0	56	31	49	0.18	0.28
	10–15	98	0.9	52	27	45	0.16	0.26
	15–20	184	1.7	98	73	84	0.42	0.49
	20–25	143	1.4	76	51	66	0.29	0.38
						Average	0.26	0.35
	Composition of the mixed gas (Nos. 1-4): Ar: 0.95% : N_2: 53.0% : CH_4: 45.9% N_2/Ar: 55.8							
XII	0–10	171	1.3	80	55	90	0.32	0.52
	10–15	44	0.4	29	4	15	0.02	0.09
	15–20	59	0.6	38	13	20	0.08	0.12
	20–25	75	0.8	49	24	25	0.14	0.14
	25–30	52	0.6	34	9	17	0.05	0.10
						Average	0.12	0.19
	Composition of No. 1-gas: Ar: 0.77%; N_2: 46.6%; CH_4: 52.7%; N_2/Ar: 60.5							
	Composition of the mixed gas (Nos. 2-5): Ar: 1.05%: N_2: 65.2%: CH_4: 33.8%: N_2/Ar: 62.1							

[a] Formed N_2 was obtained as the difference between the observed N_2 and 25.0 ml · 100 g^{-1} which was obtained as an average N_2 contained in the initial samples.

[b] Formation rate of N_2 was calculated from the Form. N_2 (ml · 100 g^{-1}) by dividing by 173 (185-12) (days).

[c] Formation rate of CH_4 was calculated from the observed CH_4 (ml · 100 g^{-1}) by dividing by 173 (days).

Calculation of these formation rates was carried out on the assumption that the redoxipotential of the incubated samples was too high for formation of N_2 and CH_4 in the first 12 days of the incubation period.

Table 47 Average evolution rates of gases from the bottom sediments. (T. Koyama)

	Station I (Open water area)	**Station XI** (*Lepironia* swamp)	**Station XII** (*Utricularia* and *Lepironia* swamp)
Area (m^2)	11,000	47,000	12,700
Weight of dry sediment ($\times 10^6$ g)[a]	800	3,760	1,020
(Average evolution rate)			
CO_2-C mg/day/100 g dry Sed.	0.54	0.94	0.54
CH_4-C mg/day/100 g dry Sed.	0.04	0.19	0.10
N_2-N mg/day/100 g dry Sed.	0.06	0.33	0.15
CO_2-C kg/day/whole area[b]	4.7	36	5.5
CH_4-C kg/day/whole area	0.4	7.2	1.0
N_2-N kg/day/whole area	0.4	12	1.5
CO_2-C mg/m^2/day	430	770	430
CH_4-C mg/m^2/day	40	150	80
N_2-N mg/m^2/day	40	260	120
Total C/N[c]	11.8	3.8	4.3

[a] Weight of dry sediment of the core samples of 1 m in depth was calculated from the average density of 1.0, the average water content of 92% and area of each location.

[b] Average evolution rate from the whole area of each location was calculated on the assumption that the formation rates of the gases in the core samples of 1 m in depth are similar to those in the upper layers (0-30 cm) and that the formation ates are remarkably low in the deeper layers than 1 m in depth.

[c] Total C = CO_2-C + CH_4-C.

Table 48 Composition of gases from the sediment collected from the surface waters of Tasek Bera (13 Aug. 1973). (Oxygen was previously removed from the gas.) (T. Koyama and O. Sato)

	$ml \cdot m^{-2} \cdot day^{-1}$	%
N_2	16.8	53.8
Ar	0.3	0.91
CH_4	14.1	45.3
N_2/Ar		59

ml · 706 cm^{-1}) which were localised were not trapped by the apparatus, and
2) some of the gases produced in the sediments dissolved directly in the water at the mud-water interface.

3.5.3 *The distribution and dynamics of calcium* (T. Mizuno and M. Yoshimi)

The waters of Tasek Bera are soft and characterised by low calcium content and also the lack of molluscs although other organisms are quite abundant (see later). The vast amount of litter that falls into the swamp belies the amount of calcium that should be present in the waters. This section deals with the distribution of calcium in all the components both living and dead within the swamp system. From this the movement of calcium from one compartment to another is deduced.

The principal components that were considered were the water, the sediments (detritus), algae, macrophytes, invertebrates and fish. Samples collected were initially air-dried followed by drying in the oven at 60°C for 48 hours and cooled in a dessicator. The samples were than ground into a powder. A portion of the powder was weighed, ashed in a pot and the resultant ash was dissolved in 2 ml 6 N HCl solution. The solution was made up to 50 ml with distilled water and was neutralised with 6 N NaOH. The calcium content was than determined using the EDTA titrimetric method. (APHA-AWWA-WPCF 1971).

The calcium ion content in the water was 0.41 mg · l^{-1}. Ten principal plant types were analysed (Table 49). *Cryptocoryne griffithii* contained the highest concentration of calcium (9.09 mg · g dry wt^{-1}), and the dominant plants *Eugenia* sp., *Pandanus helicopus*, *Utricularia flexuosa* and *Lepironia articulata* contained 7.91, 7.51, 4.98 and 0.73 mg · g dry wt^{-1} respectively. These high concentrations could provide a source of calcium to the water when they die. The concentration of calcium in the leaves was consistently higher than in the roots of specific plants (Table 49). This indicates that there is little mobility of calcium from the leaves to the roots prior to dehiscence of the leaves. The calcium content was higher in larger size detrital particles when compared with the finer particles (Table 50). This could be due to te fact that finer particles tend to be more decomposed and also that their surface area to volume ratio is greater, enhancing leaching.

The average calcium content of 19 fish species analysed was 45.6 mg · g dry wt^{-1} with a range of 69.4 mg · g dry wt^{-1} and 22.2 mg · g dry wt^{-1} for *Ophicephalus striatus* and *Fluta alba* respectively. These values compare well with

Table 49 Calcium content in the principal plants (mg/g dry wt). (T. Mizuno and M. Yoshimi)

Lepironia articulata		*Nymphoides indica*	
living leaf	0.73	living leaf	4.72
dead leaf	0.72	root	2.41
root	0.65	*Pandanus helicopus*	
Eleocharis ochrostachys		living leaf	7.51
living leaf	1.92	dead leaf	8.55
root	0.79	living stem	4.00
Utricularia flexuosa	4.98	dead stem	3.44
Cryptocoryne griffithii	9.09	root	2.09
Nitella sp.	4.01	*Eugenia* sp. (Jambu)	
Blyxa echinosperma	4.04	living leaf	7.91
Attached algae	4.20	dead leaf	10.60

Table 50 Calcium content in detritus (mg/g dry wt). (T. Mizuno and M. Yoshimi)

No.	Rough 3 mm	Fine < 1-3 mm	No.	Rough 3 mm	Fine < 1-3 mm
1	2.46	–	11	1.40	–
2	2.74	–	12	0.68	–
3	1.83	1.65	13	1.90	–
4	2.56	1.85	14	1.60	–
5	0.68	0.82	15	1.85	–
6	0.83	0.89	16	0.90	–
7	1.00	0.89	17	0.92	0.59
8	1.29	–	18	1.11	–
9	1.69	–	19	1.21	–
10	3.04	–	20	1.29	–
Average				1.58	1.12

the same species from other localities (Table 51).

Among the invertebrates the shrimps contained the highest concentration of calcium with an average of 54.8 mg·g dry wt^{-1}, while their discarded shells after ecdysis contained 219.0 mg·g dry wt^{-1} (Table 52). Insect larvae such as Odonata nymphs and chironomid larvae contained substantially lower calcium content averaging 3.3 mg·g dry wt^{-1}

The distribution of calcium is summarised in Fig. 65. Among the plants three categories are distinguishable. *P. helicopus, Eugenia* sp. and *C. griffithii* belong to the same category with high calcium content. Attached

Table 51 Calcium content in fishes. (T. Mizuno and M. Yoshimi)

Species	No. indivs. sampled	Body length (mm)	Dry weight (g)	Ca mg/g dry wt	
				Lake Bera	River Marang
Ophlicephalus striatus	1	138	14.75	69.4	
Nandas nebulosus	2	46	0.99	63.2	
Lusiocephalus palture	1	84	1.46	61.6	
Trichopsis vittatus	2	38	0.84	58.3	55.3
Rasbora heteromorpha	11	15	0.22	57.4	65.1
R. dorsiocellata	8	13	0.11	54.6	
Puntius tetrazona partipentazona	3	38	2.45	56.4	
Notopterus notopterus	1	183	23.28	51.0	
Trichogaster leeri	1	52	1.86	46.5	
Cyclocheilichthys opogon (young)	1	86	7.03	45.5	
Cyclocheilichthys apogon (young)	1	44	0.90	43.5	
Pristolepis fasciatus	1	60	4.44	43.9	
Puntius fasciatus	1	60	1.23	42.5	
Oxygaster oxygastroides	1	52	0.51	35.6	53.9
Hemirhamphodon pogonognathus	22	29	0.49	31.9	31.5
Kryptopterus apogon	1	82	1.13	29.5	
K. macrocephalus	1	54	0.49	25.6	
Osteochylus hasselti	1	99	14.22	29.0	
Fluta alba	1	260	3.20	22.2	
Average				45.6	

Table 52 Calcium content in shrimps and insects. (T. Mizuno and M. Yoshimi)

Aquatic animals	Size	No. indivs.	Total dry wt.	Ca-ion mg/g dry wt.	Average mg/g dry wt.
Shrimps	large	1	0.67	52.9	
	intermediate	2	0.39	56.0	54.8
	small	11	0.53	55.6	
Exuvae of shrimps		1	0.04	219.0	
Insects Odonata nymphs		14	0.09	3.17	3.3
Chironomid larvae		338	0.10	3.42	

algae and *U. flexuosa* belong to the second category wile *L. articulata* and *E. ochrostachys* belong to the low calcium content category. Fish and the shrimps had high calcium content and this could be related to the fact that

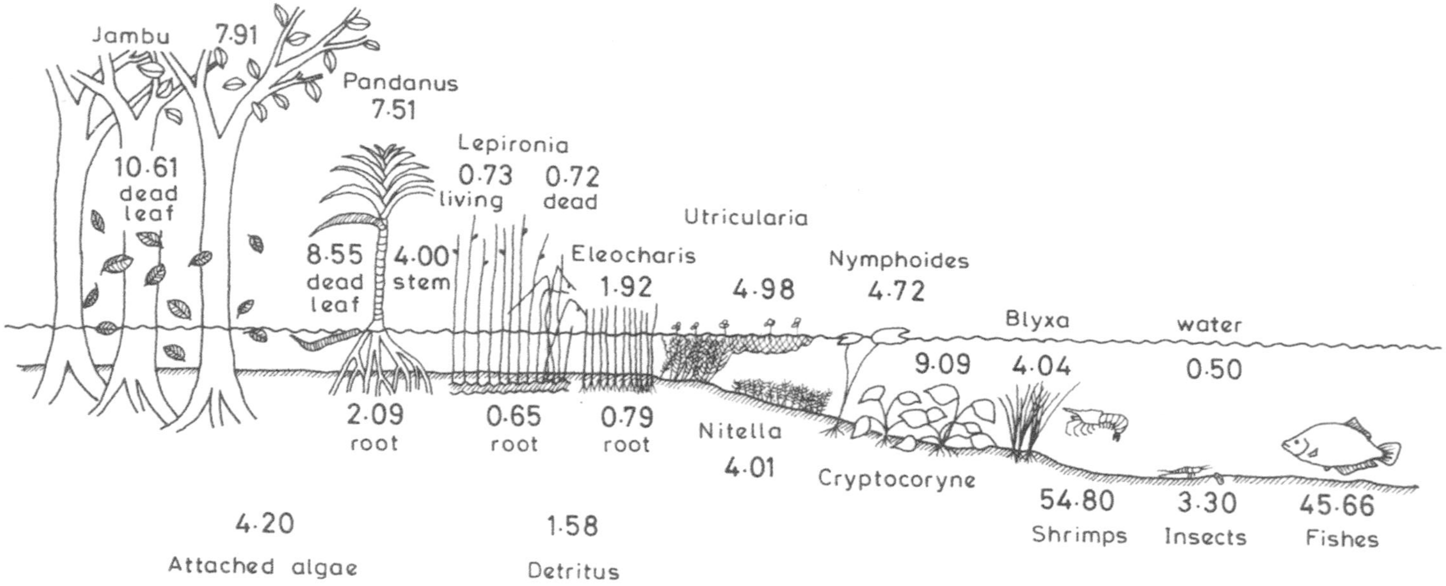

Fig. 65 Distribution of calcium in the aquatic ecosystem of Tasek Bera (Ca mg · g dry wt^{-1}). (T. Mizuno and M. Yoshimi)

they form the highest trophic level and are better at concentrating calcium. The insect larvae on the other hand did not have high calcium content despite their exoskeletons.

Rate of calcium release. Samples of known weight representing various plant species were placed in plastic cages wrapped in cloth and incubated in swamp water. At regular intervals over a year, samples were removed to determine the rate of decomposition and calcium loss.

Utricularia flexuosa which composed mainly of soft tissue decomposed almost completely within a year (Fig. 66). Plants that contained more fibrous tissue like *L. articulata* and *P. helicopus* decomposed more slowly losing about 32–33% of their initial weights in a year. The decomposition rate of *Eugenia* sp. and *E. ochrostachys* were 50% and 69% respectively over the year.

Calcium was leached out very rapidly in *U. flexuosa* and *E. ochrostachys*, i.e. 99.9% and 97.5% was lost within half a year respectively. *P. helicopus* and *Eugenia* sp. lost 94.5% and 62.3% in a year respectively. This could explain the low concentrations of calcium in detritus.

Stomach content of fishes. The food habits of the dominant fish were qualitatively and quantitatively analysed. Only three species, *Osteochilus hasselti* (Valenciennes), *O. spilurus* Bleeker and *Labiobarbus festiva* (Haeckel) were herbivorous and the others were carnivorous (Table 53). An attempt on quantitative analysis of the gut content of the herbivorous fish on a second occasion was not successful as their guts were empty. The gut of carnivorous fish contained larvae, nymphs and pupae of insects, shrimps, zooplankton and young fish (Table 53). The stomachs of *Notopterus notopterus* (Pallas) and *Pristolepis fasciatus* (Bleeker), both of which are nocturnal, contained fish scales, shrimps and insect larvae. *Puntius fasciatus* (Bleeker) and *P. tetrazona partipentazona* (Fowler) fed selectively on chironomid larvae, statoblasts of moss animalcules (Bryozoa), zooplankton and small terrestrial animals.

The average calcium ion content of inland waters in the world is estimated at 22.02 mg·l^{-1} (Livingstone 1963). However, a wide range does occur. In Africa the calcium content ranged from 51.70 mg·l^{-1} in the Blue Nile (Talling and Rzoska 1967) to 1.20 mg·l^{-1} in Lake Tumba (Beadle 1974). Most of the inland waters of Africa have high calcium content. In the Amazon basin, waters ranged from very poor concentrations (0.02–0.2 mg·l^{-1}) in blackwaters, to 0–19.5 mg·l^{-1} in clear water, to 2.9–6.16 mg·l^{-1} in the Devonian strips of the lower Amazon basin, and 3.1–204.0 mg·l^{-1} in the Carboniferous

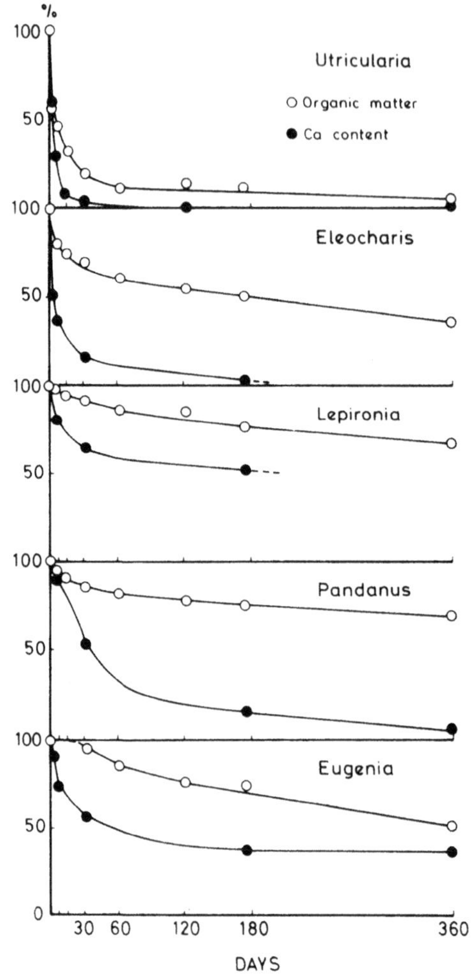

Fig. 66 Comparison between the decomposition rate of organic matter and the dissolution rate of calcium in the aquatic plants and litter. (O. Sato)

strips of the lower Amazonia (Sioli 1968; Sioli et al. 1969). In Thailand the average calcium content was 19.8 $mg \cdot l^{-1}$ while most of the water samples in Japan contained less than 19.9 $mg \cdot l^{-1}$ averaging at 9.5 $mg \cdot l^{-1}$ (Kobayashi 1958). In Malaysia, Johnson (1967) reported the calcium content to be poor and comparable with the results of Tasek Bera.

In general calcium dissolves slowly during decomposition processes (Alliwill 1968), but results from experiments in Tasek Bera seem to indicate that the calcium in litter dissolves rapidly in acid-soft waters. Since calcium is poor in the water itself, coupled with the fact that acid waters do not deposit

Table 53 Stomach content of fishes in the swamp in Tasek Bera. (Shiraishi et al. 1972).

Types of food	A	B	C	D	E	F
Terrestrial small animals						
Acarina (ticks)	35	5				
Araneae (spiders)	1				1	
Ant						1
Other insects		3				
Aquatic insects						
Chironomidae	553	139	13	124	168	18
Ephemeroptera	9		6	36	168	12
Trichoptera	10	9			3	
Odonata			3			
Hemiptera			2			2
Other aquatic animals						
Shrimps	1			38	10	2
Bryozoa	3485	190				
Tubificidae	1					
Hydracarina	3					
Fish scales	2			20	10	1
Zooplankters						
Rhizopoda	130					
Rotatoria	43					
Brachiopoda	283		4			1
Copepoda	275		1		5	3
Phytoplankters						
Cyanophyceae	980					
Bacillariophyceae	101					
Chlorophyceae	10	10				
No. of individuals examined	181	96	41	16	22	20

A = *Puntius fasciatus;* B = *P. tetrazona partipentazona;* C = *Cyclocheilichthys apogon;* D = *Notopterus notopterus;* E = *Pristolepis fasciatus;* F = *Krytopterus macrocephalus.*

calcium carbonate, it is inferred that it is rapidly taken up by the plants and animals soon after its release. Although the processes of calcium absorption by plants and animals were not studied, the fact that most of the fishes are carnivorous suggests that they obtain calcium from food such as insects, shrimps and young fish which contain fair amounts of calcium.

Calcium intake could also occur through the gills and fins of fishes as has been demonstrated by Mashiko and Jozuka (1962) and Jozuka et al. (1963) who examined the process of calcium absorption of crucian carp in fresh-waters and *Duymaria flagellifera* in the sea. However, no experimental work has been done to determine if the amount of calcium absorbed through the gills and fins is greater than the amount obtained through its food.

It is really not known whether the absence of molluscs in Tasek Bera is due to poor calcium concentrations in the water. However, it could be postulated that perhaps the concentration of calcium is below the threshold requirements of successful shell construction.

Based on assumptions the dynamic process of calcium in Tasek Bera can be constructed (Fig. 67). An estimate of about 459.0 $mg \cdot m^{-2} \cdot y^{-1}$ of calcium is recycled in the system from death of aquatic plants which comprises about 60% of the total calcium found in the grasses and leaves of ***P. helicopus***. In the swamp-forst where *Eugenia* spp. is considered dominant, about 5000 $mg \cdot m^{-2} \cdot y^{-1}$ (62.3% of calcium contained in litter) passes through the water. Other sources of calcium in the swamp-forest are from detritus, animal faeces, dead animals, castings of invertebrates, etc.

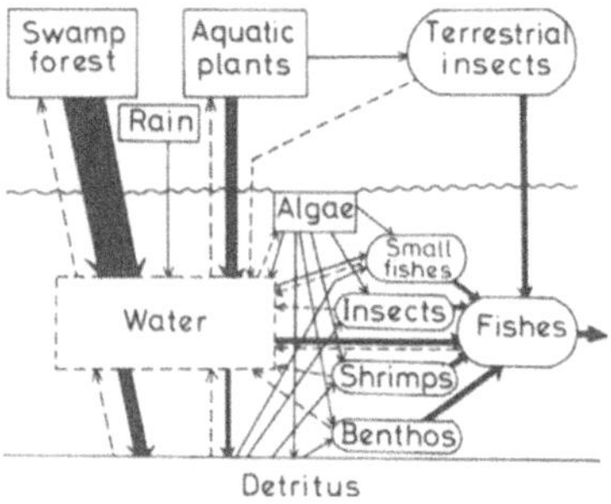

Fig. 67 A simplified model of calcium flow in the study area at Tasek Bera. (T. Mizuno and M. Yoshimi)

Since the swamp-forest composes the major proportion of Tasek Bera (see Table 1) the main source of calcium should be from the litter of the forest. The 5000 $mg \cdot m^{-2} \cdot y^{-1}$ which contrasts greatly with the amount in the water, must therefore be taken in by plants, algae and secondary producers which are then transferred to the next higher trophic level.

Figure 67 shows the movement of calcium in this swamp on the basis of the data obtained. However there are still some unsolved problems.

4. Decomposition

O. Sato, Y. Tezuka and T. Koyama

Biological aspects of decomposition have been reviewed in two valuable volumes by Dickinson and Pugh (1974). Most of the studies on the decomposition of plant litter have been done on terrestrial systems in temperate regions. Very few studies have been carried out in fresh-water systems, especially dystrophic lakes and freshwater swamps like Tasek Bera.

Decomposition refers to weight losses in dry litter due to mineralization and solubilization of the organic matter, and particulate losses. The primary and main consumers of litter in the swamp appear to be micro-organisms. Unfortunately microbial studies on litter decomposition in aquatic ecosystems in the tropics have been rather scarce. The decomposition of organic matter to carbon dioxide and oxygen consumption by respiration have been discussed in the previous chapter.

The main energy source in the aquatic environment in the swamp-forest as a whole is the litter falling from the canopy, since most of the photosynthetically active radiation is intercepted by trees before reaching the water surface. Furthermore, the colour of the water markedly reduces light penetration, particularly in the red region of the spectrum. Studies on litter-fall, litter accumulation and litter decomposition will give a better understanding of the dynamics of the swamp ecosystem. The decomposition of litter has a marked influence on the water quality, such as dissolved oxygen content, nutrient content, hydrogen ion concentration and organic matter content of the tropical swamp-forest.

4.1 Distribution and activity of microbes (Y. Tezuka)

As described in previous chapters, Tasek Bera is a highly complicated eco

Furtado, J. I. and Mori, S., eds., Tasek Bera. ISBN 90 6193 100 2.

system with unique physiographic features. Although several microbiological studies have been carried out in freshwater ecosystems in temperate regions, such studies are rather scarce in tropical regions. One of the few such studies is that of Schmidt (1969), which describes the vertical distribution of bacteria and algae in a tropical lake, Lago do Castanhe which is connected to the Amazon River. Lago do Castanhe resembles Tasek Bera in that the water level fluctuates significantly throughout the year. However, it is a harmonic lake whereas Tasek Bera is a dystrophic one. Schmidt's (1969) data show that the number of saprophytic bacteria in the Lago do Castanhe was low (less than $500 \cdot ml^{-1}$) in contrast to the relatively large numer of algae in that lake. This study on the microbes of Tasek Bera is important and interesting because of the paucity of such studies in tropical freshwater ecosystems.

4.1.1 *Population density and standing crop of bacteria and fungi* (Y. Tezuka, R. P. Lim and J. I. Furtado)

Bacteria. Bacterial density in water was determined by two methods. One was to count the number of viable, heterotrophic bacteria by the pour plate method, using a dilute (1/10 strength) nutrient agar. This medium is one of the best for measuring the number of viable, heterotrophic bacteria in freshwater. The plates were incubated at room temperature (25-30°C) for 7 days.

The other method was to count the number of bacterial colonies collected in net plankton, using an Utermohl inverted microscope. The counting of total bacteria in water under a microscope was difficult because of the presence of a large number of non-bacterial, colloidal particles in the water samples. In the case of mud, only viable, heterotrophic bacteria were counted by he same method used in water.

The number of viable, heterotrophic bacteria at Station I is shown in Fig. 68, where it can be seen that the surface water is poorer in bacteria than the lower layers of the water body. Generally, the water at a depth of 1 m is richest in bacteria. The number of bacteria at Stations V, XI, and XII is shown in Table 54. No great difference in the number of bacteria is seen among the surface waters at Stations I, V, XI, and XII. However, bottom waters at Stations XI and XII contain about one order of magnitude more bacteria than their surface waters. The water of Tasek Bera is poor in bacteria and comparable to the oligotrophic lakes in Japan (Tezuka and Kawai 1975). This may be due to the inaccessibility of organic matter to the bacteria due to prevailing environmental conditions.

The bacterial density in the net plankton is low, averaging 74.3 colonies $\cdot l^{-1}$

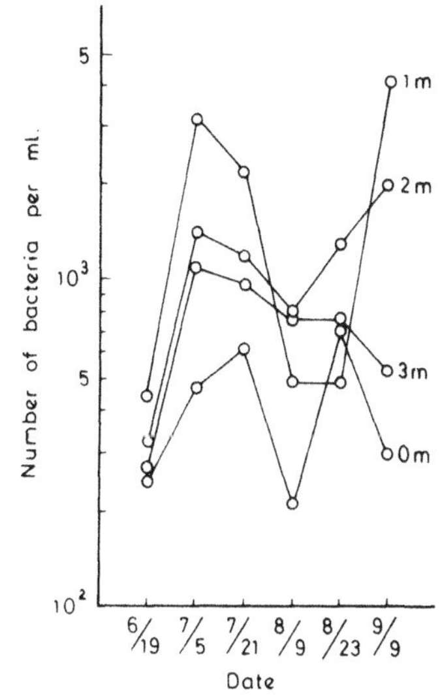

Fig. 68 The density of viable, heterotrophic bacteria at depths of 0, 1, 2 and 3 m in the water body at Station I (open water), observed in 1973. (Y. Tezuka)

Table 54 Mean, minimum and maximum densities of viable, heterotrophic bacteria in waters at Stations V, XI and XII. Observations were made on five occasions from 5 July to 9 September 1973. (Y. Tezuka)

Station		Density (cells · ml^{-1})		
		Mean	Minimum	Maximum
V	(0 m)	4.7×10^2	2.2×10^2	6.8×10^2
XI	(0 m)	6.9×10^2	6.2×10^2	9.8×10^2
XI	(0.6–0.9 m)	4.5×10^3	1.6×10^3	6.7×10^3
XII	(0 m)	4.8×10^2	2.5×10^2	6.3×10^2
XII	(0.6–0.9 m)	6.0×10^3	2.4×10^3	1.1×10^4

with a maximum of 695.3 colonies · l^{-1} (October, Station II) and a minimum of 2.3 colonies · l^{-1} (December, Station II) (Table 55). The *Pandanus-Utricularia* sublittoral has the highest density of bacteria while the swamp-forest stream channel (Station V) appears to have the lowest density. There is seasonal fluctuation in bacterial density with an abundance of bacteria immediately after the monsoonal rains, especially the northeast monsoon. This seasonality may be due to the flushing of bacteria from the forest floor of the

Table 55 Mean, minimum and maximum densities of bacteria in the net plankton at Stations I-V in Tasek Bera (1971–1972). (R. P. Lim and J. I. Furtado)

Stations	Density (No. of colonies · litre^{-1})		
	Mean	Minimum	Maximum
Limnetic, man-modified			
Station I	64.4	6.1	332.9
(Main channel)		(Dec. 1971)	(March 1971)
Sublittoral, man-modified			
Station II	121.6	2.3	695.3
(Pandanus-Utricularia)		(Dec. 1971)	(Oct. 1971)
Station III	60.1	19.6	186.9
(Utricularia)		(Oct. 1971)	(Aug. 1971)
Littoral, man-modified			
Station IV	67.3	2.9	233.9
(Lepironia)		(Nov. 1971)	(March 1971)
Limnetic, swamp-forest			
Station V	55.3	3.3	123.5
(Main channel)		(Oct. 1971)	(March 1971)
Overall	74.3	2.3	695.3
		(Dec. 1971)	(Oct. 1971)

watershed into the swamp by run-off. The bacterial species have not been identified but in gross morphology they resemble species inhabiting seepage waters especially those inhabiting the forest floor (Ratnasabapathy, pers. com.). Their occurrence and source are still unclear and need further detailed study.

The reason for the low standing crop of viable, heterotrophic bacteria in Tasek Bera may be the low organic matter content available for the nutrition of bacteria. To verify this premise, water samples were enriched with glucose and incubated. The results (Table 56) indicate that glucose does not stimulate bacterial growth. This suggests that the factor governing the low standing crop of heterotrophic bacteria is not necessarily restricted to the lack of available carbon source. In a subsequent experiment both glucose and mineral nutrients (nitrogen and phosphorus) were added to water samples and incubated. From this experiment, it is evident that nitrogen and/or phosphorus is limiting the growth of heterotrophic bacteria in Tasek Bera (Table 56). The low levels of the standing crop of both bacteria and phytoplankton in the open water indicate the low mineral nutrient content of the water.

Table 56 Effect of nutrient additions to the limnetic water on the growth of heterotrophic bacteria. Sample waters were incubated at 25-30°C in the laboratory after nutrients were added. (Y. Tezuka)

I. Effect of glucose
Sample water was collected on 8 August 1973 at Station I (0 m depth). Glucose was added at a concentration of 10 mg · l^{-1}.

Incubation time (days)	0	1	2	4	7
No. of bacteria · ml^{-1}	4.7×10^3	3.2×10^3	4.5×10^3	4.0×10^3	5.7×10^3

II. Effect of glucose, nitrogen and phosphorus
Sample water was collected on 23 August 1973 at Station I (0 m depth). Concentrations of nutrients added were: glucose, 20 mg · l^{-1}; $(NH_4)_2SO_4$, 10 mg · l^{-1}; KH_2PO_4, 10 mg · l^{-1}.

Incubation time (days)	0	1	2	4	7
No. of bacteria · ml^{-1}	8.7×10^4	3.6×10^6	2.2×10^6	1.9×10^5	4.6×10^5

Table 57 Mean, minimum and maximum densities of viable, heterotrophic bacteria in muds at Stations I, XI and XII. Observations were made on six occasions from 18 June to 7 September 1973. (Y. Tezuka)

Station		Density (cells · g^{-1} wet or dry mud)		
		Mean	Minimum	Maximum
I	Wet	4.3×10^4	3.0×10^4	7.6×10^4
	Dry	5.5×10^5	2.9×10^5	1.1×10^6
XI	Wet	3.0×10^5	1.2×10^5	3.9×10^5
	Dry	3.4×10^6	1.5×10^6	4.3×10^6
XII	Wet	3.1×10^5	1.5×10^5	4.4×10^6
	Dry	3.2×10^6	1.4×10^6	4.3×10^6

Table 57 shows that the bottom muds at Stations XI and XII contain more bacteria than that at Station I. The ash content of the mud at Stations I, XI and XII were determined twice. Mean values of 46.6, 50.8 and 54.2% of dry weight of mud were obtained for Stations I, XI and XII, respectively, showing a slightly higher organic matter content at Station I. This could imply that the bottom at Station I contains organic matter which is more resistant to bacterial decomposition than that of the other two stations. Nevertheless it is obvious that the bottom mud of Tasek Bera is poor in bacteria. Hogetsu et al. (1954) reports that peat at a depth of 0-50 cm in Ozegahara moor in the highlands of Central Japan contains heterotrophic bacteria at densities between 1.1 × 10^6 and 1.7 × 10^7 cells·g^{-1} fresh peat. The pH of the peat is between 4.2 to 4.8. Even the surface muds of many Japanese lakes with less organic matter content contain much more heterotrophic bacteria (Tezuka and Kawai 1975). The fact that Tasek Bera has such a low bacteria content in both the water and mud poses an interesting problem.

Fungi. Since bacteria proliferate under weak alkaline conditions while fungi thrive under weak acidic conditions, it was expected that Tasek Bera would support an abundant growth of fungi as its water is weakly acidic and has a high organic matter content. Hence the number of fungi was determined by the following two methods. One was by plate counts using Martin's medium (Martin 1950) which has been widely used for the determination of soil fungi. The plates were incubated at room temperature (25–30°C) for 3 to 5 days. The other method was to count fungal spores collected in the net plankton using an Utermohl inverted microscope.

The number of viable fungi in the water (Table 58) ranges from 4 to 160 ml^{-1}, although more are present in the bottom muds. As in the case of bacteria, there is a difference in the abundance of fungi in the mud at Station I when compared with those at Stations II, XI and XII. It is not possible to offer a comparison of these data with that of any other region in the world since virtually no research in this field has been done. Nevertheless, it may be suggested that bottom muds are the sites where decomposition of organic matter occurs most actively because much more bacteria and fungi are present in the bottom muds than in the overlying water.

The density of fungal spores in the net plankton is low, averaging 120.2 spores·l^{-1} (minimum 0 and maximum 1,961.6 spores·l^{-1}) (Table 59). The highest fungal spore density is in the modified and forested limnetic and the *Lepironia* littoral (Stations I, V and IV). Their abundance is of a short duration, appearing in the January (northeast monsoon) floods, reaching a peak in the subsequent dry season (March) and declining after the dry period. This short duration of fungal abundance in the net plankton suggests that they may be tychoplankton, most probably from the forest floor of the watershed. They are thus probably associated with decomposition processes and their short lifespan in the plankton may be attributed to changes in water conditions towards the end of the dry season and the decomposition processes.

4.1.2 *Microbial contribution to decomposition in water and mud* (Y. Tezuka)

Unfortunately, there is no simple and accurate technique for assessing *in situ*

Table 58 Mean, minimum and maximum densities of viable fungi in water and mud. Observations were made on five occasions from 19 June to 9 September 1973. (Y. Tezuka)

Station		Density (No. of fungi · ml^{-1})		
		Mean	Minimum	Maximum
1. Water (No. · ml^{-1})				
	(depth m)			
I	0	21	6	57
	1	21	10	43
	2	24	6	59
	3	52	8	160
V	0	22	4	54
XI	0	31	5	57
	0.6–0.9	76	23	145
XII	0	28	12	51
	0.6–0.9	153	4	54
2. Mud (No. g wet or dry mud^{-1})				
I	Wet	4.0×10^3	8.0×10^1	1.7×10^4
	Dry	5.3×10^4	8.9×10^2	2.3×10^5
XI	Wet	1.9×10^4	4.5×10^3	3.0×10^4
	Dry	2.0×10^5	5.5×10^4	3.1×10^5
XII	Wet	2.7×10^4	6.3×10^3	4.8×10^4
	Dry	9.1×10^5	6.4×10^4	2.8×10^6

the microbial contribution to the decomposition of organic matter in aquatic environments. Hence, only a qualitative study of the decomposition of organic matter was attempted.

Water. To determine if the organic matter in the limnetic water of Tasek Bera is microbiologically labile, the following two experiments were carried out.

The oxygen uptake of limnetic water incubated in the laboratory was measured. The results (Table 60) show that more than 2.5 mg · l^{-1} of oxygen is consumed by the water during 30 days of incubation. This suggests that the organic matter in the water is microbiologically labile. However, the second experiment, which measured changes in the organic carbon content of the limnetic water during its incubation in the laboratory, does not verify this inference. The concentration of organic carbon (Table 61) was almost con-

Table 59 Mean, minimum and maximum densities of fungal spores in the net plankton at Stations I-V in Tasek Bera (1971–1972). (R. P. Lim and J. I. Furtado)

Stations	Density (No. of spores · litre^{-1})		
	Mean	Minimum	Maximum
Limnetic, man-modified			
Station I (Main channel)	215.9	0.0 (June 1971, Oct. 1971, Dec. 1971, Jan. 1972)	1961.6 (March 1971)
Sublittoral, man-modified			
Station II (*Pandanus-Utricularia*)	58.6	0.0 (Jan. 1971, June 1971, Oct. 1971, Dec. 1971)	422.5 (March 1971)
Station III (*Utricularia*)	6.0	0.0 (Jan. 1971, June 1971, Oct.-Nov. 1971)	36.8 (April 1971)
Littoral, man-modified			
Station IV (*Lepironia*)	116.8	0.0 (Jan. 1971, June 1971, Oct. 1971)	1244.2 (March 1971)
Limnetic, swamp-forest			
Station V (Main channel)	203.5	0.0 (Jan. 1971, Jan. 1972)	1948.9 (March 1971)
Overall	120.2	0.0 (Jan. 1971, June 1971, Oct.-Dec. 1971, Jan. 1972)	1961.6 (March 1971)

Table 60 Oxygen consumption of the surface water collected at Station I on 22 July 1973. The water sample was incubated at 25-27°C in the dark. (Y. Tezuka)

Incubation time (days)	0	5	10	20	30
Oxygen consumed (mg · l^{-1})	0	1.26	1.91	2.53	2.62

stant for 60 days, suggesting the stability of the organic matter in the water. Therefore nothing conclusive can be drawn from these experiments, since the experiments were carried out only once. However if duplicates of these

Table 61 Changes in total organic carbon (TOC) and microbes of the swamp water incubated at 25-27°C in the dark. The water sample was collected at 0 m depth of Station I on 19 June 1973. (Y. Tezuka)

Incubation time (days)	0	5	15	31	60
TOC ($mg \cdot l^{-1}$)	9.1	9.1	9.8	8.5	9.4
Bacteria (number $\cdot ml^{-1}$)	2.5×10^2	5.5×10^3	1.6×10^4	2.2×10^4	7.4×10^2
Fungi (number $\cdot ml^{-1}$)	11	6	12		

Table 62 Oxygen consumption of mud. The samples were collected on 9 August 1973. Wet mud was suspended in the mineral water and incubated at 25-27°C in the dark. (Y. Tezuka)

Station	Oxygen consumption	
	$mg\ O_2 \cdot kg$ wet $mud^{-1} \cdot week^{-1}$	$mg\ O_2 \cdot kg$ dry $mud^{-1} \cdot week^{-1}$
I	34	450
XI	198	2,050
XII	120	1,220

experiments were carried out, it may be possible to postulate that the rather active oxygen uptake of the water is not concerned with the actual breakdown of organic compounds, but with the oxygenation of organic compounds, e.g. aromatic compounds of the humic substances. The data on the stability of colouring substances produced from litter (see later) strongly suggests that the organic matter contained in limnetic water is stable and resistant to microbial attack, i.e. the rate of organic matter decomposition within the water is fairly low. Low standing crop of microbes also supports this supposition.

Mud. A single experiment to measure the oxygen consumption of mud suspended in nutrient-enriched distilled water was conducted. Oxygen consumption of the mud collected from Station I is much lower than that of mud from Stations XI and XII (Table 62). The results are consistent with the abundance of microbes in the respective mud samples at the various stations. Values of 120 and 198 $mg\ O_2 \cdot kg$ wet wt of $mud^{-1} \cdot week^{-1}$ of oxygen consumption for mud collected from Stations XI and XII are very much lower than those with less organic matter content from harmonic lakes in Japan (Hayashi, pers. comm.). It is therefore suggested that the sediments of Tasek

Bera decompose at a fairly low rate. Additional information to support this view was obtained by incubating dried samples of *Lepironia* immersed in nutrient-enriched water, where only 6% of organic matter mineralized after 30 days incubation at 25–30°C.

4.2 Decomposition of plants (O. Sato)

Decomposition of allochthonous matter in aquatic ecosystems has been studied mainly to understand the nutrient cycle and the factors governing an aquatic ecosystem like oxygen consumers and nutrient suppliers (Willoughby 1974). Allochthonous matter, an important source of energy in any ecosystem, plays a more important role in swamp-forests than in any other aquatic ecosystem. Very little information on production and decomposition of litter in tropical swamps is available (Klinge and Rodrigues 1968a, 1968b). This study is aimed at estimating the production and decomposition rates of litter in the *Eugenia* swamp forest and also to study the changes in the litter quality during the process of decomposition.

4.2.1 *Litter production by forest trees*

Litter can by classified into two categories for convenience, i.e. fine litter which can be collected in litter traps and coarse (wood) litter. The amount of fine litter, i.e. leaves, twigs, barks, fruits, etc., can be estimated more accurately by means of litter traps than that of wood litter. In this study only the production and decomposition rates of fine litter are considered.

Ten litter traps (1 m^2 each) were placed with the rims 1.5–2 m above the ground, depending on the fluctuations in water level during the wet season, in a 20 × 60 m^2 plot. Litter fall was collected once or twice a month. The litter samples were separated into three categories i.e. leaves, twigs and others, and dried in an oven at 85°C for 48 hours. Each of these was then weighed.

Figure 69 shows the pattern of litter-fall in the traps over a period of 5 months, together with the water level fluctuations, an indication of climatic conditions. Sixty-six percent of the litter consisted of leaves while most of the remainder was twigs. The rate of litter fall was 1.7 $g \cdot m^{-2} \cdot day^{-1}$ during the wet season. Twig litter fall increased markedly during the dry season. John (1973) studied litter fall in a moist semi-deciduous forest in Ghana, tropical Africa and related the major peak of litter fall to the short, dry season. In

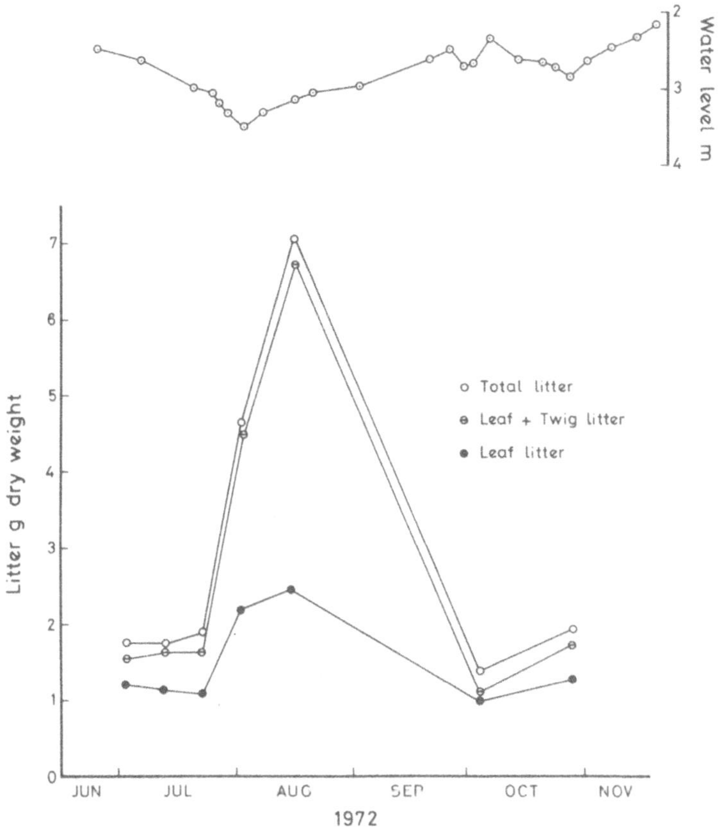

Fig. 69 Litter-fall and water level in Tasek Bera. (O. Sato)

tropical rain forests, litter fall has been considered to be more or less continuous throughout the year. However, some researchers have observed that there was a tendency for extensive litter fall during or shortly after relatively dry periods, even in humid equatorial forests. The peak of litter fall in this swamp-forest can also be related to a short, dry season from June to August.

It was difficult to estimate the average annual litter fall from the data for five months since the litter fall rate varied seasonally. Hence, the annual litter fall was estimated in two ways. One was to extrapolate from the average daily litter fall. This gave an over-estimate of the annual litter fall because the dry season, during which more litter falls, occupied half of the observation period whereas in actual fact, it occupies only a quarter of the year. In another estimation of litter fall, the litter fall rates in the dry and wet seasons were estimated separately. The annual litter fall was then calculated, based on the assumption that a year has a dry season of three months, with the remaining nine months making up the wet season. The results are shown in Table 63 where row A is based on the first estimation while row B is the latter calculation. The true annual litter production is probably a value between the two values obtained, because during the observation period of 33 months from August 1970 to March 1973, only a very short dry period was observed from March to April, judging from fluctuations in the water level in the swamp.

The estimates of leaf litter production correspond to the lower range of values of litter production in moist semi-deciduous and moist evergreen forests in West Africa (John 1973). However, total litter fall values are higher than those of moist semi-deciduous forests (John 1973). The total litter production of the *Eugenia* swamp forest is somewhat lower than that of a typical rain forest as described by Jensen (1974).

4.2.2 *Amount of dying plants*

Litter accumulation on the swamp forest floor in the dry season was higher than values obtained in typical rain forests in Malaysia. The typical soil profile of the swamp-forest in the study area consists of a 4–5 cm deep A_0 layer (raw humus layer), a 10–15 cm deep A horizon, which is coloured by humic substances, a red B horizon and a greyish-green clay horizon. The ground water level was about 15 cm below the soil surface in the dry season of 1972.

No termite nests or ant nests were observed in the swamp-forest. Invertebrate activity as sweepers of the litter on the ground in the tropical rain forest seem to be very limited in the dry season because of the high humidity of the forest soil. This limited role of the invertebrates in the decomposition of litter

Table 63 Dry weight of litter fall ($g^{-2} \cdot y^{-1}$). (O. Sato)

Calculation method	Leaves	Twigs	Others	Total
A	550	480	100	1,130
B	460	240	100	800

A was calculated from the average litter fall rate.
B was calculated on the assumption that the litter fall rate was constant through a year except the dry season.

implies that the disappearance of litter on the soil surface is mainly due to microbial decomposition and mineralisation.

Quantitative sampling of the A_0 layer (raw humus layer) was carried out on 28 July 1972. Five quadrats measuring 50×50 m^2 were sampled. The litter was then treated in the same manner as litter collected in litter traps. It was impossible to separate partly decayed leaf litter from small fruits, seeds, flowers and others to a satisfactory degree. Hence these were also classified together with leaf litter. The dry weights of each of the samples are given in Table 64. The amount of litter on the ground could be measured only once in the dry season, whereas it was not possible to do so in the wet season since the forest floor was submerged. The amoung of leaves and twigs are five times greater than the dry weight of the standing crop of litter on the ground in a semi-deciduous forest (John 1973).

The relationship between the mean yearly temperature and organic matter in the A_0 layer is shown in Figure 70. The accumulation of organic matter in the A_0 layer in the swamp-forest floor is 7 to 8 times greater than the values reported in other tropical forests. The rate of litter supply to the forest is rather low when compared to those of West Africa and Thailand (John 1973, and Yoda and Kira 1969, respectively). In spite of the lower litter supply rate, the accumulation of organic matter on the swamp-forest floor is greater than that in other tropical forests. The greater accumulation of organic matter on the swamp-forest suggests that the swamp water covering the forest floor inhibits the decomposition of organic matter by reducing the amount of oxygen available to the organisms.

4.2.3 *Weight-loss of dead plants through decomposition*

The decomposition rate of litter was estimated by three methods. In the first method, the decomposition rate of litter on the forest floor was obtained by dividing the mean weight of the ground litter by the annual amount of litter fall. In the second method, the decomposition rate of leaf litter was determined by the following experiment. Leaf litter (about 100 g dry wt), collected from the swamp-forest, was packed into each of ten Salan net bags. These bags were set on the forest floor on 1 August 1972. The bags were collected after 32 days, dried in an over at 85°C for 48 hours and weighed to determine weight losses in the litter. Five of these bags were then placed on the ground while the remaining five were placed at the bottom of the swamp. These ten bags were collected after 117 days, dried and weighed to determine weight losses in the litter content. In the third method, plastic cages containing

Table 64 Dry weight of the standing crop of ground litter. (O. Sato)

Quadrate number	Leaves ($kg \cdot m^{-2}$)	Twigs ($kg \cdot m^{-2}$)	Total ($kg \cdot m^{-2}$)	Leaf/Total %
1	1.94	0.32	2.26	86
2	2.22	2.77	4.99	44
3	1.29	0.53	1.82	70
4	1.46	0.54	2.00	73
5	1.18	0.50	1.68	70
Mean*	1.62 ± 0.44	0.93 ± 1.03	2.55 ± 1.38	67

* Mean + one sigma.

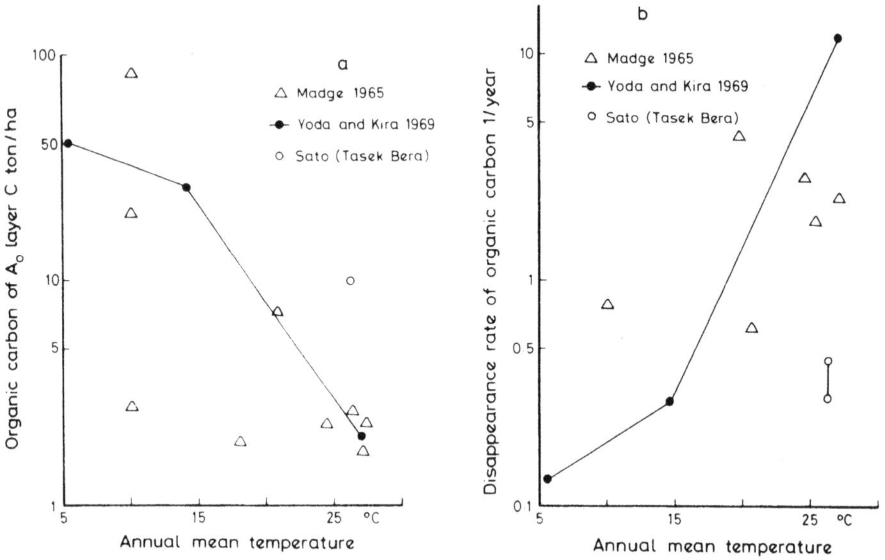

Fig. 70 Relationship between the mean yearly temperature (°C) and organic matter in the A_0 layer (tons $C \cdot ha^{-1}$). (O. Sato)

Eugenia leaves were kept in water channels of the swamp for a whole year and the weight losses were measured at specific intervals.

In the second and third methods, the decomposition rates were calculated on the assumption that the rate of decomposition of litter can be expressed by the following equation: $dm/dt = -km$, where m is the weight of the litter and k (per year) the decomposition rate coefficient. The validity of this equation for calculating the decomposition rate was confirmed by an experiment which will be discussed in the following subsection.

The decomposition rate coefficients of leaves and twigs were estimated to be 0.3–0.4 · y^{-1} and 0.3–0.5 · y^{-1} respectively, by the first method (Table 65). Twig litter appears to decompose somewhat more slowly than leaf litter. With the second method the weight loss of the litter was 4.2% for the first 32 days in the dry season and 15% for 149 days (Table 66). The decomposition rate coefficient was calculated to be 0.5 · y^{-1} for the period of 32 days and 0.4 · y^{-1} for 149 days. Decomposition proceeded more rapidly during the dry season than in the wet season. However, during the wet season, there was no

Table 65 Rate of litter disappearance (y^{-1}). (O. Sato)

Method	Leaf	Twig	Total
A	0.4	0.5	0.45
B	0.3	0.3	0.3
C	0.4-0.5		
D	0.6		

A and B were calculated from the figures in the rank A and B in Table 63 respectively.
C was determined by bag method.
D was estimated by the model experiment described in Section 4.

Table 66 Decomposition of ground litter. (O. Sato)

Sample	Starting material	Residue after 32 days	Loss of weight	Residue after 149 days	Loss of weight
	g	g	%	g	%
1	96	91	5	81	16
2	89	86	3	76	15
3	102	99	3	86	16
4	118	113	4	101	14
5	105	100	5	88	16
6	101	96	5	85	16
7	113	109	4	96	15
8	117	112	4	100	14
9	110	104	5	93	15
10	118	113	4	98	17
Mean			4.2		15

Samples 1-5 were allowed to stand on the soil through a whole period and samples 6-10 were allowed to stand on the soil during first 32 days an then in lake water for 117 days.

significant difference in the weight losses of litter placed on the ground when compared with those placed at the bottom of the swamp. The decomposition of litter on the ground in the swamp-forest was probably prevented by excess water from rain on every day of the wet season. The decomposition rate of *Eugenia* leaves as estimated by the third method was $0.6 \cdot y^{-1}$ which is the highest decomposition rate obtained in the experiments.

The slight variations obtained in the rate coefficients are negligible when compared with the differences in decomposition rates due to temperature (Fig. 70b). Here it is clearly seen that decomposition is inhibited in spite of the high temperature. Although temperature is the most important factor controlling litter decomposition it cannot be a limiting factor in the tropical swamp-forest. One of the major factors affecting litter decomposition in the swamp-forest is that development of anaerobic conditions is accelerated in the water-logged litter. However, the water-logged condition is not the sole reason for the slow decomposition of the litter on the ground. The population of soil animals in the swamp-forest is considerably lower when compared with that in the tropical rain forests where termites contribute greatly to the disappearance of ground litter.

The relationship of the decomposition rates to other biological, chemical and physical conditions will be discussed in the following subsection.

4.2.4 *Chemical changes in decomposing plants*

Of the many species of plant life, as shown in the list of flora of Tasek Bera, five dominant species, *Eugenia clavifolia, Utricularia flexuosa, Eleocharis ochrostrachys, Pandanus heliocopus* and *Lepironia articulata* were studied with respect to the physico-chemical nature of the plant materials on decomposition.

Living leaves of *Eugenia clavifolia*, whole *Utricularia flexuosa* plants, and living stems of *Eleocharis ochrostrachys*, killed by submersion in hot water (80°C) for 3 minutes and dried, were used for this experiment. Freshly shed leaves of *Pandanus heliocopus* and standing dead stems of *Lepironia articulata* were collected and dried. The leaves of *Pandanus*, stems of *Lepironia* and *Eleocharis* were cut to 5 cm lengths. Each species of the plant material was mixed thoroughly, placed in aliquots of 5 g into plastic cages and enveloped in Salan cloth. The *Eugenia* leaves and the *Utricularia* plants were treated in the same manner.

The plastic cages were grouped into eleven samples, each sample consisting of three cages of each leaf type, giving a total of fifteen cages per sample. All the cages except for 15, which were used for chemical analyses to determine the initial nutrient status, were hung at a depth of 25 cm from bamboo rafts floating in the water. At intervals of 1, 2, 4, 8, 16, 32, 64, 123, 178 and 380 days a set of cages was lifted from the water. The plant material remaining in the cage was dried at 85°C for 48 hours, weighed and milled to a fine powder.

The carbon, nitrogen and hydrogen content of the plant residues were measured with a CHN-corder MT-IS Yanako (manufactured by Yanagimoto Mfg. Co. Ltd).

The carbohydrate content of the plant residues was estimated employing the following technique. 0.5 ml of 72% sulphuric acid was added to about 20 mg of the powdered plant material and kept overnight at room temperature. 12 ml of water was then added to the mixture and the resulting solution heated on a water bath for 2 hours. The hydrolysate was then cooled in running water and filtered through a glass-fibre filter. The carbohydrate content of the hydrolysate was determined by the phenol sulphuric acid method (Handa 1966), using suitable dilutions of the filtrate.

Figure 71 shows the logarithms of the percentage weights of the plant materials suspended in swamp water, plotted as a function of time. In the first

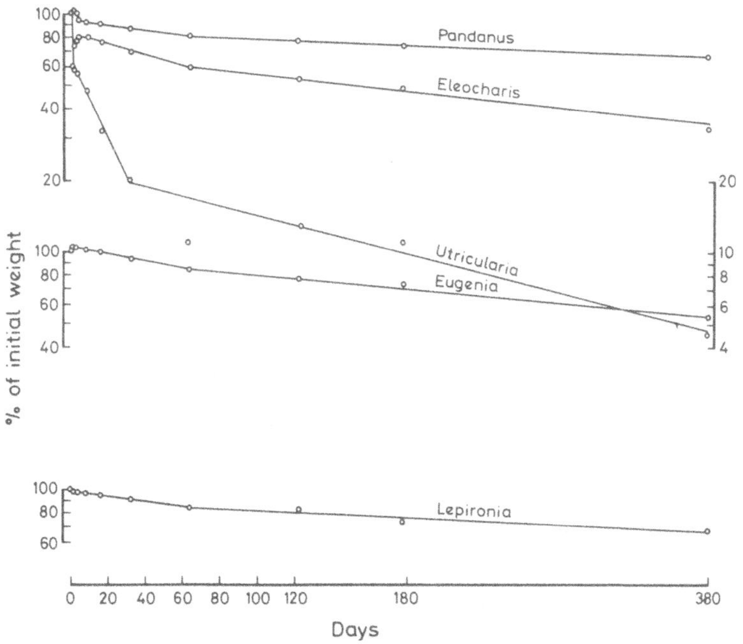

Fig. 71 Percentage weight loss of plant material suspended in swamp water. (O. Sato)

day, 30% of *Eleocharis* stems and 40% of *Utricularia* disappeared by leaching. However, the weights of *Pandanus* and *Eugenia* increased during the initial four days. This increase in weight was also observed from the second to the eighth day. This increase in total organic matter of *Pandanus* and *Eugenia* during the initial days may be due to the growth of bacteria and fungi on the surface of the plants.

The decomposition process of organic matter can be divided into two stages – an early stage of 2 months where decomposition is rapid, probably due to leaching of soluble organic matter and the decomposition of readily decomposable organic matter by micro-organisms such as fats, starch and proteins and at a later stage where the decomposition rate is relatively slower.

Forty percent of *Utricularia* disappeared after one day. After one year only 2% of the original organic matter remained in the cage. In the case of *Eleocharis* 40% of the organic matter remained behind after one year. *Eugenia* leaves, *Lepironia* stems and *Pandanus* leaves decomposed at a fairly steady rate during the early and later stages except for the slight weight increase on the first day. They have similar trends of rapid weight loss in the first two months followed by a slower rate in the later period. About 50% of the initial weight of *Eugenia* and 30% of *Lepironia* and *Pandanus* disappeared in one year.

Iversen (1973) showed that autumn-shed beech leaves placed in net bags in a Danish spring brook (water temperature 1–12.8°C) lost 90% of their weight in one year. Mathews and Kowaleczewski (1969) studied the decomposition of three types of leaf litter i.e. oak (*Quercus robur* L.), sycamore (*Acer pseudoplatanus* L.) and willow (*Salix* spp.) and found that the oak litter disappeared more slowly than the other two species, but by the end of a year, almost all the litter disappeared. In field experiments, Kaushik and Hynes (1968) exposed leaf discs in a river and found percentages of ash-free weight losses after six months of exposure of *Ulmus, Acer, Quercus, Alnus* and *Fagus* to be 68.9, 56.8, 23.1, 21.8 and 13.0, respectively.

Although these experiments were conducted at lower temperatures than the present study, almost all the litter of beech, oak, sycamore and willow disappeared within a year. On the other hand, about 50% or more of the organic matter of *Eugenia, Pandanus* and *Lepironia* remained after decomposition in Tasek Bera (water temperature: 23-30°C). The higher rate of decomposition of litter at lower temperatures may be due to the more fragile quality of the leaf material. Also, the slower rate of decomposition in the tropical swamp, Tasek Bera, may be attributed to poor quality of its water – low oxygen content, low nutrient concentration and low pH which is not conducive to much microbial activity.

Since the decomposition curves are linear (Fig. 71) the decomposition rate of organic matter in the cages can be expressed by the following equation: $dm/dt = -km$, where m is the weight of plant residues and k is the decomposition coefficient, which is similar to that described by Olson (1963) for decomposition rates of terrestrial litter. Hence this equation allows the calculation of the rate of decomposition and also the time taken to lose 95% (3/k).

The differences in the decomposition rates during the initial stages of decomposition, which is affected by many factors such as leaching, preparation of samples and growth of microorganisms, is difficult to explain. However, there was a general trend of rapid loss of plant material during the initial stages and more slowly during the later stages.

It appears that losses in dry weight are affected strongly by solubilization of organic matter during the initial stages. Therefore, true decomposition is represented by the subsequent decay of the organic matter. The true decomposition parameters can be calculated from the slope of the straight lines in the later stages. The disappearance (decomposition) coefficients at the later stages are listed in Table 67. Here the differences in the rates of weight loss of five plant species may be caused by the various physical and chemical nature of the plants. *Pandanus* and *Lepironia* can be classified as rigid and tough plants, *Eugenia* and *Eleocharis* as intermediate and *Utricularia* as fragile. The tougher plants naturally have the lowest decomposition rate.

The C and N content and C/N ratio of the plant tissues as a function of the exposure time is given in Fig. 72. There are small variations in the C content of the plant residues throughout the decomposition period. The variations in the N content are also slight except in the case of *Utricularia* and *Eleocharis* which initially had vere low nitrogen content. However the variations in the

Table 67 Disappearance coefficient and time required for the loss of 95% of the original plant material for *Utricularia*, *Eleocharis*, *Eugenia*, *Lepironia* and *Pandanus*. (O. Sato)

Sample	Disappearance coefficient		95% weight loss $3/k_2$ per year
	Initial stage k_1 per year	Later stage k_2 per year	
Utricularia	13.5	1.16	2.6
Eleocharis	0.83	0.68	4.4
Eugenia	1.43	0.58	5.2
Lepironia	0.98	0.27	11.1
Pandanus	0.55	0.18	16.7

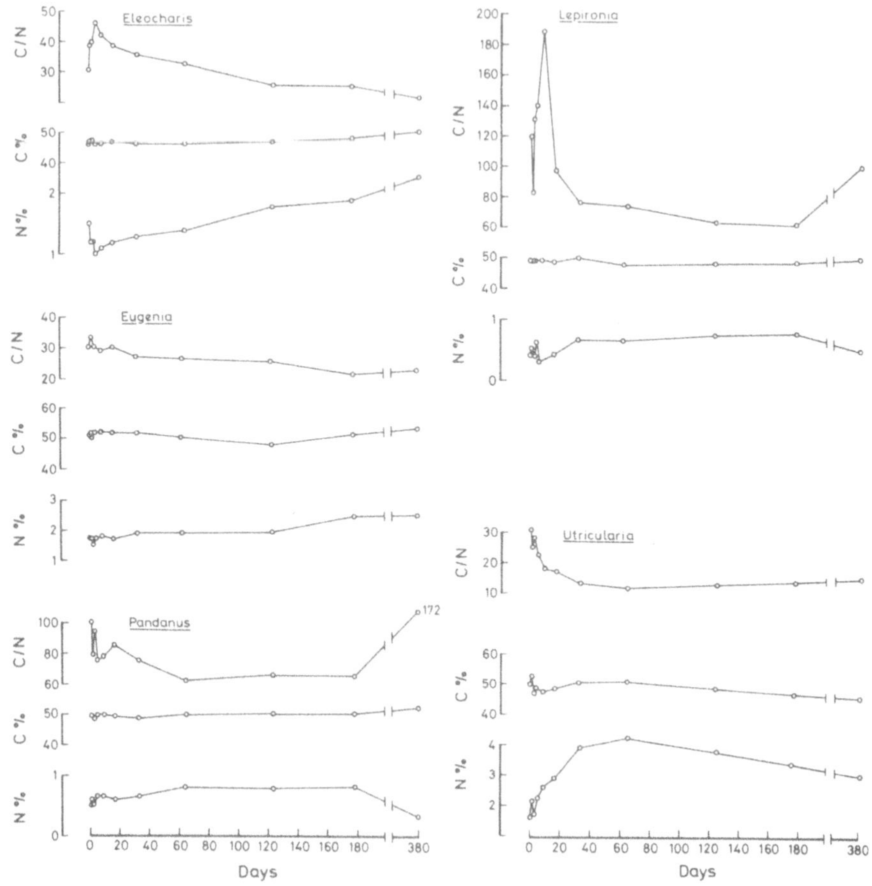

Fig. 72 Carbon and nitrogen content (%) and C/N ratio of various plant tissues as a function of exposure time (days). (O. Sato)

C/N ratio are remarkable (Fig. 72). The N content of *Utricularia* increased rapidly initially, its C/N ratio reaching a value of 10 at the end of this initial stage. In the later stages of decomposition, the C/N ratio of *Utricularia* remained fairly constant at about 10. *Eleocharis* and *Eugenia*, decomposing at slower rates showed an increase in N content throughout the experimental period, but there was a steady decrease in the C/N ratio. Although the N content increased in *Pandanus* and *Lepironia*, their changes were less marked.

Figure 73 shows the percentage changes in the N content of plant samples in the cages as a function of exposure time. During the initial 16 days, the N

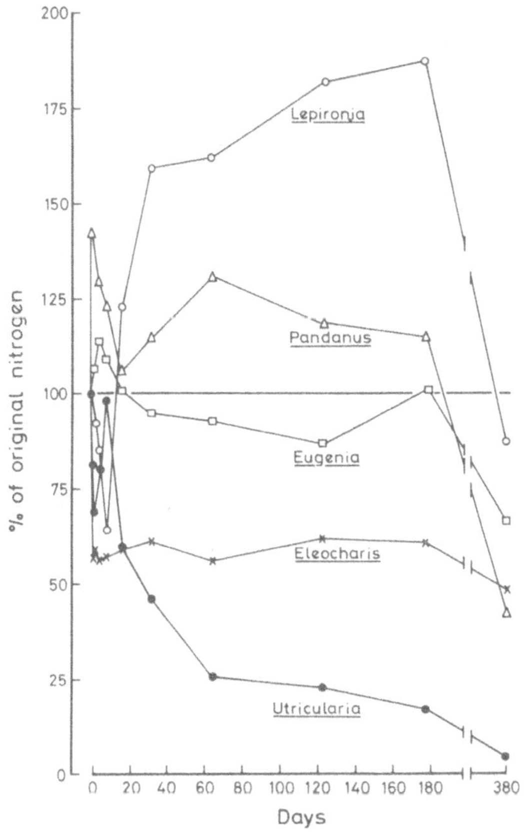

Fig. 73 Percentage changes in nitrogen content of various plant samples as a function of exposure time. (O. Sato)

content in the plant samples show complex variations caused by leaching of N-compounds into the swamp water and microbial growth incorporating N-compounds from the swamp water. The content of *Lepironia and Pandanus* whose original N content was low, increased with time for six months, while that of *Utricularia* and *Eugenia* decreased with time. About 40% of the N content of *Eleocharis* was lost by leaching on the first day, after which the N content remained fairly constant. From these studies, *Utricularia* and *Eugenia* appear to act as the N sources of the swamp during their decomposition while *Lepironia* and *Pandanus* act as the N consumers in the early stages of their decomposition. The N content of mud in the *Lepironia* swamp was 1–2% of the organic matter and higher than those of *Lepironia* and *Pandanus*. This indicates the consumption/absorption of N compounds in the swamp water by *Lepironia* plant residues in the early stages of decomposition.

The changes occurring in the chemical composition of the plant residues are given in Fig. 74. Protein content was calculated from the N content on the assumption that all nitrogen is proteinaceous and that the N content of protein is 16%. The original carbohydrate content of *Utricularia* was about 50% while that of the final sample was only 5%. This decrease in carbohydrate may mean that the carbohydrate in the *Utricularia* is available for the activities of bacteria and fungi. Ten percent of the total weight of *Eleocharis* is water soluble and this disappeared in the first day. The increase in carbohydrate content during the first eight days might be due to microbial growth in the sample surfaces. The chemical composition of the other three species changed very slowly with time. The percentage of carbohydrate in the samples decreased slowly while those of lignin and other components increased slightly with time.

The relationship between the decomposition rate and the N content of the samples after 64 days at which time the later phase of disappearance begins is seen in Fig. 75a. It is seen that the plants with higher protein content have higher disappearance rates than those with lower protein content.

Figure 75b shows the relationships between the disappearance rate and the carbohydrate content of the plant residues after decomposition for one year. It is observed that plant samples with higher carbohydrate content have lower disappearance rates. Therefore the amount of stable carbohydrates like cellulose is closely related to the stability of plant materials in the swamp water.

The role of invertebrates in the decomposition of litter on the forest floor has been widely recognised. Mathews and Kowaleczewski (1969) showed that animals do not play as great a role in aquatic systems. Salan bags which were used in this experiment excluded most of the invertebrate fauna and therefore, the disappearance of the organic matter was due only to microbial decomposition and physical leaching. The biomass of invertebrate fauna in Tasek Bera is small, especially in the swamp mud because of the extremely low oxygen content of the swamp bottom. In view of Mathews and Kowaleczewski's (1969) work and the small invertebrate biomass of Tasek Bera, it is reasonable to postulate that the role of animals in the decomposition of plant materials at the bottom of the swamp is negligible.

4.3 Origin of black water
(O. Sato)

Yellow to brown colouring material is present in most river and lake waters although their concentrations may not be sufficient to be visible to the naked

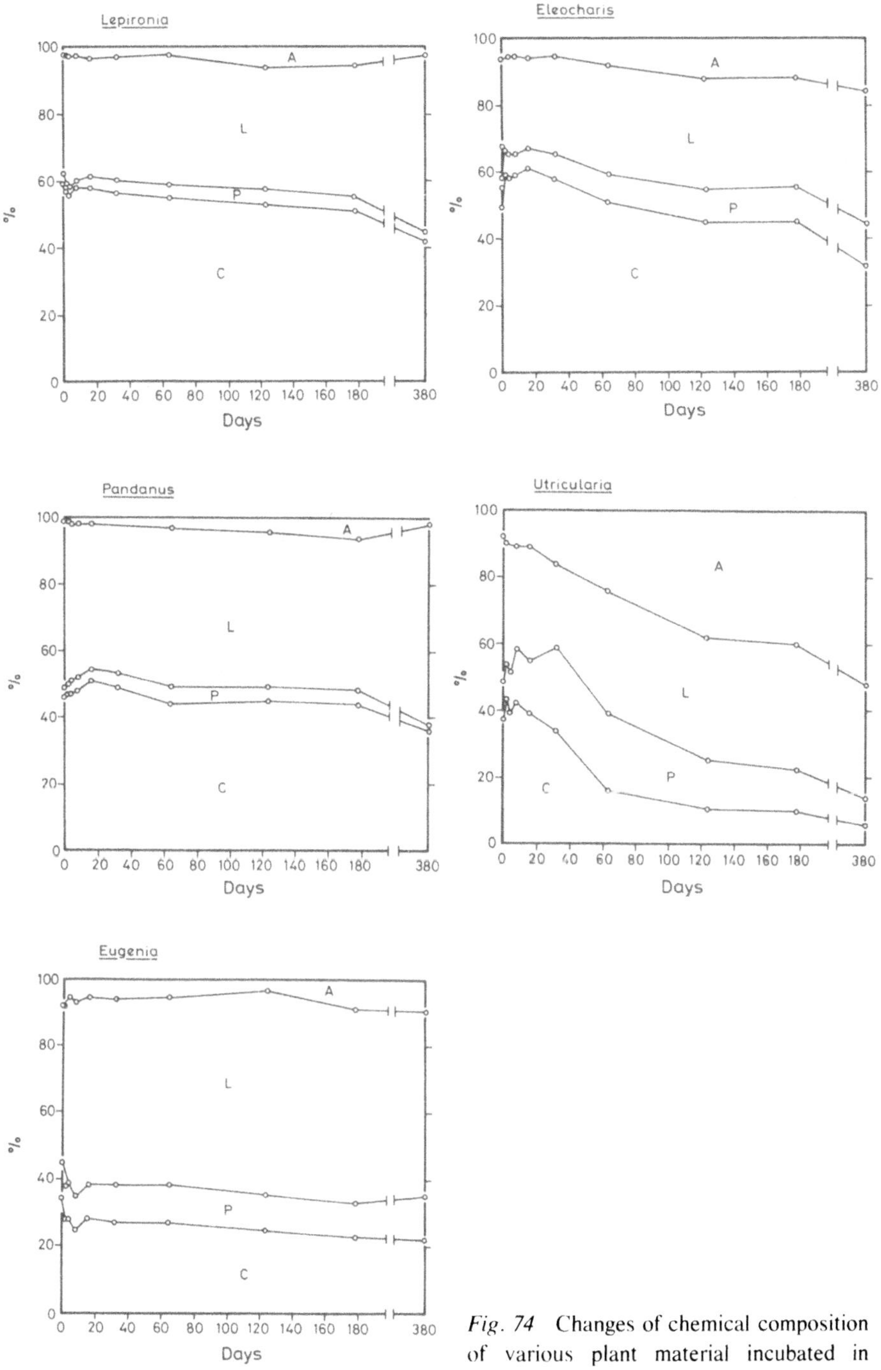

Fig. 74 Changes of chemical composition of various plant material incubated in swamp water. (O. Sato)

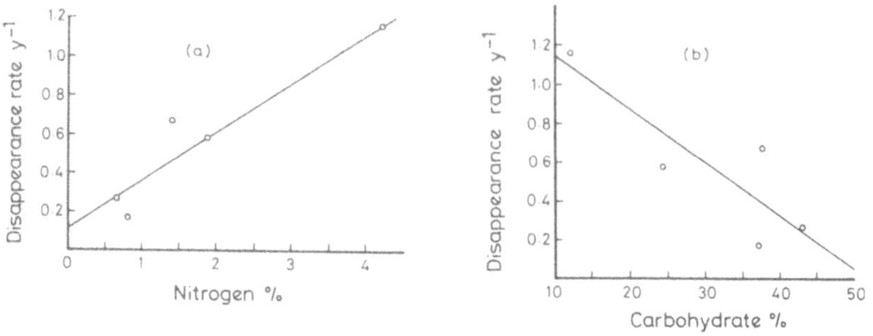

Fig. 75 Relationship of disappearance rates of plant material to (a) nitrogen content of 64-day-old samples and (b) carbohydrate content of 380-day-old samples. (O. Sato).

eye. Information regarding the colouring organic material in black water, i.e. their origin and genesis, physical and chemical properties and their physical roles is very scarce, especially of tropical black waters despite their wide distribution.

The increasing demand for water supplies for public consumption has accelerated studies on organic matter in black water. Technologists have interested themselves in removing the organic colouring material from black water such that the water can be made available for public consumption (Black et al. 1963). Others (Shapiro 1957; Black 1963; Christman and Gassemi 1966; Midwood and Felbeck 1968; Fotiyev 1968; Christman and Minear 1971) studied the chemical nature of colouring material in water by applying soil science techniques developed for characteriszation of soil humic substances. They concluded that the yellow substances which separated from surface waters are fluvic acids and aromatic poly-hydroxy-methoxy carboxylic acids. Their polycondensate nature has been recognised by using Sephadex gel filtration techniques (Christman and Minear 1971).

Limnologists have been interested in their influences on primary production such as (a) their effects on light attenuation, (b) chelating effects on algal growth (Shnitzer 1971), (c) effects on organic phosphorus compounds (Christman and Minear 1971), and (d) effects of some unknown growth factors (Prakash and Rashid 1973).

The geochemical roles of colouring material have been the subject of much speculation although information on the subject is limited. On the basis of their physical and chemical properties it is considered that colouring material plays a prominent role in controlling the concentration of metals in waters (Manskaya and Drozdova 1968; Shapiro 1964).

Yellow to brown colouring material are released during the degradation of organisms. Sioli (1964, 1969) reported that the waters of Rio Negro in the Brazilian Amazon Basin became coloured while passing through tropical podzol in the catchment area. Colour in Tasek Bera, however, does not seem to be derived from soil organic matter but from peaty sediments in the *Lepironia* swamp or litter on the swamp-forest floor since tropical podzolic soil has not been observed in the catchment area of Tasek Bera.

The question which arises at this pont is, which is the main source of the colouring material – peaty sediments or litter? The distribution of colouring material in Tasek Bera was studied experimentally. Their physical and chemical properties were also studied.

4.3.1 *Distribution of colouring material*

Water samples from Tasek Bera were filtered through glassfilters and the filtrates used for measuring light absorption at 350 μm with a Hitachi 101 spectrophotometer using a 1 cm light path cell. The amount of permanganate consumption (COD) of the filtrates was determined according to APHA-AWWA-WPCF (1971). pH of the swamp waters was measured *in situ*.

The relationship between light absorption at 350 μm and COD of swamp waters from Tasek Bera and a small pond in the forest suggests that the light absorption of the water is mainly due to organic matter.

The light absorption of water from Station I ranged from 0.13 to 0.2 throughout the year. These small fluctuations in colour intensity may explain the negligibly small ranges in the daily and annual variations of the pH in Tasek Bera. This indicates that the main buffering factor in the water is organic acid because there was no correlation between the pH and carbon dioxide concentration of the water. During the dry season beginning in June and ending in August of 1972, there was no rain in the catchment area of Tasek Bera. During this period, the *Lepironia* swamp and most of the swamp-forest dried up although the open water region still contained water. Light absorption (350 μm) in the open water during this season was 0.2 which was similar to observed values during the other seasons.

The swamp was flooded in the wet season (December to January). Light absorption of the water measured in December was 0.9. In October, when the water level in the swamp was gradually rising, the colour intensity was rather low. In fact, on October 21 the lowest level of colour intensity was observed. This may be due to the fact that there was a heavy shower (67 mm in a few hours) on October 16, though the water level returned to its original level after 11 days. The pH was also higher than that observed at the beginning of the month. The low concentration of colouring matter and high pH observed five days after the heavy rain may be considered to have resulted from a low supply rate of organic acids.

Light absorbances of water samples (at 350 μm) from Yoda's Plot and Stations XIX and XX which are adjacent to the swamp-forest were low when compared with those of the *Lepironia* swamp.

To understand more fully the changes in the concentrations of colouring matter in the upper to lower streams of the swamp, extensive observations were conducted on 6 October 1972. However, no clear differences were observed in the colour intensities of any of the regions studied (Fig. 76).

4.3.2 *Chemical properties of the colouring material*

Studies on the physical and chemical nature of colouring materials of lake and river waters in temperate regions indicate the colouring material to be some

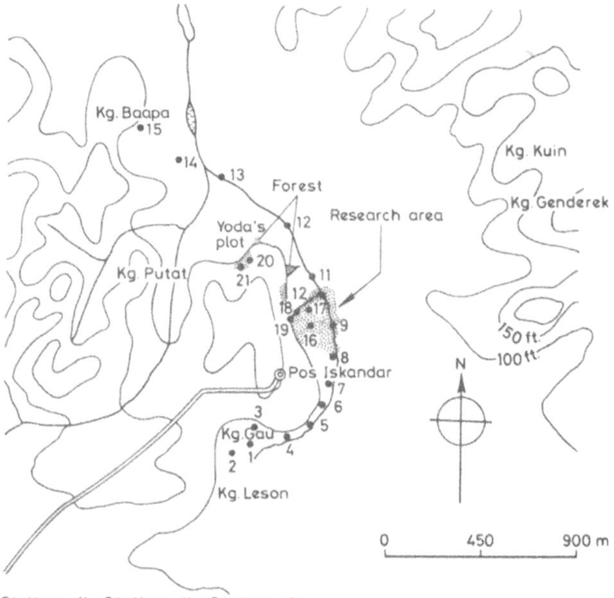

Station	K	Station	K	Station	K
1	.185	8	.163	15	.195
2	.200	9	.174	16	.167
3	.180	10	.165	17	.164
4	.173	11	.163	18	.140
5	.170	12	.168	19	.123
6	.170	13	.160	20	.118----Yoda's plot
7	.162	14	.157	21	.122----Yoda's plot

Fig. 76 Distribution of colour intensities of water in the study area. (O. Sato)

kind of humic substance similar to fluvic acids. However, since the colouring material in the waters of Tasek Bera could be coagulated by the addition of small amounts of Ca^{++} or Cu^{++}, it is suspected that their physical and chemical nature differ from that of soil fluvic acids.

$CaCl_2$ solution was added in increasing amounts into a series of test tubes from a microburette. The volumes of the solutions in the test tubes were made up to 1 ml with distilled water. 5 ml of the sample water was then introduced into each of the tubes, the contents carefully shaken and allowed to stand at room temperature for 9 hours. The amount of $CaCl_2$ solution required for the complete precipitation of the colouring matter was recorded.

The extraction of the colouring matter was conducted by the following coagulation method. To freshly collected water from Tasek Bera (60 l), $CuSO_4$ $5H_2O$ was added to coagulate the colouring matter. The amount of divalent ions required for the coagulation of the colouring matter was predetermined experimentally. The coagulated matter was allowed to stand at room temperature overnight. The precipitate collected in a plastic bottle, the supernatant having been siphoned off, was dissolved using sodium hydroxyde, keeping the pH at 11. The alkaline solution was filtered through a Millipore HA filter to separate suspended substances from the amorphous metal hydroxides. The filtrate was immediately acidified to pH 2, using hydrochloric acid. The solution was desalted on Dowex 50. The desalted solution was then concentrated under reduced pressure in a rotating evaporator. The concentrated sample was transferred to a cellophane tube and dialysed overnight with distilled water. Part of the colouring matter passed through the cellophane. The solutions of colouring matter inside the cellophane tube, fraction A, and that outside the tube, fraction B, were concentrated under reduced pressure in a rotating evaporator. The concentrated colouring matter was oven dried at 85°C for 48 hours and used for analysis.

Coagulation of the colouring matter occurred at 4 milli-equivalents of Ca ions and complete precipitation occurred at 6 milliequivalents. According to the classification based on the solubility of humic substances, water soluble colouring matter was classified as fluvic acids. However, fluvic acids from the soil are more stable with $CaCl_2$ and they show very little precipitation even with large amounts of $CaCl_2$ (Kononova 1966). Since the method for the measurement of coagulation used in this experiment differed from that proposed by Kononova (1961), the results here cannot be compared strictly with the results obtained by soil scientists. The high sensitivity of the colouring matter with Ca ions indicates that the material differs from soil fluvic acids.

In a preliminary experiment Ca ions were used to compare the results with data presented by soil scientists. To prepare a sample of organic matter Cu ions were used to coagulate the colouring matter as above because of its action as a sterilizer.

Ultra-violet and visible spectra were observed using the Hitachi Perkin-Elmer 139 spectrophotometer. The infra-red spectrum was recorded using a Perkin-Elmer Model 457 grating infra-red spectrophotometer. The carbon, nitrogen and hydrogen contents of the colouring matter ware measured with a CHN recorder MT-1S Yanaco (manufactured by Yanagimoto Mfg. Co. Ltd.).

Figure 77 shows the infra-red absorption spectrum of fraction A and fraction B. No significant difference is observed between the two spectra. These spectra also closely resemble those of soil humic and fluvic acids (Shnitzer and Khan 1972), organic matter of river waters (Midwood and Feldbeck 1968) and yellow organic acid (Shapiro 1957). The prominent bands in the infra-red spectrum are assigned as follows:

3400 Hydrogen-bonded OH and NH
2940 Aliphatic C—H stretching
1720 C=O stretching
1620–1640 Aromatic C=O stretching or stretching of hydrogen-bonded C=O of quinons.
1400 Aliphatic C—H deformation and OH stretching

The pattern of the spectrum is also very similar to those of fluvic acids and immatured humic acids like the B and Rp types described by Kumada (1974).

Figure 78 shows the ultra-violet and visible spectra of fractions A and B. Table 68 shows the spectrophotometric properties of colouring matter including humic acids from rotting plant material (Kumada 1974). The absorp-

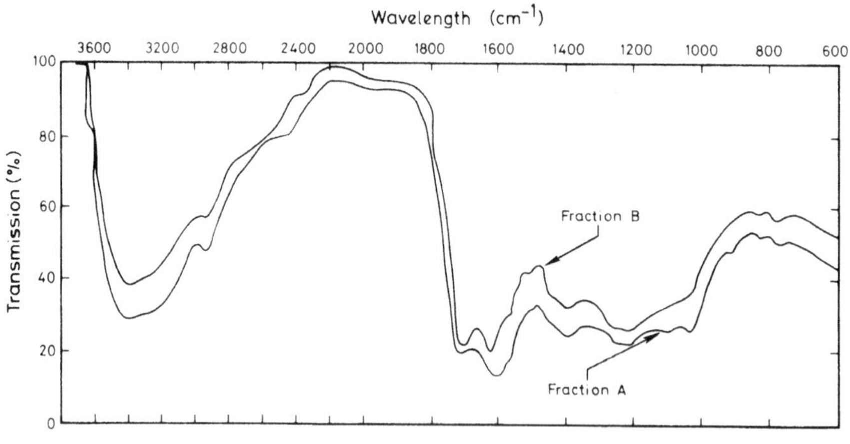

Fig. 77 Infra-red spectra of colouring material, fractions A and B. (O. Sato)

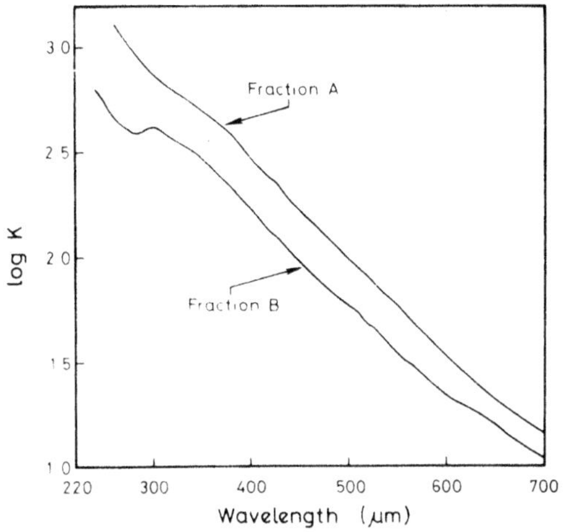

Fig. 78 Ultra-violet and visible spectra of colouring material, fractions A and B. (O. Sato)

Table 68 Spectrophotometric properties of colouring materials and humic acids. (O. Sato)

Sample	Δ log K[a]	K_{600}[b]	RF[c]	Type of H.A.
Fraction A	0.87	0.118	32.0	Rp
Fraction B	0.94	0.104	26.1	Rp
Pinus densiflora RW[d]	0.99	0.032	8.8	
Cryptomeria japonica RW	0.95	0.039	11.6	
Fagus crenata RW	0.77	0.039	18.0	Rp
Quercus variabilis RL[e]	0.76	0.141	31.8	
Cryptomeria japonica RL	0.74	0.129	32.1	
Fagus crenata RL	0.76	0.147	34.8	
Sochiken soil	0.61		76	B
Kuragari soil	0.63		61	
Tanemori soil	0.46		144	A
Miyahara soil	0.58		90	

[a] $\Delta \log K = \log K_{400}/K_{600}$, where K is absorption coefficient at 400 or 600 μm.

[b] K_{600} is defined as the absorption coefficient at 600 μm of 0.01% solution of humic acid in 0.1% NaOH.

[c] $RF = K_{600} \times 1{,}000/(\text{ml of } 0.1\ N\ KMnO_4$ consumed by 30 ml of the humic acid solution).

[d] RW: Rotten wood.

[e] RL: Rotten leaves.

tion spectrum of crude colouring matter from Tasek Bera shows no maximum and minimum peaks in the ultra-violet and visible regions, as observed also in yellow organic acids and soil humic substances. However, a very weak absorption maximum for fraction A and a very weak absorption discontinuity for B was observed around 300μm (Fig. 78).

Kumada and his co-workers (1967) proposed a classification diagram (Fig. 79) for soil humic acids in which humic acids were classified on the basis of the nature of their colours, RF and $\triangle$ log K, where

- RF — is the intensity of light absorption at 600 μm per unit amount of humic acid.
- $\triangle$ log K — is the slope of the absorption curve of the alkaline humic solution, where the logarithms of absorbance are plotted against wavelength.

Several properties of humic acids, like elementary composition, infra-red spectra, functional groups, intrinsic viscosity, cation exchange capacity, the crystalline nature, etc. vary in parallel with their humic types (Kumada 1974). Kumada (1974) also confirmed that a decrease in values of $\triangle$ log K indicates progression in humification of humic acids, the increase in cation exchange capacity and amorphous carbon-like structure recognizable by X-ray diffraction, and the decrease of the following values: the ratio of hydrolysable nitrogen to total N, the ratio of 0.1 N $KMnO_4$ consumed by chromic acid in alkaline solution to that consumed in acid solution and a minimal concentra-

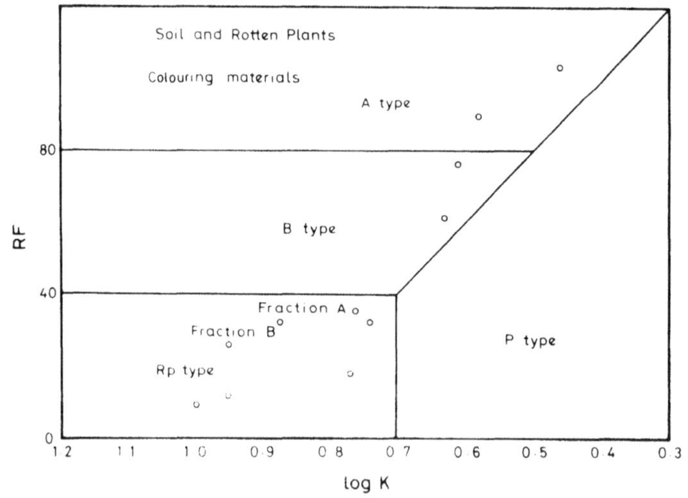

Fig. 79 Classification diagram of humic acid. (O. Sato)

tion of electrolyte required for the complete coagulation of a definite amount of humic acid in a given time. These findings show that the value of $\Delta \log K$ is very useful in predicting the chemical and physical nature of humic substances.

The $\Delta \log K$, K_{600} and RF of fractions A and B are listed in Table 68 together with those of humic substances extracted from rotting plant material and various soils (Kumada 1974). RF and $\Delta \log K$ values of fractions A and B, plotted on Kumada's classification diagram is shown in Figure 79. The values of $\Delta \log K$ and RF of fractions A and B coincide with those of the Rp type humic acids from rotting leaves. This suggests that the colouring matter of the waters of Tasek Bera can be classified as the Rp type humic acid. The results, therefore, confirm the suggestion in the previous section that the colouring matter originated from litter on the forest floor. The data on the elementary composition shown in Table 69 also indicate that the colouring matter in the waters of Tasek Bera is due to immature humic acids like the Rp type humic acid extracted from rotting plant material. The colouring substance is certainly not fluvic acid, since fluvic acid cannot be coagulated with small amounts of Ca ions and its oxygen content is higher than that of the colouring matter in Tasek Bera.

Table 69 Composition of elements in colouring material and Rp type humic acid measured by Kumada.

Sample	C%	N%	H%	C/N
Pinus densiflora RW[a]	59.4	5.36	0.54	109.1
Cryptomeria japonica RW	59.4	5.58	1.41	42.2
Fagus crenata RW	56.1	5.68	3.79	14.8
Cryptomeria japonica RL[b]	59.8	5.99	2.63	22.8
Quercus variabilis RL	55.1	5.12	3.75	14.7
Fagus crenata RL	57.8	5.07	2.43	23.8
Fraction A	55.0	4.56	2.75	20.0
Fraction B	59.6	4.10	2.47	24.1

[a] RW: Rotten wood.
[b] RL: Rotten leaves.

4.3.3 *Role of depositing mud*

Mud-core samples were taken from the *Lepironia* swamp for this study. These samples were frozen in polyethylene bags and stored in deep freeze

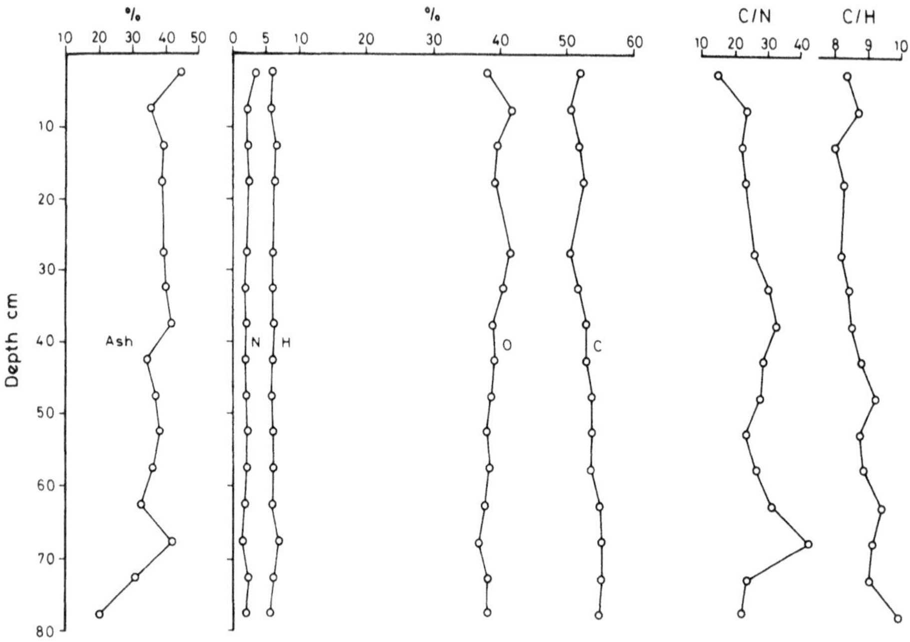

Fig. 80 Composition of elements of core I from the *Lepironia* swamp. (O. Sato)

before analysis. The thawed samples were filtered through glass fibre filters. The filtrate was stored in a freezer for analysis of light absorption, pH, total carbon and carbohydrate carbon. Mud remaining on the glass fibre filters was dried, weighed and powdered for elemental analysis.

To estimate the rate of formation of organic colouring matter from the swamp mud, 50 ml of wet mud, collected from Station I was mixed with 450 ml of distilled water and incubated at 25–30°C for 42 days. The optical density of the filtrate from the mixture was measured with a Hitachi 101 spectrophotometer. The dissolved organic carbon content of the filtrate was determined by measuring the weight of $BaCO_3$ produced from carbon dioxide released through chromic acid oxidation of organic carbon in the presence of $BaCl_2$.

The vertical distribution of the elemental composition of two core samples is shown in Figs. 80 and 81. Analytical results of the interstitial water are listed in Table 70 together with the changes in the dissolved organic matter and colour intensity.

The light absorption of the interstitial water was slightly higher than that of the swamp water. The Δ log K values of the colouring matter from the interstitial water are similar to those of the colouring matter from the swamp

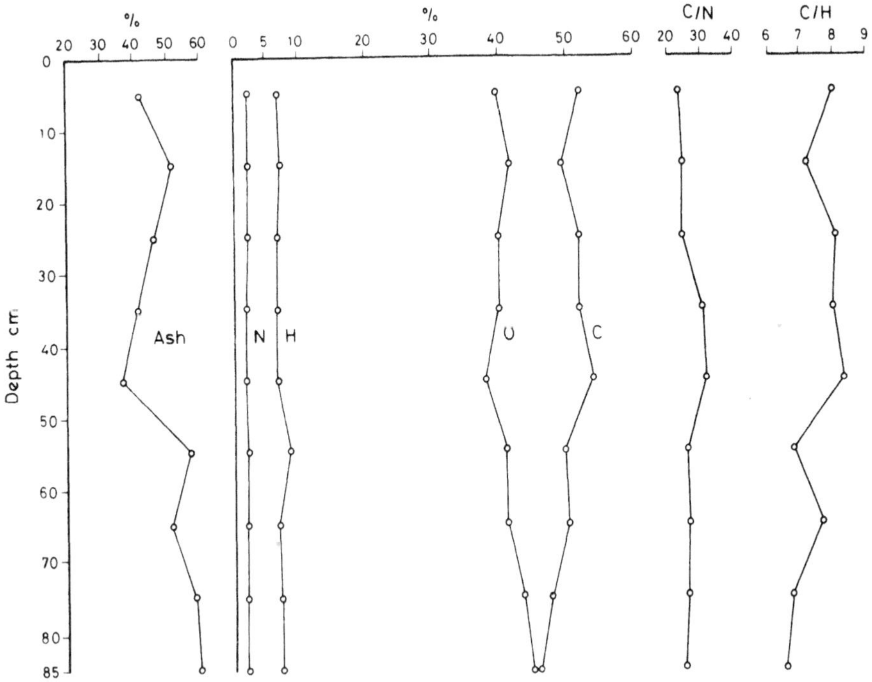

Fig. 81 Composition of elements of core II from the *Lepironia* swamp. (O. Sato)

Table 70 Some properties of interstitial water of core samples from the *Lepironia* swamp. (O. Sato)

Depth (cm)	pH	K[a] at 350 μm	Δ log K	DCC[b] (μg/l)	DOC[c] (μg/l)	DCC/DOC (%)
0–10	5.20	0.177	0.647	7.22	26.1	26.9
10–20	5.20	0.175	0.736	9.51	50.0	19.0
20–30	5.20	0.233	0.735	8.40	36.3	23.1
30–40	5.30	0.269	0.842	8.31	29.3	28.4
40–50	5.00	0.275	0.778	10.62	34.0	31.2
50–60	5.20	0.270	0.977	10.21	34.7	30.3
60–70	5.03	0.333	0.827	10.07	28.1	36.0
70–80	5.13	0.242	0.778	12.47	31.4	39.8
80–90	5.20	0.259	0.833	13.02	38.2	40.4

[a] K at 350 μm is absorbancy at 350 μm of sample.
[b] DCC: Dissolved carbohydrate carbon.
[c] DOC: Dissolved organic carbon.

water. The pH of the two waters was also similar, ranging from 5.0 to 5.3. These facts suggest that the swamp mud contributed to the colouring matter of the swamp water.

The increase in dissolved carbohydrates in the interstitial water with core depth suggests that some of the carbohydrates in the swamp water are derived from the mud by hydrolysis of polysaccharides. This may explain why plant tissues in the deeper parts of the mud have become easily fragmented.

The ash content of core Sample I is higher than that of core Smaple II. This difference might be due to differences in the sampling sites. The ash content of core Sample I tends to increase with depth while that of core Sample II decreases slightly with depth. This indicates that the condition of the deposited muds varies from time to time and from place to place. Hence, the rate of decomposition of organic matter to ash ratio cannot be calculated on the basis of changes of organic matter to ash ratio. However, there was no clear increase in ash content with core depth in the samples. This indicates that the decomposition rate of organic matter is very low in the swamp mud.

The results of the experiments which were done to estimate the production rate of colouring matter in the *Lepironia* swamp mud and from swamp mud of the open water region are given in Table 71. This indicates that the production rate of colouring material in the mud was very low. The low rate of decomposition of organic matter to CO_2 and CH_4 was confirmed by the study of the gaseous metabolism of swamp mud conducted by Koyama (described in the previous subsection).

From these results it can be concluded that the contribution of the swamp mud to the colouring of the swamp water is negligibly small when compared with that of litter in the swamp-forest.

4.3.4 *Colouring of water in Tasek Bera by dissolved plant material from the swamp-forest and macrophytes in the Lepironia reed swamp*
(O. Sato and Y. Tezuka)

Fresh leaves of two species of *Eugenia* (*E. clavifora* and *E.* sp.), fresh *Utricularia* sp., fresh stems of *Eleocharis ochrostrachys*, shed leaves of *Pandanus helicopus* and dead stems of *Lepironia articulata* were prepared as described in the previous sub-section to study the colouring of the swamp water due to dissolved organic material from the swamp-forest and macrophytes in the *Lepironia* swamp. Algal detritus was collected from the bottom of the *Lepironia* swamp during the dry season. Leaf litter was collected from

Table 71 Formation of colouring substances and dissolved organic carbon (DOC) from the litter and the swamp mud.

A. Formation of colouring substances from plant tissues, ground litter and swamp mud. (Figures are measured at 350 μm) (O. Sato)

Sample	mg	First 15	Second incubation period (days)							$K \cdot l^{-1}$ gL 22
			1	2	3	4	5	13	22	
Eugenia	29	0.065	0.002	0.010	0.007	0.011	0.028	0.026	0.044	1.52
Eugenia	86	0.124	0.002	0.013	0.016	0.017	0.025	0.046	0.053	0.55
Eugenia	127	0.165	0.005	0.020	0.020	0.026	0.034	0.047	0.070	0.55
Eugenia	180	0.400	0.033	0.047	0.050	0.060	0.072	0.104	0.135	0.75
Eugenia	218	0.520	0.138	0.068	0.075	0.087	0.103	0.159	0.205	0.94
Eugenia	64	0.178	0.020	0.020				0.028	0.037	0.58
Pandanus	69	1.080	0.065	0.113	0.135	0.172	0.210	0.348	0.448	6.49
Lepironia	181	0.410	0.037	0.046	0.039	0.039	0.043	0.048	0.062	0.34
Eleocharis	91	0.368	0.016	0.032	0.038	0.050	0.057	0.093	0.120	1.32
Utricularia	210	0.600	0.165	0.273	0.312	0.322	0.322	0.378	0.400	1.90
Attached algae	100	0.750	0.103	0.263	0.293	0.338	0.380	0.490	0.430	4.30
Forest litter	1152	1.675	0.211	0.312	0.345	0.420	0.515	0.760	0.800	0.69
Forest litter	1162	2.125	0.255	0.350	0.420	0.488	0.610	0.840	0.850	0.73
Swamp mud	105	0.060	0.003				0.025	0.028	0.045	0.43
Swamp mud	413	0.051	0.002				0.023	0.026	0.052	0.13
Swamp mud	1050	0.058	0.004				0.034	0.047	0.073	0.07

B. Formation of dissolved organic carbon (DOC) and colouring matter from litter. (Y. Tezuka)

C. Formation of DOC and colouring matter from mud. (Y. Tezuka)

Incubation time (days)	0	1	3	7	15	33	60	0	42
pH	6.7	5.6	5.9	6.0	5.9		6.9	7.5	7.2
OD (350 μm)	0	0.65	1.50	2.60	3.60	4.50	4.60	0.05	0.05
DOC ($mg \cdot l^{-1}$)	0	50	93	113	153	210	225	0	4.9

$K \cdot l^{-1}$ gL: Absorbance measured at 350 μm after 22 days incubation and calculated on the basis of 1 g litter in 40 ml of water.

the swamp-forest and air dried and wet swamp mud from the *Lepironia* swamp was used in the experiment.

Each of the samples was put into 40 ml of distilled water in glass bottles and incubated at 25–30°C. After 15 days, the colour intensity of the filterates (the samples were filtered through glass fibre filters) was measured with a spectrophotometer using a 1 cm light path cell. In order to exclude the effect of colour formation during sample preparation, 40 ml of distilled water was added to each of the bottles which contained the remainder of the sample and incubated for a further 22 days. At regular intervals, the colour intensity of these filtrates was measured. The results are shown in Table 71.

A comparison of the absorbances calculated on the basis of 1 g litter in 40 ml of water shows that all the plant material except for *Pandanus* and attached algae release colouring matter at approximately the same rates. *Pandanus* and attached algae give a deeper colouration than the other plants. Samples exposed to air for a long time tend to colour the water more intensely. A comparison of the intensities of colour formation after the initial incubation of 15 days with that after 22 days shows that the colour intensity in the former was generally stronger than that in the latter. This suggests that the greater part of the colouring matter must have been released during the sample preparation. Tezuka and Kawai (1975) conducted a similar experiment in which 20 g of air-dried litter was submerged in 2 litres of distilled water and incubated for 60 days. Dissolved organic carbon and optical densities of the filtrates were measured at regular intervals during the incubation period. A similar though relatively higher rate of colouration was observed (Table 71).

Rates of colouration of the water were also studied in samples containing varying ratios of ground litter to water. Varying quantities of litter were put into 40 ml of distilled water in glass bottles and incubated for 22 days. At regular intervals light absorbances of the filtrates were measured. The results are shown in Figure 82. The light absorption is expressed on the basis of 1 g litter in 40 ml of water. The results of this experiment indicate that the amount of colouring matter increased with increasing quantities of ground litter. The maximum amount of colouring matter (calculated on the basis of 1 g of ground litter) was observed in samples containing 600 to 700 mg of litter in 40 ml of water. These results are still insufficient to estimate the rate of colour formation in the swamp because the formation of colour is affected by several factors.

The high rate of colour formation in experiments using air-dried samples suggests that humification of organic matter in air is an important step in the formation of colouring matter in the swamp. This process occurs on the swamp-forest floor in the dry season. The chemical nature of the colouring

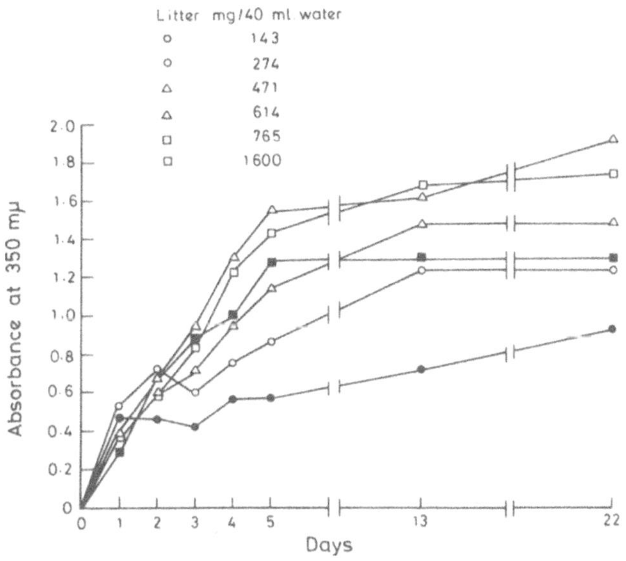

Fig. 82 Formation of colouring material from different amounts of forest litter placed in a finite volume of water. (Absorbance at 350 μm is expressed on the basis of one gram litter in 40 ml water.) (O. Sato)

matter and the humification process of the plant matter has to be studied in more detail for a better understanding of colour formation in the swamp waters.

4.3.5 *Reaction of iron to colouring matter*

Several investigators have noted that water colour increases following rainfall, suggesting that the colouring matter originated from the soil (Shapiro 1957). The black water of the Rio Negro originates from the tropical podzol which is extensively distributed in the swamp-forest area. (Sioli 1964).

However, in Tasek Bera, heavy rain tends to dilute the colour of the water. Ground-water sampled from a well near the swamp did not contain dissolved oxygen but contained ferrous ion (4.3 $mg \cdot l^{-1}$) and was colourless and clear. These facts and the absence of tropical podzol in the catchment area suggest that the water colour in the swamp did not originate from the soil. The intensity of colour of the water in the *Lepironia* swamp was rather strong whereas it was much lighter in the swamp-forest. However, there was no distinct difference in the colour intensity of the upper and lower streams in Tasek Bera.

Laboratory experiments to study the colouration process of the swamp water by plant litter and swamp mud were conducted. The results show that the colour was formed mainly through the decay of ground litter. But this conclusion seems to be in conflict with the data for colour distribution.

Brown flocculent precipitates were found on the bottom surface near the banks of the swamp-forest several times. The precipitate may be closely involved in the colouration process and reduction of colour in the swamp.

The brown precipitates were collected in a test tube. It was alkalified to pH 11 by adding NaOH. The brown suspension dissolved to a brown solution and then changed to a reddish brown precipitate. Most of this reddish brown precipitate was dissolved in hydrochloric acid to give a yellow solution. Qualitative analysis using α,α'dipyridyl confirmed that the main component of the reddish-brown precipitate was an iron compound. Solubility tests and qualitative analyses suggest the brown precipitates to be coprecipitates of ferric hydroxide and humic substances. Ferric ions in the ground water may have oxidized to ferric hydroxide by dissolved oxygen and the resulting ferric hydroxide would have coprecipitated with humic substances. This coprecipitation may have reduced the colour of the water in the swamp-forest. Since no thick layer of the brown precipitates was found anywhere in the study region, the precipitates must be presumed to be removed by the flow of water or dissolved by some chemical process.

The sedimentation rate in the swamp was found to be 0.84 g $C \cdot m^{-2} \cdot day^{-1}$ (Okino, pers. comm.). This high sedimentation rate which cannot be explained by primary production of phytoplankton, suggests that the origin of the sediments is allochthonous. The rate of oxygen consumption by the brown sediment was slow indicating that it is resistant to bacterial decomposition (Okino, pers. comm.). The similarity between the brown sediment collected by Okino and the brown flocculent precipitate adhering to algae in the *Lepironia* swamp confirm that some of the brown flocculent that precipitated to the bottom of the swamp was transported from the shore into the swamp in the form of colloidal particles.

The nature of the iron-humate association is not well understood. The concentration of iron (1.0–1.5 mg $Fe \cdot l^{-1}$) in the brown filtrate of the swamp water after filtration through a Millipore HA filter is slightly higher than in most river waters. However, the ratio of iron to total dissolved inorganic substances is clearly higher than that of normal river water anywhere in the world. The formation of a soluble metal complex with humic substances seems to play an important role in the disappearance of the brown flocculent sediments and in the formation of brown water.

A large quantity of the brown floc was collected and kept in polyethylene bottles at room temperature. After one month a reddish-brown film covered the inside walls of the bottle. This indicates that the ferric hydroxide in the floc was first reduced to soluble ferrous hydroxide by microbial or chemical reduction and then the ferrous ions were oxidized again to insoluble ferric hydroxide on the inner surface of the wall of the bottle by oxygen diffusing through the wall of the bottle. The remaining water in the bottle was deeply coloured. This phenomenon might have been caused by the soluble, brown iron-humate liberated from the brown floc by reduction of ferric hydroxide. This process also seems to play an important role in the formation of the brown water.

The α,α′dipyridyl reactive iron and acid soluble iron in the swamp water amounted to 0.8–0.6 $mg \cdot l^{-1}$ and 0.4–0.2 $mg \cdot l^{-1}$ respectively. The reaction rate of iron with α,α′dipyridyl was considerably slower than with free ferrous ions. Since the iron chelating with the colouring matter is believed to be in the ferric state (Shapiro 1964), this slow reaction rate might be the result of slow reduction and slow dissociation of the chelated ferric ions.

Acetate buffer and α,α′dipyridyl solution were added to two test tubes containing 10 ml aliquots of swamp water each. One test tube was covered with aluminium foil and the other exposed to sunlight. Colour intensities of the reaction mixtures were observed at regular intervals. The results show that irradiation by sunlight accelerated the formation of ferrous ion (Fig. 83). Acceleration of the reaction by sunlight suggests that the organic matter in the water is capable of reducing ferric to ferrous ion in the same way as oxalic acid in a system of aqueous oxalic acid and ferric ion (Beattie, Bricker and Garvin 1961).

In the measurement of primary production there was a greater consumption of oxygen in the light bottle when compared with the dark bottles as shown in Fig. 84. This suggests that oxygen consumption in the swamp water

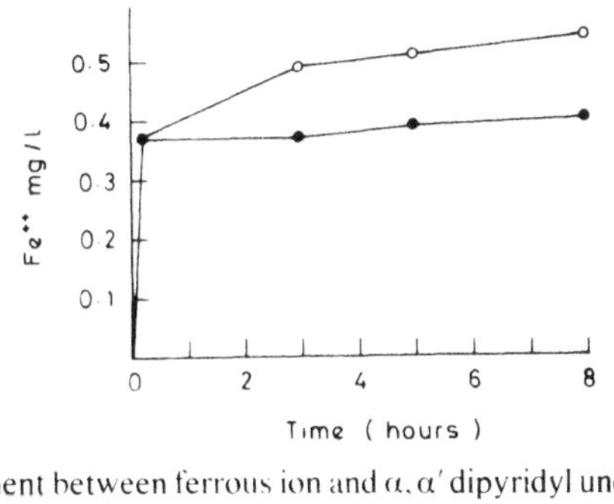

Fig. 83 Colour development between ferrous ion and α,α′ dipyridyl under light (o) and dark (•) conditions. (O. Sato)

is accelerated by a photochemical reaction. This photochemical reaction might be related to the reduction of ferric hydroxide.

Tezuka (pers. comm.), in measuring the respiration of the Tasek Bera water, found that the oxygen consumption was not accompanied by the production of carbon dioxide. These findings suggest that the chemical changes in the formation of the iron humate involve two steps: firstly, the ferric ions in the humate are reduced to ferrous ions by transforming the organic matter into another form containing more oxygen atoms, and secondly, some of the ferrous ions are oxidized to ferric hydroxide by absorbing dissolved oxygen, these chemical changes being accelerated by sunlight.

Table 72 Temperature, dissolved oxygen, total carbon dioxide, pH and ferrous ion of water in the swamp-forest measured on the 5th, 6th and 7th of November 1972. (O. Sato)

Date and time	Depth (cm)	Temp. (°C)	pH	O_2 $ml \cdot l^{-1}$ I[a]	O_2 $ml \cdot l^{-1}$ II[b]	CO_2 $ml \cdot l^{-1}$	Fe^{++} $ml \cdot l^{-2}$ I[a]	Fe^{++} $ml \cdot l^{-2}$ II[b]
5th								
13:45	5	27.8	5.15	0.69	0.70	12.2	0.47	0.44
6th								
7:00	5	26.5	5.01	0.41	0.47	11.6	0.71	
12:35	5	27.5	5.27	0.77	0.72	11.2	0.39	
15:00	5	28.0	5.05	0.72		11.6	0.44	0.41
17:30	5	26.8	5.15	0.47	0.47	12.4	0.30	0.30
7th								
7:00	5	27.0	5.12	0.41	0.45	11.7	0.69	

[a] I are the results of samples collected from the upper stream of small rapids.
[b] II are the results of samples collected from the lower stream of small rapids.

Diurnal fluctuations of temperature, dissolved oxygen concentrations, pH and ferrous ion concentration in the waters of the swamp-forest are shown in Table 72. Since a great proportion of water of Tasek Bera is shaded from direct sunlight, even in the daytime, the fluctuations in oxygen content are not great. However, the oxygen concentration was distinctly higher during the day than during the night. Ferrous ion concentration was slightly higher in the morning, decreasing gradually towards evening. It may be possible that the concentration of iron in the swamp water is higher during the night.

Hence the steps in the formation of colouring matter in the waters of Tasek Bera can be summarized as follows:

1) The organic component of the colouring matter is a brown humic substance originating from plant litter in the swamp-forest.
2) Iron is supplied from the soil as soluble ferrous ions.
3) The ferrous ions are oxidized to ferric hydroxide by oxygen near the shores of the swamp.
4) Part of the ferric hydroxide reacts with the soluble brown humic substances to form a soluble ferric humate which gives a deeper brown colouration.
5) Part of ferric humate coprecipitates with residual ferric hydroxide to form a brown floc.
6) The brown floc under anaerobic conditions releases the coprecipitated ferric humate into the water again by the reduction and solubilization of ferric hydroxide.
7) Reduction of ferric atoms in the ferric humate is accelerated by exposure to sunlight without the accompanying CO_2 formation. The ferrous ions formed are then oxidized to ferric hydroxide by absorption of dissolved oxygen.

The chemical and physical nature of the colouring matter varies greatly with the nature of iron under aerobic and anaerobic conditions. The conclusions are speculative as they are deduced from insufficient data. However, it is hoped that these findings will provide an impetus for more detailed studies.

5. Primary production

I. Ikusima and J. I. Furtado

Primary production forms an important component in Tasek Bera basically because it is a swamp with more than 75% of the area being occupied by vegetation. Primary production can be divided into two broad categories: (1) algal production and (2) macrophyte production which can further be subdivided into the non-emergents and emergents. The emergent macrophytes which are dominated by the swamp-forest form the most important component in primary productivity. In the following text, the various floral communities will be described and primary production will be dealt according to the two broad groups (i.e. algal and macrophyte) with detailed studies on the more dominant species.

5.1 Classification of primary producers (I. Ikusima)

5.1.1 *Algae*

(M. Ratnasabapathy, S. Kumano, T. Watanabe and T. Mizuno)

The algae of Tasek Bera have been well studied (Prowse 1957, 1969; Mizuno and Mori 1970; Ratnasabapathy and Kumano 1974). Prowse (1969) described one species and four varieties of desmids new to this region. Mizuno and Mori (1970) listed species belonging to different groups: Cyanophyta - 5 species; Pyrrophyta - 1; Bacillariophyta - 19; Chrysophyta - 1; Euglenophyta - 1; and Chlorophyta - 54. Ratnasabapathy and Kumano (1974) listed 168 species, varieties and forms of Desmidiales and found a general affinity of the desmid flora to the broad Indo-Malayan-North Australian phytogeographical region (cf. Krieger 1932) and its particularly close relationship with Sumatra.

Furtado, J. I. and Mori, S., eds., Tasek Bera. ISBN 90 6193 100 2.

To date a total of 328 species, varieties and forms have been recorded (Table 73) (Prowse 1957, 1969; Mizuno and Mori 1970; Ratnasabapathy and Kumano 1974; Lim, Furtado, Fernando and Skuja, unpublished). Of these, 23 species including 5 unidentified species belonged to the Cyanophyta, 5 to the Pyrrophyta, 61 including 8 unidentified species belonged to the Bacillariophyta, 7 belonged to the Chrysophyta, 24 including 3 unidentified species belonged to the Euglenophyta, 293 including 14 unidentified species belonged to the Chlorophyta, 2 species of Charophyta and 3 species of Rhodophyta.

Table 73 List of phytoplankton and epiphytes from Tasek Bera.

(P) indicates phytoplanktonic forms.
(C) indicates attached forms on *Cryptocoryne griffithii*.
(U) indicates attached forms of *Utricularia* spp.
(L) indicates attached forms on *Lepironia articulata*.

* from C. H. Fernando and H. Skuja.
** from R. P. Lim and J. I. Furtado.
*** from G. A. Prowse

CYANOPHYTA
Cyanophyceae
Chroococcales
Chroococcaceae
Chroococcus tenax (Kirch.) Hieron.
***Gloeocapsa* sp. (P)
Coelosphaerium kuetzingianum Näg. (U)
Merismopedia elegans A. Braun
M. glauca (Ehrnb.) Näg. (P) (L)
Microcystis incerta Lemm. (U)
Nostocales
Oscillatoriaceae
Oscillatoria borneti Zukal (L)
**O. chalybea* (Mert.) Gom.
O. geitleriana Elenkin
**O. princeps* Vauch.
O. pseudogeminata Schmid
O. splendida Grev. (L)
Phormidium inundatum Kütz.
P. minnesotensis (Tilden) Drouet
Microcoleus chthonoplastes Thur. (L)
Lyngbya limnetica Lemm. (L)
Nostocaceae
Anabaena sp.
Cylindrospermum sp.
Scytonemataceae
Scytonema sp.
Rivulariaceae
Rivularia sp.
Stigonematales
Stigonemataceae
Hapalosiphon flexuosus Borzi (U)
H. stuhlmannii Hieron. (L)
Stigonema panniforme (Ag.) Born. et Flah. (L)

EUGLENOPHYTA
Euglanophyceae
Euglenales
Euglenaceae
Euglena acus Ehrb.
E. fusca (Klebs) Lemm.
E. gracilis Klebs
**E. granulata* (Klebs) Lemm.
E. mutabilis Schmitz
E. oxyuris Schmarda
E. proxima Dangeard
E. sanguinea Ehrb.
**Phacus caudatus* Hubn.
P. longicauda (Ehrb.) Duj.
P. myersi Skvort.
P. tortus (Lemm.) Skvort.
P. trifacialis Prowse

Trachelomonas hispida (Perty) Stein em. Defl.
**T. intermedia* Dang.
T. klebsi Defl.
T. similis Stokes
**T. volvocina* Ehrb.
T. volvocinopsis Swir.
T. volzii Lemm.
Peranemaceae
Notosolenus lens Skuja
Peranema sp.
Petalomonas sp.
Astasiaceae
Astasia sp.

PYRROPHYTA
Cryptophyceae
Cryptomonadales
Cryptomonadaceae
Cryptomonas sp.
Dinophyceae
Gymnodiniales
Gymnodiniaceae
Gymnodinium sp.
Peridiniales
Glenodiniaceae
Glenodinium sp.
Peridiniaceae
Peridinium sp.
Dinococcales
Dinococcaceae
Tetradinium sp.

CHRYSOPHYTA
Xanthophyceae
Mischococcales
Chlorobotrydaceae
Botryococcus sudeticus Lemm.
Chrysophyceae
Ochromonadales
Synuraceae
Chrysodidymus gracilis Prowse
Ch. synuroides Prowse
**Synura uvella* Ehrb.
Dinobryaceae
Dinobryon divergens Imhof.
D. sertularia Ehrb. (U)
Monosigales
Phalansteriaceae
Phalansterium digitatum Stein

BACILLARIOPHYTA
Centrales
Coscinodiscaceae
Melosira sp.
Cyclotella meneghiniana Kütz.
Pennales
Fragilariaceae
Diatoma elongatum Ag.
D. vulgare Bory
Tabellaria fenestrata (Lyngb.) Kütz.
Asterionella sp.
Fragilaria sp.
Synedra ulna (Nitz.) Ehrb.
Eunotiaceae
**Eunotia arcus* Ehrb.
E. curvata (Kütz.) Lagerst. (U)
E. curvata var. *capitata* (Grun.) Patrick (U)
**E. diodon* Ehrb.
E. faba (Ehrb.) Grun.
**E. flexuosa* (Bréb.) Kütz.
E. flexuosa var. *eurycephala* Grun. (L)
E. gracilis (Ehr.) Rbh. (P)
E. hexaglyphis Ehrb. (U)
E. lunaris (Ehrb.) Grun. (P)
E. naegelii Migula (U)
E. paludosa Grun.
E. pectinalis (Kütz.) Rbh. (P)
E. pectinalis var. *minor* (Dillw.) Rbh. (U)
E. robusta Ralfs (P)
E. robusta var. *diadema* (Ehrb.) Ralfs (P) (L) (U)
E. sudetica O. Müll. var. *incisa* (Mayer) A. Cleve
E. serra Ehrb. (U)
E. theronii Cholnoky
Naviculaceae
Anomoeneis serians (Breb. ex Kütz.) Cleve var. *acuta* Hustd. (L) (U)
A. serians var. *brachysira* (Breb. ex Kütz.) Hustd. (L) (U)

Table 73 (Continued)

Frustulia rhomboides (Ehrb.) De Toni (P) (L) U)
F. rhomboides var. *saxonica* (Rhb.) De Toni (P) (U)
Pinnularia abaujensis (Pant.) Ross (L)
**P. gibba* Ehrb. var. *mesogomyla* (Ehrb.) Hustd.
P. luculenta A.S. (P)
P. maior (Kütz.) Rhb. (Ehrb.) (P)
P. nobilis Ehrb.
Pleurosigma sp.
Neidium iridis (Ehrb.) Cleve var. *amphigomphus* (Ehrb.) V. Heurck
Navicula sp.
Stauroneis sp.
Gomphonemataceae
**Gomphonema longiceps* Ehrb. var. *subclavatum* Grun.
Nitzschiaceae
Nitzschia sp.
Surirellaceae
Stenopterobia sp.
Surirella biseriata Bréb. (P)
**S. biseriata* var. *constricta* Grun.
S. elegans Ehrb.
S. engleri O. Müller f. *angustior* O. Müller (U)
S. linearis W. Smith (P)
S. linearis var. *constricta* (Ehrb.) Grun.
S. pseudospiralis Hustd. (P)
S. spiralis Kütz. (U)

CHLOROPHYTA
Chlorophyceae
Volvocales
Chlamydomonadaceae
Chlamydomonas sp.
Tetrasporales
Palmellaceae
Palmella mucosa Kütz. (U)
Gloeocystis gigas (Kütz.) Lager. (U)
Tetrasporaceae
Schizochlamys gelatinosa A. Br. (C)
Chlorococcales
Hydrodictyaceae
Pediastrum duplex Meyen
P. boryanum (Turp.) Menegh. var.
Oocystaceae
Chlorella luteoviridis Chod.
Ankistrodesmus falcatus (Chod.) Ralfs (U)
A. fusiformis Chod.
Dictyosphaeraceae
Westella sp.
Scenedesmaceae
Scenedesmus bijuga (Turpin) Lagerh.
**S. diagonalis* Ley
S. ecornis (Ralfs) Chod.
S. ecornis var. *disciformis* Chod.
S. perforatus Lemm.
Coelastrum sp.
Ulotrichales
Ulotrichaceae
Ulothrix cylindricum Prescott
Microsporaceae
Microspora sp.
Chaetophorales
Chaetophoraceae
Stigeoclonium sp.
Chaetophora sp. (C)
Coleochaetaceae
Coleochaete sp. (C)
Chaetosphaeridium sp. (C)
Oedogoniales
Oedogoniaceae
Oedogonium spp. (U)
Bulbochaete praereticulata Jao (L) (U)
Zygnematales
Zygnemataceae
Mougeotia spp. (P) (L) (U)
Spirogyra spp. (P) (L) (U)
Zygnema spp. (P) (L) (U)
Sirogonium sp.
Mesotaeniaceae
Cylindrocystis brebissonii Menegh.
C. crassa De Bary (P)
Netrium digitus (Ehrb.) Itzigs. et Rothe (P) (U)
**N. digitus* var. *lamellosum* (Breb.) Grönbl.
Spirotaenia sp.

Gonatozygaceae
Gonatozygon brebissonii De Bary
**G. kinahini* (Arch.) Rbh.
G. moniliformis Ehrb.
G. monotaenium De Bary
**G. monotaenium* var. *pilosellum* Nordst.
G. sudanense Grönbl. et Scott (L) (U)
Desmidiaceae
Penium cylindrus (Ehrb.) Bréb.
**P. exiguum* W. West var. *globerrimum* Grönbl.
P. margaritaceum (Ehrb.) Bréb.
P. silvae-nigrae Rabanus (P) (L) (U)
P. spirostriolatiforme West et West
P. spirostriolatum Narker
Closterium abruptum W. West var. *angustissimum* Schm.
C. acerosum (Schrank) Ehrb.
C. biclavatum Borges (L)
C. cuspidatum Bail.
C. dianae Ehrb. (P)
**C. dianae* var. *arcuatum* (Bréb.) Rbh.
C. dianae var. *minus* (Wille) Schroder (U)
C. ehrenbergii Menegh.
C. gracile Bréb. (P) (U)
C. gracile var. *striolatum* Krieg. (U)
C. infractum Messik.
C. intermedium Ralfs (P) (L) (U)
C. kuetzingii Bréb.
C. libellula Focke (P)
C. libellula var. *elongatum* (Krieg.) Scott et Prescott
C. libellula var. *interruptum* (W. et W.) Donat (P)
C. longoense Nordst. var. *crassius* Gutw.
C. navicula (Bréb.) Lutkem. var. *crassum* (W. et W.) Grönbl.
**C. macilentum* Bréb.
**C. moniliferum* Bory) Ehrb.
**C. nematodes* Josh.
C. porrectum Nordst. (U)
C. ralfsi Breb. var. *hybridum* Rbh. (U)
C. setaceum Ehrb. (P)
C. striolatum Ehrb. (P)
**C. venus* Kütz.
Roya sp.
Pleurotaenium baculoides (Roy et Biss.) Playf. (U)
****P. burmense* (Josh.) Kütz. var. *elegans* Prowse (P)
P. coronatum (Bréb.) Rbh. var. *nodulosum* (Bréb.) W. West
P. ehrenbergii (Bréb.) De Bary
P. ehrenbergii var. *undulatum* Schaarschm. (U)
P. elatum (Turner) Borge var. *subundulatum* Hirano (P)
P. kayei (Arch.) Rbh. (U)
P. minutum (Ralfs) Delp.
P. minutum var. *crassum* (W. West) Krieg.
P. nodosum (Bail.) Lund. var. *gutwinskii* Krieg. (P) (U)
P. ovatum Nordst.
P. ovatum var. *inermius* Möbius
P. subcoronulatum (Turn.) West et West
P. trabecula (Ehrb.) Näg.
P. trabecula var. *rectum* (Delp.) West et West
P. tridentulum (Wolle) W. West
P. tridentulum var. *gracile* Korster
P. undatum Scott et Prescott (U)
P. verrucosum (Bail.) Lund. var. *validum* Scott et Gronbl. (P) (U)
P. verrucosum var. *bulbosum* Krieg.
Docidium baculum Bréb.
Triploceras gracile Bail.
T. gracile f. *curvatum* Scott et Prescott
T. gracile var. *undulatum* Scott et Prescott (P) (L) (U)
T. splendens Prowse (U)
Tetmemorus brebissonii (Menegh.) Ralfs
T. brebissonii var. *tenuissima* Möbius (U)
T. laevis (Kütz.) Ralfs (P)
Euastrum acanthophorum Turn. f. *minus* Scott et Prescott (P) (U)
E. ampullaceum Ralfs
E. ansatum Ehrb. (P) (U)
E. ansatum var. *dideltiforme* Ducell.
E. ansatum var. *pyxidatum* Delp.

E. binale (Turpin) Ehrb. var. *sectum* Turner (P)
E. cuneatum Jenner
E. denticulatum (Kirchen.) Gay
E. didelta Ralfs var. *bengalicum* Lagerh.
E. gnathophorum West et West
E. gnathophorum var. *bulbosum* Scott et Prescott
E. humerosum Ralfs (P)
E. insulare (Wittr.) Roy var. *silesiacum* Grönbl. (P)
E. longicolle Nordst.
E. longicolle var. *capitatum* West et West f. *minus* Scott et Prescott (P) (U)
E. moebii (Borge) Scott et Prescott var. *tetrachastriforme* West et West f. *latum* Scott et Prescott (U)
E. rostratum Ralfs
E. rostratum var. *bioculatum* Scott et Prescott (P) (U)
E. sinuosum Lenorm. var. *reductum* West et West (P) (U)
E. sinuosum var. *scrobiculatum* Nordst.
**E. sublobatum* Bréb. (P)
Ichthyocercus longispinus (Borge) Krieg. (P) (U)
Ichthyodontum sachlanii Scott et Prescott var. *parorthium* Scott et Prescott
Micrasterias alata Wall.
M. alata f. *tumida* Prowse (P) (U)
M. anomala Turn.
M. anomala var. *kalimantana* Scott et Prescott
M. anomala var. *sumatrana* Scott et Prescott (P) (U)
M. apiculata (Ehrb.) Mengh. var. *lacerata* Turn.
M. ceratofera Josh. (P) (U)
M. foliacea Bail. (P) (L) (U)
M. foliacea var. *quadrinflata* Scott et Prescott
M. foliacea var. *spinosa* Prowse (P) (U)
**M. fimbriata* Ralfs var. *elephanta* Wolle
M. lux Josh.
M. lux var. *sachlanii* Scott et Prescott (P) (U)
**M. mahabuleshwarensis* Hobs.
M. mahabuleshwarensis var. *chauliodon* Scott et Prescott (P) (U)
M. mahabuleshwarensis var. *surculifera* Lagerh. (U)
M. pinnatifida (Kütz.) Ralfs
M. quadridentata (Nordst.) Grönbl. f. *indonesiensis* Scott et Prescott
M. radians Turn.
**M. rotata* (Grev.) Ralfs
**M. sol* (Ehrb.) Kütz. var. *ornata* Nordst.
M. thomasiana Arch. var. *evoluta* Krieg. (P) (U)
M. thomasiana var. *notata* (Nordst.) Grönbl. (U)
M. torreyi Bail var. *doveri* (Biswas) Krieg. (P) (U)
M. torreyi var. *crameri* (Bern.) Krieg. (U)
M. torreyi var. *curvata* Krieg.
Cosmarium alpecta Roy et Biss.
C. armatum Joshua var. *verrucosum* Hirano (L)
C. askenasyi Schm.
C. askenasyi f. *latum* Scott et Prescott
C. auriculatum Riensch (P)
C. binerve Lund. var. *subangulatum* Scott et Prescott
C. ceylanicum W. et West f. *minus* Scott et Prescott (U)
C. circulare Reinsch f. *minor* West et West
C. contractum Kirchn. var. *pachydermum* Scott et Prescott (U)
C. crassangulatum Borge
C. decoratum W. et West var. *dentiferum* West et West
C. denticulatum Borge var. *rotundatum* Borge
C. diplosporum (Lund.) Lütkem.
C. globosum Bulnh. var. *wollei* West et West
C. lagerheimianum (Turn.) Scott et Prescott (U)
C. malleum Krieg.

Table 73 (Continued)

C. marinvernianum (Racib.) Schmidle (P)
C. maximum (Borge) West
C. moniliforme (Turp.) Ralfs var. *indentatum* Scott et Grönbl. (P)
C. nudum (Turn.) Gutw.
C. obsoletum (Hantz.) Reinsch var. *sitvens* Gutw. (P) (I)
C. portianum Arch. (P)
C. quadrifarium Lund.
C. reniforme (Ralfs) Arch. var. *elevatum* West et West (P)
C. sexnotatum Gutw. (P)
C. spinuliferum West et West
C. subexcavatum West et West
**C. subturgidum* (Turn.) Schmidle
C. trilobulata Reinsch (P)
C. trilobulata var. *printzii* Messik (P)
**C. westi* Bernard
C. zonatum Lund. var. *pyriforme* Scott et Prescott
Xanthidium acanthophorum Nordst. var. *raciborskii* Gutw.
X. antilopaeum (Bréb.) Kütz. var. *laeve* Schm. f. *longispinum* Scott et Prescott
X. armatum (Bréb.) Rbh.
X. armatum var. *anguligerum* Krieg. (U)
X. burkillii West et West
X. kalimantanum Scott et Prescott (U)
X. lepidum West et West
X. perissacanthum Scott et Prescott
X. superbum Elfv. var. *centricornis* Prowse (P) (L) (U)
Arthrodesmus arcuatus Josh. var. *minus* Scott et Prescott
A. arcuatus var. *incrassatus* Scott et Prescott
A. convergens Ehrb.
A. curvatus Turn. var. *incrassatus* Scott et Prescott
A. gibberulus Josh.
A. impar (Jacob.) Grönbl.
A. octocornis Ehrb.
A. sachlanii Scott et Prescott (P) (U)
A. spechtii Scott et Prescott
A. subvalidus Grönbl. (P) (U)
A. sumatranus Scott et Prescott
Cosmocladium sp.
***Staurastrum anisacanthum* Scott et Prescott
S. arthrodesmiforme Behre
***S. cerastes* Lund.
S. cerastes var. *coronatum* Krieg. f. *inflatum* Scott et Prescott (P) (L)
S. columbetoides West et West (P)
S. connatum (Lund.) Roy et Biss.
S. corniculatum Lund. var. *variabile* Nordst. (U)
S. curvatum W. West
S. cuspidatum Bréb.
S. cuspidatum var. *divergens* Nordst.
S. dejectum Bréb.
S. ensiferum Turner
S. exporrectum Scott et Prescott
***S. forficulatum* Lund. (P)
S. freemani West et West (P)
S. freemani var. *triquetrum* West et West
S. furcatum (Ehrb.) Bréb. (P)
S. gracile Ralfs (P)
S. gracile var. *ornatum* Krieg. (P)
***S. limneticum* Schm. (P)
S. limneticum var. *burmense* West et West
***S. longispinum* (Bail.) Arch. var. *bidentatum* (Wittr.) West (P)
S. megacanthum Lund.
S. megacanthum var. *orientale* Prescott (P)
S. megacanthum var. *scoticum* West et West (P)
S. modestum Behre
S. octodontum Skuja
S. orbiculare Ralfs var. *depressum* Roy et Biss. (L)
S. orbiculare var. *denticulatum* Nordst.
***S. prionotum* Scott et Prescott (P)
S. punctulatum Bréb.
***S. saltans* Josh. (P)
S. saltans var. *polycharax* Scott et Prescott (U)
S. saltans var. *sumatranum* Scott et

Prescott
S. sexangulare Lund. (P)
S. sexangulare var. *attenuatum* Turn.
***S. sexangulare* var. *bidentatum* Gutw. f. *crassum* Scott et Prescott (P)
S. sexangulare var. *productum* Nordst. (P)
S. sexangulare var. *subglabrum* West et West (U)
S. smithii (G. M. Smith) Teiling
S. stauroton Scott et Prescott
S. subauriculatum West et West var. *truncatum* West et West
S. subgracillimum West et West
S. subsaltans West et West
S. subsaltans var. *indonesianum* Scott et Prescott (P) (U)
**S. subtrifurcatum* West et West var. *major* West et West
S. tetracerum Ralfs (U)
***S. thienemanii* Krieg.
S. thienemanii f. *triradiatum* Scott et Prescott
S. tohopekaligense Wolle
S. tohopekaligense f. *minus* (Turn.) Scott et Prescott (U)
S. triforcipatum West et West
S. trissacanthum Scott et Prescott var. *dissacanthum* Scott et Prescott
S. wandae Raciborski var. *brevispinum* Grönbl.
S. wildmanii Gutw. (P) (W)
S. wildmanii var. *majus* (West et West) Scott et Prescott (P) (U)
S. zahlbruckneri Lütkem. var. *mamillatum* West et West
***S. zonatum* Börges.
Sphaerozosma granulatum Roy et Biss.
Onychonema laeve Nordst. var. *micracanthum* Nordt.
Spondylosium nitens (Wal.) Arch. var. *triangulare* Turn. f. *javanicum* Gutw.
S. moniliforme Lund.
S. planum (Wolle) West et West (U)
Hyalotheca dissiliens (Smith) Bréb. (P)
H. dissiliens var. *hians* Wolle (P)
H. inflata Scott et Grönbl.
H. mucosa (Dillw.) Ehrb. (P)
H. undulata Nordst (P)
H. undulata var. *perundulata* Grönbl.
Gorenbladia neglecta (Racib.) Teiling (L)
G. neglecta var. *elongata* Scott et Grönbl.
Desmidium aptogonum Bréb. var. *tetragonum* West et West (P) (U)
D. baileyi (Ralfs) Nordst.
D. baileyi f. *longiprocessum* Scott et Prescott (P) (U)
D. baileyi f. *tetragonum* Nordst.
D. bengalicum Turn. (P)
D. coarctatum Nordst.
D. grevillei (Kütz.) De Bary (P) (L) (U)
D. graciliceps (Nordst.) Lagerh. (P)
D. quadratum Nordst.
D. suboccidentale Scott et Prescott
D. swartzii Ag. (P)
**D. swartzii* var. *quadrangulatum* (Rolp.) Roy
Bambusina brebissonii Kütz. (P) (L) (U)
Phymatodocis irregularis Schm.
P. irregularis var. *intermedia* Gutw.
P. nordstedtiana Wolle
P. nordstedtiana var. *triangularis* Prowse
Streptonema quadrangularis Prowse
S. trilobatum Wall.

CHAROPHYTA
Charophyceae
Charales
Charceae
Chara sp.
Nitella sp.

RHODOPHYTA
Rhodophyceae
Florideophycideae
Nemalionales
Batrachospermaceae
Batrachospermum tortuosum Kumano (C)
B. cylindrocellulare Kumano (C)
B. beraense Kumano (C)

Amongst the various taxa, Chlorophyta, Bacillariophyta and Cyanophyta are especially conspicuous in Tasek Bera. Chlorophyta is the most dominant order with the Zygnematales being the most abundant and diverse in terms of species representing about 65% of all taxa listed (269 species, varieties and forms) (Table 73). The great abundance and diversity of desmids is striking and possibly nowhere else in Malaysia would one find such a rich assemblage.

The tabulation of Scott and Prescott (1961) for the desmid taxa of different regions according to various authors has been partially modified and expanded to include the Philippines (Behre 1956) and Tasek Bera (Table 74). The data for Tasek Bera represent collections for a single location whereas in other countries collections were made in several localities. Some of the interesting features noted are (1) the desmid flora of Tasek Bera is rich in species and comparable to the other localities listed; all nine filamentous genera are represented and most of the rare genera and species are represented; (2) in Tasek Bera the genera represented by the most number of species are *Staurastrum*, *Cosmarium* and *Euastrum* which is similar to the algal composition of Indonesia and Arnhem Land.

Some species, varieties and forms of desmids are unique to Tasek Bera. These are *Triploceras splendens* Prowse, *Micrasterias foliacea* Bail. var. *spinosa* Prowse, *M. alata* Wall. f. *tumida* Prowse and *Xanthidium superbum* Elfv. var. *centricornis* Prowse. A few of the desmids of Tasek Bera are also found in Sri Lanka, India, Burma and Japan. Some of the desmids of Tasek Bera extend to Thailand and Cambodia. Many of them are common to northern Australia. Several of the desmids are also common both to Tasek Bera and Indonesia and there is a close affinity of the desmid flora of Tasek Bera to that of Sumatra (e.g. the swamp of Menggela and eastern Sumatra). In general the results indicate that the desmid flora of Tasek Bera belongs to the broad phytogeographical area designated as the Indo-Malayan-North Australian region proposed by Krieger (1932).

The algae of Tasek Bera occur as free floating, loosely entangled or attached forms amongst the aquatic macrophytes such as *Lepironia articulata, Utricularia* spp., *Eleocharis*, sp. and *Cryptocoryne griffithii*. In some regions of the swamp where light penetrates to the bottom, an algal community has developed in the benthos.

Phytoplankton. Phytoplanktonic algae are common in static pools, among the reeds and submerged aquatic macrophytes but are rare in open water where there is a current and in stream channels especially in the swamp-forest. Amongst the phytoplankton, diatoms such as *Tabellaria fenestrata, Eunotia gracilis, E. lunaris, E. robusta, Frustulia rhomboides* and *Pinnularia*

Table 74 Summary of distribution of Desmidiales in Tasek Bera and neighbouring regions (modified from Scott and Prescott 1961) (S. Kumano).

Genera	Scott & Prescott Indonesia		West & West Sri Lanka		Skuja Burma		Scott & Prescott Arnhem Land		Behre Philippines		Tasek Bera*	
	Taxa	Percent	Taxa	Percent	Taxa	Percent	Taxa	Percent	Taxa	Percent	Taxa	Percent
Spirotaenia	1	0.2									1	0.4
Mesotaenium					1	0.3	1	0.4				
Cylindrocystis	2	0.4	3	1.1					1	0.8	2	0.7
Metrium	2	0.4			1	0.3					2	0.7
Conatozygon	3	0.6	2	0.7	1	0.3	2	0.8			6	2.2
Penium	3	0.6	16	5.8	1	0.3	1	0.4			6	2.2
Closterium	40	7.6	29	10.4	33	11.0	18	7.1	25	20.0	27	10.1
Roya											1	0.4
Pleurotaenium	33	6.3	18	6.5	14	4.7	13	5.1	4	3.0	20	7.5
Docidium	1	0.2	2	0.7			1	0.4			1	0.4
Triploceras	2	0.4	1	0.4	1	0.3	3	1.2			4	1.5
Tetmemorus	2	0.4	1	0.4			3	1.2	1	0.8	3	1.1
Ichthyocercus	1	0.2									1	0.4
Ichthyodontum	2	0.4									1	0.4
Euastrum	61	11.6	34	12.2	18	16.6	37	14.6	10	8.0	22	8.3
Euastridium	1	0.2										
Micrasterias	37	7.0	8	2.8	7	2.3	20	7.9	10	8.0	26	9.8
Cosmarium	111	21.1	69	24.7	140	46.8	49	19.3	37	29.0	32	12.0
Cosmocladium					1	0.3					1	0.4
Arthrodesmus	23	4.4	7	1.8	6	2.0	13	5.1	3	2.0	11	4.1
Xanthidium	23	4.4	10	3.6	7	2.3	7	2.7	3	2.0	9	3.3
Staurastrum	145	28.5	62	22.6	57	19.0	69	27.2	29	23.0	59	22.1
Sphaerozosma	1	0.2	1	0.4	1	0.3	2	0.8			1	0.4
Spondylosium	4	0.8	1	0.4	3	1.0					3	1.1
Onychonema	4	0.8	2	0.7	1	0.3	3	1.2			1	0.4
Hyalotheca	4	0.8	5	1.8	3	1.0	2	0.8	3	2.0	6	2.3
Groenbladia	3	0.6									2	0.7
Phymatodocis	2	0.4	1	0.4	1	0.3					4	1.5
Bambusina	3	0.6	1	0.4	1	0.3	1	0.4			1	0.4
Desmidium	11	2.1	7	2.5	2	0.7	8	3.1	1	0.8	12	4.5
Streptonema	1	0.2					1	0.4			2	0.7
Total	526		278		299		254		127		267	

major are common. Desmids such as *Cosmarium moniliforme, Closterium dianae, C. gracile, C. libellula, Hyalotheca dissiliens, H. undulata, Micrasterias foliacea* were most common (Table 73).

Epiphytic algae. The majority of the algae are found adhering to the aquatic macrophytes. Epiphytic forms, especially diatoms, loosely adhering desmids and euglenoids may conspicuously discolour the undisturbed exposed wet mud banks of the littoral region and creeks. In similar localities scattered pockets of macroscopic Charophyta are prominent. Some of the interconnecting channels support large almost pure cultures of red algae, *Batrachospermum* spp. and fragile colonies of the palmelloid green algae, *Schizochlamys* sp. which are found attached to *Cryptocoryne griffithii* (Table 73).

Amongst the forms epiphytic on *Utricularia* spp. diatoms such as *Frustulia rhomboides* var. *saxonica, Eunotia naegelii* and the green alga *Bulbochaete praereticulata* are most common. Several species of *Zygnemetales, Closterium, Cosmarium, Xanthidium, Stautrastrum* and *Desmidium* also occur (Table 73).

Common epiphytic forms on *Lepironia articulata* are the blue-green algae *Stigonema panniforme* and *Hapalosiphon stuhlmanni* and the green algae *Bulbochaete praereticulata* (Table 73).

Benthic algae (T. Watanabe). The algal community in the sediments is similar to periphytic species occurring on aquatic macrophytes. Blue-green algae such as *Stigonema panniforme* and *Hapaloshipon stuhlmannii* and green algae like *Bulbochaete praereticulata* were dominant.

5.1.2 *Macrophytes*
(I. Ikusima and E. Soepadmo)

A total of seven submerged aquatic macrophytes have been recorded in Tasek Bera (Table 75). Of these *Utricularia flexuosa* and *Cryptocoryne griffithii* are the most dominant primary producers. These two plants are usually associated with the less dominant submerged macrophytes like *Blyxa echinosperma, Hydrilla verticillata* and *Potamogeton malaianus*.

A total of 12 emergent macrophytes were collected in the littoral region outside of the swamp-forest (Table 75). Of these only *Lepironia articulata, Pandanus helicopus* and to a lesser extent *Eleocharis ochrostachys* formed

Table 75 Macrophytes. (E. Soepadmo and I. Ikusima)

I.	**Aquatic plants**	
	a. Submerged	*Blyxa echinosperma* (Clarke) Hook f.
		Cryptocoryne griffithii Schott
		Hydrilla verticillata (L.) Royle
		Potamogeton malaianus Miq.
		*Scirpus confervoides** Poir.
		Utricularia aurea Lour.
		Utricularia flexuosa Vahl.
	b. Floating-leaves	*Nymphoides indica* (L.) Kuntze
		Ploiarium alternifolium Melch.
	c. Emergent	*Dendrobium aloifolium* (Bl.) Rchb.
		Elaeocarpus sp.
		Eleocharis ochrostachys Stead.
		Eriocaulon sexangulare L.
		Flagellaria indica L.
		Gynotroches axillaris Bl.
		Lepironia articulata (Retz.) Domin.
		Ludwigia prostrata Roxb.
		Lycopodium cernuum L.
		Nepenthes gracilis Korth.
		Pandanus helicopus Kurz
		Scirpus mucronatus L.
II.	**Terrestrial plants**	
	a. In forest edge of the lake	
	Trees or shrubs:	*Artocarpus scortechinii* Hook. f.
		Baeckia frutescens L.
		Castanopsis javanica (Bl.) DC.
		Eugenia spp. (5 species)
		Fragraea fragrans Roxb.
		Garcinia nigrolineata Planch. ex T. Anders.
		Gardenia carinata Wall.
		Ixonanthes reticulata Jack
		Ixora spp.
		Koompassia malaccensis Maing. ex Benth.
		Lithocarpus elegans (Bl.) Hatus. ex Soepadmo
		Lithocarpus wallichianus (Lindl. ex Hance) Rchd.
		Ormosia bancana (Miq.) Merr.
		Pittosporum sp.
		Rhodamnia cinerea Jack
		Sindora coriacca Maing. ex Prain
		Tristania whiteana Griff.
	Woody climbers:	*Agelaea trinervis* (Llanus) Merr.
		Bauhinia sp.

	Cnetis palala (Lour.) Merr.
	Entada sp.
	Gnetum latifolium Bl.
	Gnetum gnemonodoies Brongn.
Ferns and allies:	*Bolbitis* sp.
	Lycopodium cernuum L.
	Lygodium spp.
	Taenitis sp.
	Tectaria sp.
b. Inland forest surrounding the swamp	
	Baccaurea parvifolia (M.A.) M.A.
	Barringtonia spicata Bl.
	Canarium littorale Bl.
	C. patentinervium Miq.
	Castanopsis inermis (Lindl. ex Wall.) Benth et Hook. f.
	C. javanica (Bl.) DC
	C. rhamnifolia (Miq.) A.DC
	C. schefferiana Hance
	Clerodendron breviflorum Ridl.
	Cratoxylon arborescens (Vahl.) Bl.
	Dillenia reticulata King
	Diptercarpus costulatus v. Sl.
	Dysoxylum cauliflorum Hiern.
	Elateriospermum tapos Bl.
	Fibraurea chloroleuca Miers. (woody climber)
	Gironniera parvifolia Planch.
	G. subaequalis Planch.
	Grewia blattaefolia Corner
	Gynotroches axillaris Bl.
	Payena sp.
	Parinarium corymbosum (Bl.) Miq.
	Nauclea sp.
	Rinorea sp.
	Shorea leprosula Miq.

* A new record from Mainland Asia: previously known only from Cuba, Florida, Guyana, Sri Lanka and Madagascar (Stone 1971).

the dominant components of macrophyte community. *L. articulata* particularly is dominant, occupying about 6.2 ha or 78% of the total study area. As such, detailed studies of this species were conducted and the results will be discussed later.

5.2 Population density and standing crop (I. Ikusima, R. P. Lim and J. I. Furtado)

5.2.1 *Phytoplankton* (T. Okino, R. P. Lim, J. I. Furtado, S. Kumano and T. Mizuno)

Phytoplankton were sampled at Stations I - V (see above), and the amount of chlorophyll was measured at Stations I, V, XI, XII, and XIII. Each species of the phytoplankton was counted whenever identifiable, otherwise phytoplankton counts were pooled into broad taxa, using an Utermohl's inverted microscope (see earlier).

The phytoplankton density is the dominant component of the plankton, averaging 343.1 cells$\cdot l^{-1}$ for Stations I-V, with a minimum of 11.8 at Station V (January 1972) and a maximum of 6903.9 cells$\cdot l^{-1}$ at Station II (October 1971) (Table 76). The *Pandanus-Utricularia* (Station II, mean 738.4 cells$\cdot l^{-1}$) and the *Utricularia* (Station III, mean 440.4 cells$\cdot l^{-1}$) sublittoral are the best developed in phytoplankton, probably because of their ecological complexity being at the transition between the limnetic and the littoral, and because of the development of surface aquatic macrophytes especially *Utricularia* which permit the establishment and growth of epiphytic species. This probably influences the corresponding development of zooplankton (see later).

All habitats (Stations I-V) exhibit at least one peak abundance of phytoplankton in the early northeast monsoon (September-December) (Tables 76 and 77; Fig. 84), and may be dominated by the marked development of phytoplankton populations in the sublittoral with *Utricularia* (Stations II and III). This dominant peak in phytoplankton abundance could be due to (1) increasing nutrients through influx from the watershed favouring the temporal development of certain taxa, (2) the presence of abundant surface aquatic macrophytes for algal epiphytes, and (3) the translocation of phytoplankton from the littoral and sublittoral, especially the *Utricularia*, to the limnetic by water flow caused by monsoonal rainfall. The particular taxa which contribute to this abundance are the Chlorophyta which dominate the phytoplankton community (see later) and, to a lesser extent, the Bacillariophyta.

In addition (Table 77; Fig. 84), the sublittoral *Utricularia* (Station III) exhibits a second peak abundance of phytoplankton in April-May (southwest monsoon) due to (1) the same factors that determine the major abundance (see above), (2) the abundance of diatoms (see later), and (3) the growth of the *Utricularia* mats in the surface water (see later). The modified limnetic (Station I) also exhibits a second smaller peak in phytoplankton abundance in May-June (end of the southwest monsoon), but this is due to the Pyrrophyta

Table 76 Phytoplankton: Mean, minimum and maximum population densities of phytoplankton at all stations (I-V) in Tasek Bera (1971–1972). (R. P. Lim and J. I. Furtado)

Stations	Population density (No. of cells · litre^{-1})		
	Mean	Minimum	Maximum
Open water region			
Station I	198.3	11.9 (Jan. 1972)	835.6 (Sept. 1971)
Station II	739.4	22.9 (Jan. 1971)	6903.9 (Oct. 1971)
Littoral region			
Station III	440.4	18.8 (Jan. 1972)	2051.5 (Oct. 1971)
Station IV	213.0	15.5 (Jan. 1972)	786.5 (Sept. 1971)
Swamp-forest region			
Station V	124.4	11.8 (Jan. 1972)	481.5 (Oct. 1971)
Overall	343.1	11.8 (Jan. 1972)	6903.9 (Oct. 1971)

Table 77 Phytoplankton: Mean, minimum and maximum population densities of phytoplankton at different months in Tasek Bera (1971–1972). (R. P. Lim and J. I. Furtado)

Date	Population density (no. of cells · litre^{-1})		
	Mean	Minimum	Maximum
11 January 1971	31.6	19.2 (St. IV)	63.5 (St. I)
14 February 1971	39.3	24.9 (St. III)	57.7 (St. IV)
12 March 1971	55.7	19.6 (St. I)	109.3 (St. IV)
17 April 1971	385.9	68.3 (St. V)	1447.8 (St. III)
12 May 1971	128.7	75.3 (St. III)	219.2 (St. II)
8 June 1971	252.0	164.4 (St. III)	437.3 (St. IV)
4 July 1971	106.2	46.8 (St. I)	213.1 (St. V)
9 August 1971	291.4	193.4 (St. III)	389.3 (St. I)
8 September 1971	591.6	23.7 (St. V)	835.6 (St. I)
6 October 1971	2011.9	293.7 (St. I)	6903.9 (St. II)
7 November 1971	118.7	66.6 (St. V)	206.1 (St. III)
7 December 1971	435.6	94.1 (St. IV)	899.8 (St. III)
9 January 1972	19.7	11.8 (St. V)	40.5 (St. II)
Overall	893.7	11.8 (St. V)	6903.9 (St. II)

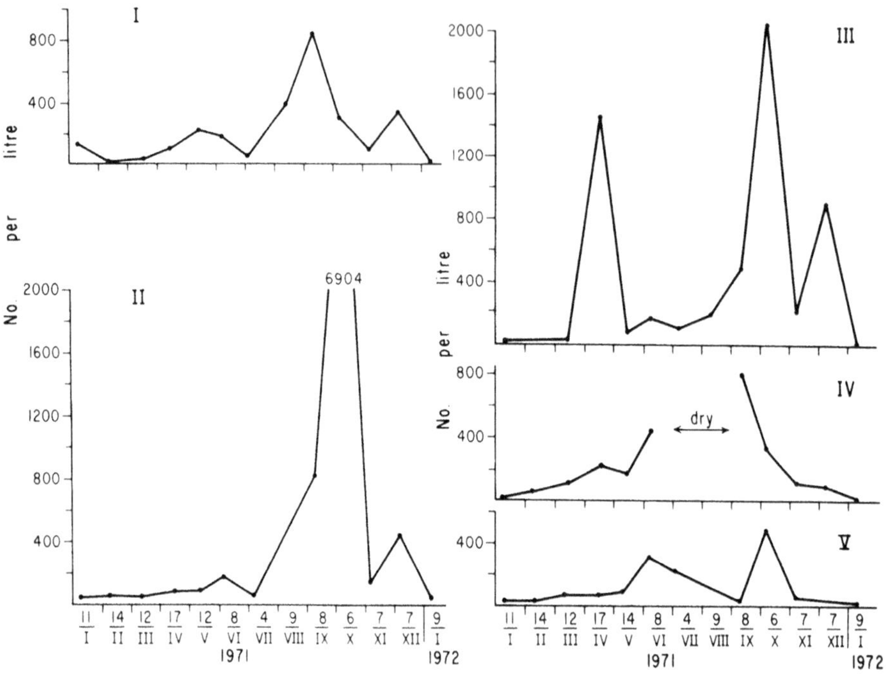

Fig. 84 Seasonal changes in standing crop of net phytoplankton. (R. P. Lim and J. I. Furtado)

(see later) and probably to initial decomposition of allochthonous matter.

5.2.1.1 *Phytoplankton association* (R. P. Lim and J. I. Furtado). Seven main orders of phytoplankton occur in Tasek Bera, namely the Cyanophyta, Rhodophyta, Bacillariophyta, Pyrrophyta, Euglenophyta, Chlorophyta and Chrysophyta, with the last vere rarely encountered. Among the various orders the Rhodophyta, Pyrrophyta and Euglenophyta are very poor, the Cyanophyta are poor, whereas the Bacillariophyta and Chlorophyta are rich and very rich, respectively, in species diversity as well as density (Tables 78 and 81)

The diversity of the phytoplankton at Stations I-III in Tasek Bera is fairly uniform averaging 30.2 species$\cdot l^{-1}$, with a minimum of 8 at Station III (February 1971) and a maximum of 58 species$\cdot l^{-1}$ at Station IV (May 1971) (Table 79). This diversity is maximal during the wet seasons (April-June 1971 and October-December 1971) with an artefact due to analysis in August 1971 (Table 80). However, this bimodal abundance in species is not due to any singular habitat of the swamp but to the contribution of the different habitats at different times of the year, and is thus dependent on various combinations

Table 78 Phytoplankton taxa: Relative abundance of various phytoplankton taxa in the standing crop (% number of individuals per litre) at different months at Tasek Bera (1971–1972) (x = 25-49%, xx = 56-74% and xxx = 75% abundance). (R. P. Lim and J. I. Furtado)

Date	Cyanophyta	Rhodophyta	Bacillario-phyta	Euglenophyta Pyrrophyta	Chlorophyta
11 Jan. 1971					xxx
14 Febr. 1971	xxx		x		xx
12 Mar. 1971	x		x		x
17 Apr. 1971			xxx		x
12 May 1971			xx	xxx	x
8 June 1971				xx	xx
4 July 1971			x		xx
9 Aug. 1971			xx		xxx
8 Sept. 1971			x		xxx
6 Oct. 1971			x		xxx
7 Nov. 1971			x		xxx
7 Dec. 1971					xxx
9 Jan. 1972			x		xx

Table 79 Phytoplankton species: Mean, minimum and maximum numbers of phytoplankton species in the standing crop at all stations (I-V) in Tasek Bera (1971–1972). (R. P. Lim and J. I. Furtado)

Stations	Species diversity (No. of spp. · litre^{-1})		
	Mean	Minimum	Maximum
Open water region			
Station I	29.1	12 (Febr. 1971)	45 (Apr. 1971)
Station II	32.2	16 (Febr. 1971)	53 (Oct. 1971)
Littoral region			
Station III	29.0	8 (Febr. 1971)	41 (July 1971)
Station IV	30.8	18 (Jan. 1972)	58 (May 1971)
Swamp-forest region			
Station V	29.7	18 (Sept. 1971)	54 (July 1971)
Overall	30.2	8 (Febr. 1971)	58 (May 1971)

Table 80 Phytoplankton species: Mean, minimum and maximum diversity of species in the phytoplankton standing crop at different months in Tasek Bera (1971–1972).

Date	Species diversity (No. of spp. · litre^{-1})		
	Mean	Minimum	Maximum
11 January 1971	23.0	18	38 (St. I)
14 February 1971	16.6	8	25 (St. V)
12 March 1971	22.8	15	29 (St. IV)
17 April 1971	34.8	19	45 (St. I)
12 May 1971	40.8	32	58 (St. IV)
8 June 1971	39.2	33	49 (St. IV)
4 July 1971	31.8	29	54 (St. V)
9 August 1971*	38.5	37	40 (St. I)
8 September 1971	26.8	18	41 (St. II)
6 October 1971	34.6	23	53 (St. II)
7 November 1971	28.4	26	32 (St. III)
7 December 1971	31.3	24	39 (St. III)
9 January 1972	20.6	18	29 (St. II)
Overall 29.9	8	58	

* Only two stations were analysed.

of environmental conditions favouring the abundance of different taxa in space and time. The April-June (southwest monsoon) species abundance is due to the modified limnetic and *Lepironia* littoral regions (Stations I and IV), whereas the October-December (northeast monsoon) one is due to the *Pandanus-Utricularia* and *Utricularia* sublittoral region (Stations II and III), being dominated by the *Utricularia* association.

Fluctuations in the diversity of phytoplankton species (Table 80) is dominated by the Chlorophyta (Table 81) which form at least 60% of the average species composition. The Bacillariophyta and Cyanophyta are less important, while the Rhodophyta, Pyrrophyta and Euglenophyta are almost unimportant in contributing to these fluctuations. The April-June abundance in species is due not only to the dominant Chlorophyta, but also to the less dominant Bacillariophyta and almost insignificant Pyrrophyta and Cyanophyta. The latter three taxa are also numerically abundant in this season, thus making them relatively important even though their numerical composition may be overshadowed by the Chlorophyta (Table 82). The October-December abundance in species composition is due to the diversity of Chlorophyta and Bacillariophyta (Table 78), both of which are also numerically abundant in this monsoonal season (Tables 78 and 82). The Chlorophyta form at least 60%

Table 81 Phytoplankton taxa: Mean, minimum and maximum number of species of various phytoplankton taxa in the standing crop at Tasek Bera (1971–1972). (R. P. Lim and J. I. Furtado)

Taxa	Species diversity (No. of spp. · litre^{-1})		
	Mean	Minimum	Maximum
Cyanophyta	2.0	1	7 (St. III, May 1971)
Rhodophyta	0.6	0	2 (St. II, Febr. 1971; St. V, Sept. 1971)
Bacillariophyta	6.6	1	12 (St. V, July 1971; St. I-IV, Apr. 1971)
Pyrrophyta	0.6	0	3 (St. I, May 1971)
Euglenophyta	0.2	0	2 (St. I, June and Aug. 1971)
Chlorophyta	20.3	1	44 (St. IV, May-June 1971)
Total	30.3	3	70

Table 82 Phytoplankton taxa: Mean, minimum and maximum population density of various phytoplankton taxa in Tasek Bera (1971–1972). (R. P. Lim and J. I. Furtado)

Taxa	Population density (No. of cells · litre^{-1})		
	Mean	Minimum	Maximum
Cyanophyta	11.8	0	114.5 (St. III, Apr. 1971)
Rhodophyta	0.7	0	8.2 (St. II, Oct. 1971; St. V, July 1971)
Bacillariophyta	52.4	1.2	932.5 (St. II, Oct. 1971)
Pyrrophyta	11.0	0	198.4 (St. I, May 1971)
Euglenophyta	0.3	0	3.7 (St. V, June 1971)
Chlorophyta	267.5	0.4	5881.4 (St. II, Oct. 1971)
Total	343.7	1.6	7138.7

of the average numerical composition of the phytoplankton.

Thus the various orders of phytoplankton may be ranked in decreasing species diversity: Chlorophyta > Bacillariophyta > Cyanophyta > Rhodophyta + Pyrrophyta > Euglenophyta (Table 81), and in decreasing numerical abundance: Chlorophyta > Bacillariophyta > Cyanophyta + Pyrrophyta > Rhodophyta + Euglenophyta (Table 82). The dominance of the Chlorophyta is due to the rich diversity of desmids (Table 73). Such diversity in, and dominance of the Chlorophyta is uncommon in the pelagic of lakes in the temperate as well as the tropical regions (cf. Hutchinson 1967, Ho 1970 and Arumugam 1971), and is due probably to the role of the extensive sublittoral and littoral in affording a complexity of biotopes and niches for speciation and colonisation.

5.2.1.2 *Cyanophyta* (R. P. Lim and J. I. Furtado) The Cyanophyta is relatively poorly represented at Tasek Bera. It is the third most important taxa (Tables 81 and 82) in species diversity with an average of 2.0 species $\cdot l^{-1}$ (maximum 7 and minimum 1) and in density with an average of 11.8 cells $\cdot l^{-1}$ (maximum 114 and minimum zero). It flourishes best in the *Utricularia* sublittoral (Station III) followed by the *Pandanus-Utricularia* sublittoral (Station II) probably because of the surface *Utricularia* mats (Tables 83 and 84).

The blue-green algae exhibit a unimodal seasonal fluctuation in density in the limnetic region (Station I and V), and a bimodal fluctuation in the *Pandanus-Utricularia* and *Utricularia* sublittoral (Stations II and III) and the *Lepironia* littoral regions (Station IV) (Figs. 85a-e: A). The bimodal pattern is best developed in the sublittoral, again due to the *Utricularia* association. Peak density occurs in the dry season (February-April 1971) extending somewhat into the southwest monsoonal season, and in the northeast monsoonal season (September-December 1971). This is probably due to a differential response by species to nutrient influx from the watershed in the monsoonal seasons, to nutrient mineralization by decomposing allochthonous organic matter, and to macrophytic substrate abundance for the development of epiphytic forms.

Peak density of Cyanophyta differs slightly from its maximum species diversity which is in May 1971 at the end of the southwest monsoon. The species most common throughout the year are *Fisherella reptans* and *Oscillatoria* sp.

5.2.1.3 *Rhodophyta* (R. P. Lim and J. I. Furtado). The Rhodophyta is poorly represented at Tasek Bera. Its is relatively uncommon in the phyto-

Table 83 Cyanophyta: Mean, minimum and maximum Cyanophyta species in the phytoplankton standing crop at all stations (I-V) In Tasek Bera (1971-1972). (R. P. Lim and J. I. Furtado)

Stations	Species diversity (No. of spp. · litre^{-1})		
	Mean	Minimum	Maximum
Open water region			
Station I	1.7	1	2
Station II	1.7	1	3
Littoral region			
Station III	2.6	1	7 (May 1971)
Station IV	2.0	1	4 (May 1971)
Swamp-forest region			
Station V	2.0	1	3
Overall	2.0	1	7 (May 1971)

Table 84 Cyanophyta: Mean, minimum and maximum population density of Cyanophyta at all stations (I-V) in Tasek Bera (1971–1972). (R. P. Lim and J. I. Furtado)

Stations	Population density (No. of cells · litre^{-1})		
	Mean	Minimum	Maximum
Open water region			
Station I	6.6	1.2 (Sept., Dec. 1971)	17.6 (April 1971)
Station II	16.1	1.2 (Jan. 1971)	81.8 (Oct. 1971)
Littoral region			
Station III	17.2	0 (Oct. 1971)	114.5 (April 1971)
Station IV	9.1	0.8 (Jan. 1971)	29.5 (Febr. 1971)
Swamp-forest region			
Station V	9.8	0.4 (Sept. 1971)	38.5 (March 1971)
Overall	11.8	0 (Oct. 1971)	114.5 (April 1971)

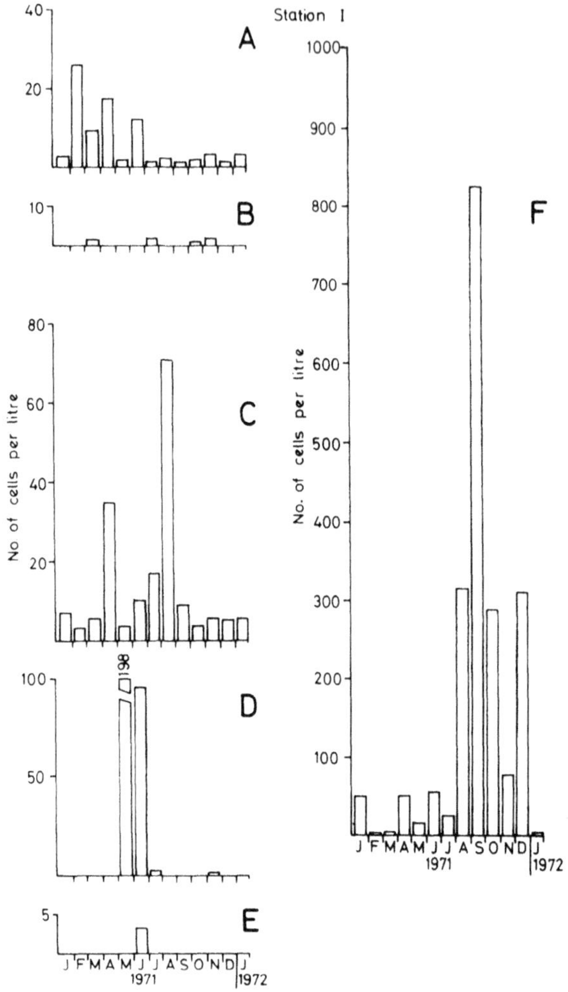

Fig. 85a Standing crop of phytoplankton at Station I in Tasek Bera. A: Cyanophyta; B: Rhodophyta; C: Bacillariophyta; D: Pyrrophyta; E: Euglenophyta; F: Chlorophyta. (R. P. Lim and J. I. Furtado)

plankton, occurring only for a few months with an average diversity of 0.6 species$\cdot l^{-1}$ (maximum 2 and minimum zero), and diversity of 0.7 cells$\cdot l^{-1}$ (maximum 8.2 and minimum zero) (Tables 85 and 86). It flourishes best both numerically and in species diversity at the *Pandanus-Utricularia* sublittoral (Station II) and the swamp-forest limnetic (Station V) regions (Tables 85 and 86). *Batrachospermum* spp., an attached algae, constitutes the major portion of this taxa in the phytoplankton. It exhibits a unimodal density fluctuation

Table 85 Rhodophyta: Mean, minimum and maximum Rhodophyta species in the phytoplankton standing crop at various stations (I-V) in Tasek Bera (1971–1972). (R. P. Lim and J. I. Furtado)

Stations	Species diversity (No. of spp. · litre^{-1})		
	Mean	Minimum	Maximum
Open water region			
Station I	0.4	0	1
Station II	0.8	0	2 (Febr. 1971)
Littoral region			
Station III	0.5	0	1
Station IV	0.5	0	1
Swamp-forest region			
Station V	0.7	0	2 (Sept. 1971)
Overall	0.6	0	2

Table 86 Rhodophyta: Mean, minimum and maximum population density of Rhodophyta at various stations (I-V) in Tasek Bera (1971–1972). (R. P. Lim and J. I. Furtado)

Stations	Population density (No. of cells · litre^{-1})		
	Mean	Minimum	Maximum
Open water region			
Station I	0.2	0	0.8 (July 1971)
Station II	1.1	0	8.2 (Oct. 1971)
Littoral region			
Station III	0.5	0	1.2 (Nov. 1971)
Station IV	0.4	0	1.6 (Nov.-Dec. 1971)
Swamp-forest region			
Station V	1.3	0	8.2 (July 1971
Overall	0.7	0	8.2 (July, Oct. 1971)

which is not very marked (Figs. 85a-e: B). The abundance in July 1971 (dry season) at the forest limnetic region (Station V) is probably due to its growth and development in these channels, while its abundance in October 1971 (northeast monsoon) at the *Pandanus-Utricularia* sublittoral region (Station II) is due to its dislodgement and translocation by heavy rains.

5.2.1.4 *Bacillariophyta* (R. P. Lim and J. I. Furtado). The Bacillariophyta is a fairly common taxa in Tasek Bera. It is the second largest taxa of the

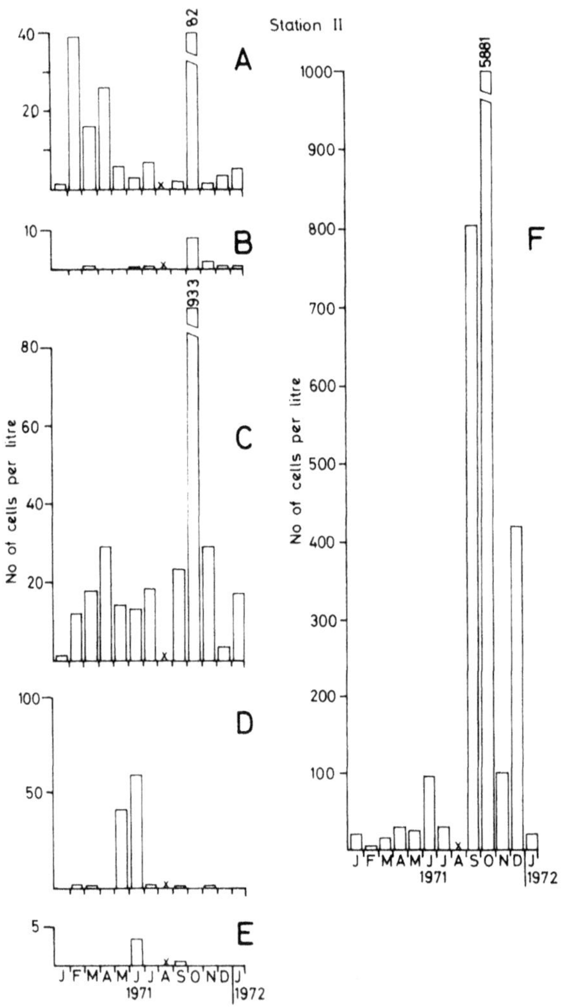

Fig. 85b Standing crop of phytoplankton at Station II in Tasek Bera. A: Cyanophyta; B: Rhodophyta; C: Bacillariophyta; D: Pyrrophyta; E: Euglenophyta; F: Chlorophyta. (R. P. Lim and J. I. Furtado)

phytoplankton both in species and numerical abundance, with an average diversity of 6.6 species $\cdot l^{-1}$ (maximum 12 and minimum 1) and density of 52.4 cells $\cdot l^{-1}$ (maximum 932.5 and minimum 1.2) (Tables 81 and 82). In species diversity this taxa is equally well developed in all habitats of the swamp (Table 87). However, in numerical abundance, it is best developed in the *Pandanus-Utricularia* and *Utricularia* sublittoral regions (Stations II and III) (Table 88).

The species composition of diatoms fluctuates somewhat with two periods of peak abundance, April 1971 (southwest monsoon) and July-August 1971 (subsequent dry season) (Table 89). This fluctuation may be due to the colonization by and establishment of different species as environmental and nutrient conditions change in the swamp.

The density of diatoms fluctuates bimodally with peak abundance in April-May 1971 (southwest monsoon) and in September-December 1971 (northeast monsoon) (Figs. 85a-e: C), with some lag in-between seasons. This coincides only partly with the fluctuations in species diversity, although the sublittoral *Utricularia* (Station III) exhibits a number of density peaks during August-December probably due to micro-climatic changes. The littoral *Lepironia* shows only one major diatom peak density in the southwest monsoon, and its poor diatom density in the northeast monsoon may be due to the drying up of the littoral in August (dry season). Since the sublittoral is best developed for six taxa, the fluctuations in the limnetic region are probably influenced by it as in ponds (Fritsch 1931). The environmental conditions favouring numerical abundance are grossly similar to those favouring species fluctuations.

Four species of diatoms are common at Stations I-V in Tasek Bera throughout the whole year (Tables 90-93). Populations of *Fragellaria* spp. are best developed in the sublittoral *Utricularia* (Station III), probably due to the influence of *Utricularia* as a substrate (Table 90). Its peak population density is during the southwest monsoon in the sublittoral region (April 1971), with a minor increase in the forest limnetic region in October 1971 due to rain. Populations of *Eunotia gracilis* are best developed in the sublittoral region especially in the *Utricularia* mats (Station III), and probably influence the adjacent limnetic regions in the monsoons. (Table 91). It shows a major population increase during the southwest monsoon (April 1971) in the sublittoral *Utricularia* region, and a minor one during the northeast monsoon (October 1971) in the *Pandanus-Utricularia* sublittoral region. *Eunotia robusta* populations are best developed in the sublittoral Utricularia region (Station III) (Table 92), and show a major population peak during the southwest monsoon (April 1971). *Frustulia rhomboides* populations are best developed in the sublittoral *Utricularia* and the littoral *Lepironia* regions (Stations

Table 87 Bacillariophyta: Mean, minimum and maximum Bacillariophyta species in the phytoplankton standing crop at various stations (I-V) in Tasek Bera (1971–1972). (R. P. Lim and J. I. Furtado)

Stations	Species diversity (No. of spp. · litre^{-1})		
	Mean	Minimum	Maximum
Open water region			
Station I	6.2	4	11 (April 1971)
Station II	6.5	1	10 (April 1971)
Littoral region			
Station III	7.0	3	11 (April, Aug. 1971)
Station IV	6.2	3	11 (Febr. 1971)
Swamp-forest region			
Station V	7.1	3	12 (July 1971)
Overall	6.6	1	12 (July 1971)

Table 88 Bacillariophyta: Mean, minimum and maximum population density of Bacillariophyta at various stations (I-V) in Tasek Bera (1971–1972). (R. P. Lim and J. I. Furtado)

Stations	Population density (No. of cells · litre^{-1})		
	Mean	Minimum	Maximum
Open water region			
Station I	14.1	2.9 (Febr. 1971)	35.2 (April 1971)
Station II	92.5	1.2 (Jan. 1971)	932.5 (Oct. 1971)
Littoral region			
Station III	84.8	2.1 (Jan. 1971)	756.7 (April 1971)
Station IV	42.9	1.6 (Jan. 1971)	174.6 (April 1971)
Swamp-forest region			
Station V	27.9	2.5 (Jan. 1971)	73.2 (July 1971)
Overall	52.4	1.2 (Jan. 1971)	932.5 (Oct. 1971)

Table 89 Bacillariophyta: Mean, minimum and maximum species diversity of Bacillariophyta in the phytoplankton standing crop at Tasek Bera (1971–1972). (R. P. Lim and J. I. Furtado)

Date		Species diversity (No. of spp. · litre^{-1})		
		Mean	Minimum	Maximum
11 January	1971	3.2	1	6 (St. I)
14 February	1971	7.4	6	11 (St. IV)
12 March	1971	6.6	6	8 (St. V)
17 April	1971	9.2	7	11 (St. I and III)
12 May	1971	7.6	6	9 (St. V)
8 June	1971	5.4	5	7 (St. IV)
4 July	1971	9.5	8	12 (St. V)
9 August	1971*	9.5	8	11 (St. III)
8 September	1971	6.2	4	9 (St. III)
6 October	1971	6.4	4	9 (St. V)
7 November	1971	5.2	3	7 (St. V)
7 December	1971	5.3	5	8 (St. V)
9 January	1972	5.6	5	6

* Only two stations analysed.

Table 90 *Fragellaria:* Mean, minimum and maximum population density of *Fragellaria* spp. at various stations (I-V) in Tasek Bera (1971–1972). (R. P. Lim and J. I. Furtado)

Stations	Population density (No. of cells · litre^{-1})		
	Mean	Minimum	Maximum
Open water region			
Station I	1.8	0	4.5 (July 1971)
Station II	3.7	0.4	9.8 (April 1971)
Littoral region			
Station III	16.2	0.4	173.3 (April 1971)
Station IV	1.9	0.4	4.5 (Febr. 1971)
Swamp-forest region			
Station V	6.9	0	22.5 (Oct. 1971)
Overall	6.1	0	173.3

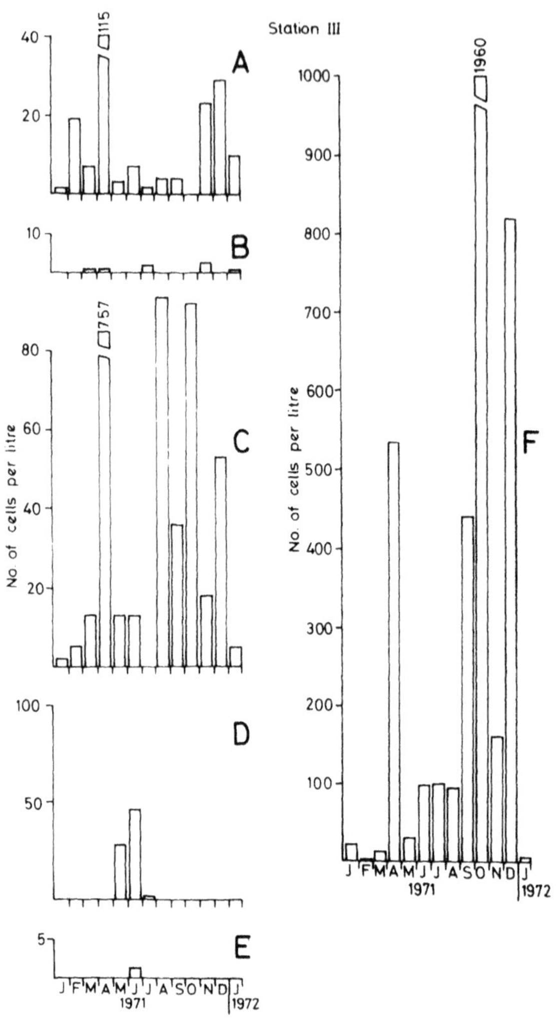

Fig. 85c Standing crop of phytoplankton at Station III in Tasek Bera. A: Cyanophyta; B: Rhodophyta; C: Bacillariophyta; D: Pyrrophyta; E: Euglenophyta; F: Chlorophyta. (R. P. Lim and J. I. Furtado)

III and IV) (Table 93). It shows a major population peak during the southwest monsoon (April-May 1971). Thus, these common diatom species flourish numerically in the southwest monsoon (April-May) probably because of favourable nutrient and environmental conditions. Their predominance in the sublittoral and littoral regions especially the *Utricularia* suggests that their population development depends on the development of this macrophytic substrate. These sublittoral and littoral populations probably influence the

Table 91 Eunotia gracilis: Mean, minimum and maximum population density of *Eunotia gracilis* at various stations (I-V) in Tasek Bera (1971–1972). (R. P. Lim and J. I. Furtado)

Stations	Population density (No. of cells · litre^{-1})		
	Mean	Minimum	Maximum
Open water region			
Station I	2.8	0.2	10.2 (April 1971)
Station II	7.1	0.4	49.1 (Oct. 1971)
Littoral region			
Station III	13.8	0	132.0 (April 1972)
Station IV	2.8	1.6	6.1 (June 1971)
Swamp-forest region			
Station V	5.1	0.2	12.7 (Oct. 1971)
Overall	6.3	0	132.0

Table 92 Eunotia robusta: Mean, minimum and maximum population density of *Eunotia robusta* at various stations (I-V) in Tasek Bera (1971–1972). (R. P. Lim and J. I. Furtado)

Stations	Population density (No. of cells · litre^{-1})		
	Mean	Minimum	Maximum
Open water region			
Station I	3.4	0	31.9 (Aug. 1971)
Station II	3.0	0	13.1 (Oct. 1971)
Littoral region			
Station III	10.5	0	90.8 (April 1971)
Station IV	1.2	0	3.3 (Febr. 1971)
Swamp-forest region			
Station V	2.7	0.4	9.8 (July 1971)
Overall	4.2	0	90.8

Table 93 *Frustulia rhomboides:* Mean, minimum and maximum population density of *Frustulia rhomboides* at various stations (I-V) in Tasek Bera (1971–1972). (R. P. Lim and J. I. Furtado)

Stations	Population density (No. of cells · litre⁻¹)		
	Mean	Minimum	Maximum
Open water region			
Station I	3.2	0	8.2 (April 1971)
Station II	6.4	1.2	21.7 (Oct. 1971)
Littoral region			
Station III	22.8	1.2	198.0 (April 1971)
Station IV	20.4	0	100.2 (May 1971)
Swamp-forest region			
Station V	8.5	0	32.3 (July 1971)
Overall	12.3	0	198.0

limnetic populations, accounting for lags in limnetic population fluctuations with the interaction of environmental factors.

The fairly common diatom species *Tabellaria fenestrata, Eunotia lunaris, E. pectinalis, Pinnularia luculenta, Surirella biseriata, S. linearis,* and *S. pseudospiralis* are mostly best developed in the sublittoral, especially the *Utricularia* region (Table 94). Most of these species exhibit a peak population density in April 1971 (southwest monsoon) exept for *T. fenestrata* and *E. lunaris* were population peaks occur in August 1971 (dry season). Although these species occur sporadically depending on favourable nutrient and environmental conditions, they contribute to the secondary dominance of the diatoms in the phytoplankton, as well as to the flux of diatoms from the sublittoral and littoral regions to the limnetic region.

5.2.1.5 *Pyrrophyta and Euglenophyta* (R. P. Lim and J. I. Furtado). The Pyrrophyta and Euglenophyta are poor in Tasek Bera. They are very poorly represented in the phytoplankton in species diversity with an average of 0.6 species · l^{-1} (maximum 3 and minimum zero) for the Pyrrophyta, and of 0.2 species · l^{-1} (maximum 2 and minimum zero) for the Euglenophyta (Table 81). In numerical abundance, the Pyrrophyta is of quarternary importance to the

Table 94 Common Bacillariophyta: The maximum population density of selected fairly common Bacillariophyta species at designated stations in Tasek Bera (1971–1972). (R. P. Lim and J. I. Furtado)

Species	Population density (No. of cells · litre^{-1})	
Tabellaria fenestrata	13.9	(Station I, August 1971)
	18.8	(Station II, August 1971)
	6.5	(Station IV, December 1971)
	9.4	(Station V, April 1971)
Eunotia lunaris	13.9	(Station I, August 1971)
	4.1	(Station II, October 1971)
	11.5	(Station III, August 1971)
	4.5	(Station V, May 1971)
Eunotia pectinalis	1.2	(Station II, July 971)
	37.1	(Station III, April 1971)
Pinnularia luculenta	4.1	(Station I, April 1971)
	2.1	(Station II, April 1971)
	107.3	(Station III, April 1971)
	4.9	(Station IV, April 1971)
Surirella biseriata	16.5	(Station III, April 1971)
Surirella linearis	1.2	(Station II, April 1971)
	3.3	(Station V, April 1971)
Surirella pseudospiralis	4.1	(Station III, April 1971)

phytoplankton, ranking close to the Cyanophyta, with an average density of 11.0 cells · l^{-1} (maximum 198.4 and minimum zero), whereas the Euglenophyta are of least importance with an average density of 0.3 cells · l^{-1} (maximum 3.7 and minimum zero) (Table 82).

The Pyrrophyta and Euglenophyta are best developed both in species diversity and numerical abundance in the limnetic region (Tables 95-98). They occur ephemerally with peak diversity and density in May-June 1971 (the beginning of the dry season after the southwest monsoon) (Figs. 85a-e: D & E). Their brief colonisation and establishment may be due to unstable environmental and nutrient conditions, especially at the initial stages of decomposition of allochtonous organic matter deposited from the watershed during the monsoon. The Pyrrophyta mainly consist of *Ceratium* sp. while the Euglenophyta of *Phacus* sp. and *Trachilomonas* sp.

5.2.1.6 *Chlorophyta* (R. P. Lim and J. I. Furtado). The Chlorophyta is the

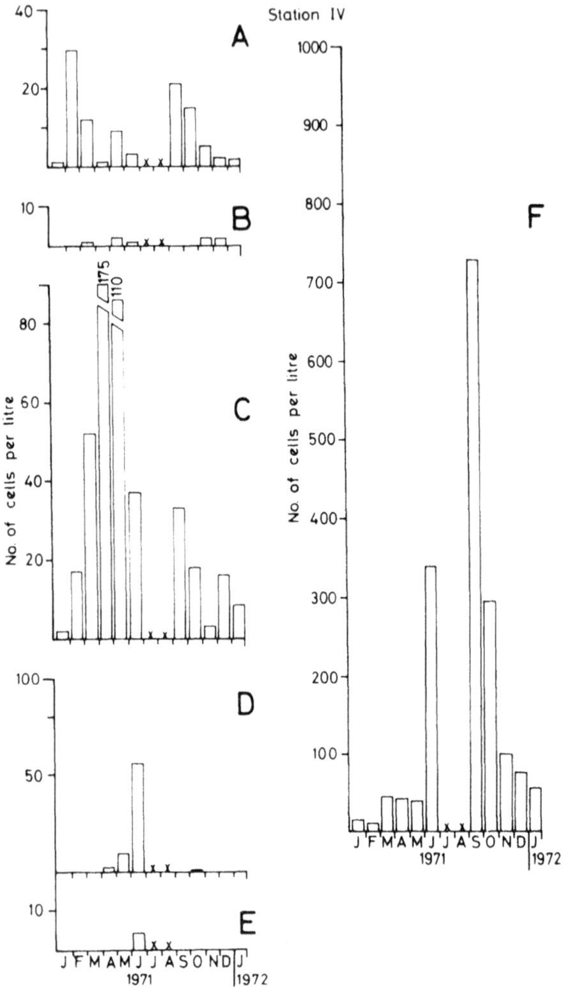

Fig. 85d Standing crop of phytoplankton at Station IV in Tasek Bera. A: Cyanophyta; B: Rhodophyta; C: Bacillariophyta; D: Pyrrophyta; E: Euglenophyta; F: Chlorophyta. (R. P. Lim and J. I. Furtado)

most common and dominant taxa in Tasek Bera. It is the largest component of the phytoplankton both in species diversity and numerical abundance, with an average diversity of 20.3 species$\cdot l^{-1}$ (maximum 44 and minimum 1) and density of 267.5 cells$\cdot l^{-1}$ (maximum 5881.4 and minimum 0.4) (Tables 81 and 82). In species diversity, this taxa is developed fairly equally and well in all habitats of the swamp, with better development in the *Pandanus-Utricularia* sublittoral (Station II) and the *Lepironia* littoral regions (Station IV) (Table 99). Whereas in numerical abundance, this taxa is best developed at the

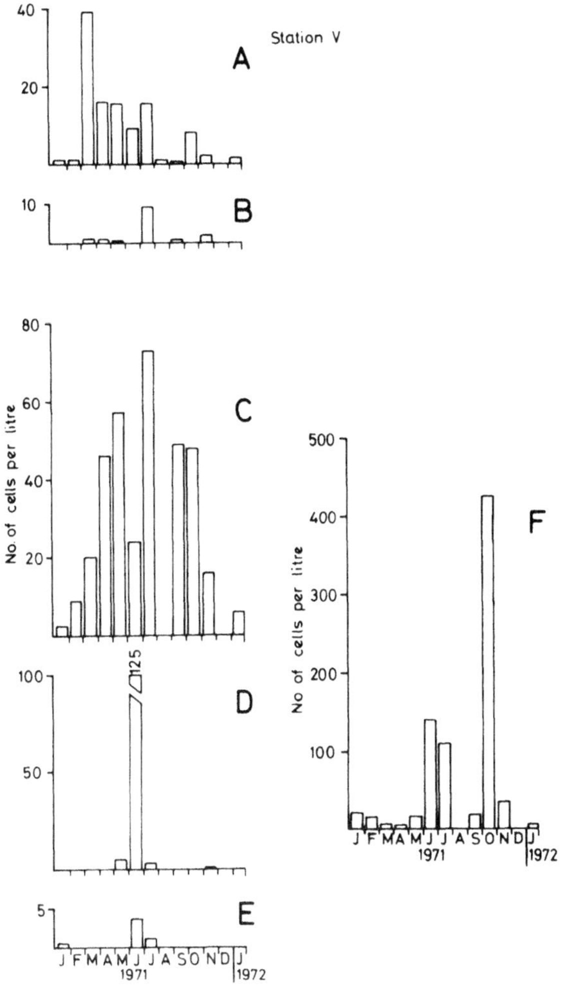

Fig. 85e Standing crop of phytoplankton at Station V in Tasek Bera. A: Cyanophyta; B: Rhodophyta; C: Bacillariophyta; D: Pyrrophyta; E: Euglenophyta; F: Chlorophyta. (R. P. Lim and J. I. Furtado)

Utricularia and *Pandanus-Utricularia* sublittoral region (Stations II and III) in comparison to the littoral and limnetic regions (Stations I, IV and V) (Table 100). The numerical abundance of the Chlorophyta, especially desmids, in the sublittoral region may be due to the ecological complexity of this transitional zone.

The species diversity of Chlorophyta fluctuates somewhat seasonally with peak abundance in the southwest monsoon season (April-June) and extending somewhat into the dry season, and in the northeast monsoon season

Table 95 Pyrrophyta: Mean, minimum and maximum diversity of Pyrrophyta species at various stations (I-V) in Tasek Bera (1971–1972). (R. P. Lim and J. I. Furtado)

Stations	Species diversity (No. of spp. · litre^{-1})		
	Mean	Minimum	Maximum
Open water region			
Station I	0.8	0	3 (May 1971)
Station II	0.7	0	2
Littoral region			
Station III	0.5	0	2
Station IV	0.4	0	2
Swamp-forest region			
Station V	0.5	0	2
Overall	0.6	0	3

Table 96 Pyrrophyta: Mean, minimum and maximum population density of Pyrrophyta at various stations (I-V) in Tasek Bera (1971–1972). (R. P. Lim and J. I. Furtado)

Stations	Population density (No. of cells · litre^{-1})		
	Mean	Minimum	Maximum
Open water region			
Station I	22.9	0	198.4 (May 1971)
Station II	8.6	0	59.3 (June 1971)
Littoral region			
Station III	5.7	0	45.8 (June 1971)
Station IV	5.9	0	54.8 (June 1971)
Swamp-forest region			
Station V	11.9	0	125.1 (June 1971)
Overall	11.0	0	198.4

Table 97 Euglenophyta: Mean, minimum and maximum diversity of Euglenophyta species at various stations (I-V) in Tasek Bera (1971–1972). (R. P. Lim and J. I. Furtado)

Stations	Species diversity (No. of spp. · litre^{-1})		
	Mean	Minimum	Maximum
Open water region			
Station I	0.4	0	2 (June, Aug. 1971)
Station II	0.2	0	1
Littoral region			
Station III	0.1	0	1
Station IV	0.2	0	1
Swamp-forest region			
Station V	0.2	0	1

Table 98 Euglenophyta: Mean, minimum and maximum population density of Euglenophyta at various stations (I-V) in Tasek Bera (1971–1972). (R. P. Lim and J. I. Furtado)

Stations	Population density (No. of cells · litre^{-1})		
	Mean	Minimum	Maximum
Open water region			
Station I	0.3	0	3.3
Station II	0.3	0	3.4 (June 1971)
Littoral region			
Station III	0.1	0	1.2
Station IV	0.2	0	2.1
Swamp-forest region			
Station V	0.5	0	3.7 (June 1971)
Overall	0.3	0	3.7

(October-December) (Table 99). This fluctuation is due to colonisation and establishment by different species as environmental and nutrient conditions change in the swamp due to influx from the watershed in the monsoons.

Table 99 Chlorophyta: Mean, minimum and maximum abundance of Chlorophyta species at various stations (I-V) in Tasek Bera (1971–1972). (R. P. Lim and J. I. Furtado)

Stations	Species diversity (No. of spp. · litre^{-1})		
	Mean	Minimum	Maximum
Open water region			
Station I	19.5	5	32 (April 1971)
Station II	22.4	6	43 (Oct. 1971)
Littoral region			
Station III	18.6	1	30 (July, Dec. 1971)
Station IV	21.6	10	44 (May-June 1971)
Swamp-forest region			
Station V	19.3	9	36 (May-June 1971)
Overall	20.3	1	44 (May-June 1971)

Table 100 Chlorophyta: Mean, minimum and maximum population density of Chlorophyta at various stations (I-V) in Tasek Bera (1971–1972). (R. P. Lim and J. I. Furtado)

Stations	Population density (No. of cells · litre^{-1})		
	Mean	Minimum	Maximum
Open water region			
Station I	156.3	1.6 (Febr. 1971)	25.4 (Sept. 1971)
Station II	621.5	4.5 (Febr. 1971)	5881.4 (Oct. 1971)
Littoral region			
Station III	332.2	0.4 (Febr. 1971)	1959.9 (Oct. 1971)
Station IV	154.5	5.5 (Jan. 1972)	732.5 (Sept. 1971)
Swamp-forest region			
Station V	72.9	4.5 (Jan. 1972)	425.4 (Oct. 1971)
Overall	267.5	0.4 (Febr. 1971)	5881.4 (Oct. 1971)

The Chlorophyta also show a bimodal seasonal fluctuation in density corresponding to the monsoons with the best development in the northeast monsoon (September-December) in the sublittoral *Utricularia,* the littoral *Lepironia* and the forest limnetic (Stations III, IV and V) (Figs. 85a-e: F). At the modified limnetic and *Pandanus-Utricularia* sublittoral regions only a unimodal density fluctuation occurs with its peak corresponding to the northeast monsoon (September-December) (Table 100). The monsoons thus influence Chlorophyta colonisation, growth and proliferation by nutrient inputs from the watershed and environmental changes, as well as by flushing these algae from one habitat to another, especially from the sublittoral *Pandanus-Utricularia* and *Utricularia* region (Stations II and III) to the littoral and

Table 101 *Oedogonium* sp.: Mean, minimum and maximum population density of *Oedogonium* sp. at various stations (I-V) in Tasek Bera (1971–1972). (R. P. Lim and J. I. Furtado)

Stations	Population density (No. of cells · litre^{-1})		
	Mean	Minimum	Maximum
Open water region			
Station I	1.09	0 (Jan. 1971, May 1971, Oct. 1971, Nov. 1971, Jan. 1972)	5.7 (Sept. 1971)
Station II	2.71	0 (Jan.-April 1971, Aug.-Nov. 1971, Jan. 1972)	29.9 (June 1971)
Littoral region			
Station III	1.92	0 (Jan.-May 1971, Oct.-Dec. 1971)	14.3 (June 1971)
Station IV	0.10	0 (Jan.-April 1971, Oct.-Nov. 1971, Jan. 1972)	9.0 (June 1971)
Swamp-forest region			
Station V	1.60	0 (Jan.-April 1971, Oct. 1971)	9.8 (June 1971)
Overall	1.40	0	29.9 (June 1971)

limnetic regions (Stations IV, V and I) (Table 100). The low Chlorophyta density in the dry season (February-March) is due probably to its displacement by flourishing bacteria and fungal spores (see earlier) which decompose allochthonous and autochthonous organic matter. These conditions may affect recolonisation by this taxa during the southwest monsoon (April-May) especially at the littoral *Lepironia*.

Twelve species of Chlorophyta are common at Stations I-V in Tasek Bera throughout most of the year (Tables 101-112). *Oedogonium* spp. (Oedogoniaceae) populations are best developed at the sublittoral *Pandanus-Utricularia* region (Stations II and III) (Table 99). They fluctuate sporadically, being most abundant at the end of the southwest monsoon in June (Table 10). Among the Zygnemataceae, *Mougeotia* spp. populations are best developed

Table 102 *Mougeotia* sp.: Mean, minimum and maximum population density of *Mougeotia* sp. at various stations (I-V) in Tasek Bera (1971–1972). (R. P. Lim and J. I. Furtado)

Stations	Population density (No. of cells · litre^{-1})		
	Mean	Minimum	Maximum
Open water region			
Station I	27.07	0 (Jan. 1971, Feb. 1971, Dec. 1971)	168.1 (Oct. 1971)
Station II	142.38	0 (Jan. 1971, Feb. 1971, April 1971, July 1971, Aug. 1971, Dec. 1971, Jan. 1972)	1620.0 (Oct. 1971)
Littoral region			
Station III	29.10	0 (Jan. 1971, Feb. 1971, April 1971, Dec. 1971)	320.6 (Oct. 1971)
Station IV	15.2	0 (Jan. 1971, Feb. 1971, April 1971, June 1971, Dec. 1971)	141.5 (Oct. 1971)
Swamp-forest region			
Station V	31.7	0 (Jan. 1971, April-May 1971, Jan. 1972)	292.4 (Oct. 1971)
Overall	49.1	0	1620.0 (Oct. 1971)

Table 103 *Spirogyra* sp.: Mean, minimum and maximum population density of *Spirogyra* sp. at various stations (I-V) in Tasek Bera (1971–1972). (R. F. Lim and J. I. Furtado)

Stations	Population density (No. of cells · litre^{-1})		
	Mean	Minimum	Maximum
Open water region			
Station I	2.87	0 (Feb.-April 1971, Jan. 1972)	7.4 (Sept. 1971)
Station II	18.42	0 (Feb.-May 1971, Aug. 1971, Jan. 1972)	204.0 (Oct. 1971)
Littoral region			
Station III	3.33	0 (Jan.-May 1971, Nov. 1971, Jan. 1972)	19.6 (Oct. 1971)
Station IV	2.50	0 (Feb.-May 1971)	14.7 (June 1971)
Swamp-forest region			
Station V	4.50	0 (Feb.-May 1971, Nov. 1971, Jan. 1972)	18.8 (Oct. 1971)
Overall	6.40	0	204.0 (Oct. 1971)

in the *Pandanus-Utricularia* sublittoral region (Station II) (Table 102), and fluctuate sporadically with peak abundance in the northeast monsoon (October) (Table 102). *Spirogyra* spp. populations are best developed in the *Pandanus-Utricularia* sublittoral region (Station II), and fluctuate sporadically with peak abundance in the northeast monsoon (October) (Table 103). *Zygnema* spp. populations have the same pattern of development and seasonal fluctuation as *Mougeotia* spp. and *Spirogyra* spp., but occur in lower numbers (Table 104). Thus, the Zygnemataceae contribute to the species diversity and numerical abundance of the Chlorophyta in the northeast monsoon, while the Oedogoniaceae contribute at the end of the southwest monsoon.

Among the Desmidaceae, *Cosmarium moniliforme* populations are best developed in the *Pandanus-Utricularia* sublittoral and modified limnetic regions (Stations II and I), and show seasonal fluctuations with peak density in August-October, corresponding to the dry season northeast monsoon

Table 104 *Zygnema* sp.: Mean, minimum and maximum population density of *Zygnema* sp. at various stations (I-V) in Tasek Bera (1970–1972). (R. P. Lim and J. I. Furtado)

Stations	Population density (No. of cells · litre^{-1})		
	Mean	Minimum	Maximum
Open water region			
Station I	0.82	0 (Jan.-March 1971, July 1971, Oct. 1971)	6.1 (Dec. 1971)
Station II	2.71	0 (Jan.-March 1971, July-Aug. 1971)	15.1 (Dec. 1971)
Littoral region			
Station III	1.18	0 (Jan.-March 1971, July 1971, Aug. 1971, Oct. 1971, Nov. 1971)	10.2 (Dec. 1971)
Station IV	0.70	0 (Jan.-April 1971, (Oct. 1971)	3.3 (June 1971)
Swamp-forest region			
Station V	8.4	0 (Jan.-March 1971, July 1971)	2.5 (June 1971)
Overall	1.2	0	15.1

transition (Table 105). *Closterium dianae* (Table 106), *C. gracile* (Table 107) and *C. libellula* (Table 108) populations are all better developed in the *Pandanus-Utricularia* sublittoral region (Station II), with the last species also abundant in the *Lepironia* littoral region (Station IV), and fluctuate with peak abundance in the sublittoral region in October (northeast monsoon) and in the littoral in March (dry season). *Hyalotheca dissiliens* (Table 109) populations are best developed in the *Pandanus-Utricularia* and *Utricularia* sublittoral regions, and fluctuate sporadically with peak abundance in the northeast monsoon (October). *Hyalotheca undulata* (Table 110) populations are more abundant and dominant than the preceding species, and show the same pattern of spatial development and temporal fluctuations. *Micrasterias foliacea* populations (Table 111) are best developed in the modified limnetic region (Station I) and show peak abundance in the northeast monsoon (November). Thus, all the common Desmidiaceae contribute to the seasonal

Table 105 Cosmarium moniliforme: Mean, minimum and maximum population density of *Cosmarium moniliforme* at various stations (I-V) in Tasek Bera (1971–1972). (R. P. Lim and J. I. Furtado)

Stations	Population density (No. of cells · litre^{-1})		
	Mean	Minimum	Maximum
Open water region			
Station I	20.49	0 (Jan.-March 1971, Dec. 1971, Jan. 1972)	258.1 (Aug. 1971)
Station II	34.65	0 (Feb. 1971, April 1971, Aug. 1971, Nov. 1971, Dec. 1971, Jan. 1972)	442.0 (Oct. 1971)
Littoral region			
Station III	1.07	0 (Jan.-May 1971, Jan. 1972)	52.4 (Oct. 1971)
Station IV	2.70	0 (Jan. 1971, Feb. 1971) April 1971, Nov. 1971-Jan. 1972)	22.5 (May 1971)
Swamp-forest region			
Station V	1.30	0 (Jan.-May 1971, Oct. 1971, Jan. 1972)	11.0 (July 1971)
Overall	12.0	0	442.0 (Oct. 1971)

fluctuation of Chlorophyta species and populations, with *Cosmarium moniliforme* and *Hyalotheca undulata* populations being dominant and with peak densities related to the northeast monsoon.

As for the filamentous green algae (Table 112), their populations are best developed in the modified limnetic and the sublittoral regions (Stations I, II and III), and show a seasonal fluctuation with peak abundance during the northeast monsoon (September-December); these contribute significantly to Chlorophyta seasonality. Less common desmid populations such as *Gymnozyga moniliformis, Staurastrum longispinum, S. megacanthum, Staurastrum* sp., *S. columbetoides, S. cerastes, S. sexangulare, Micrasterias lux, Euastrum sublobatum*, are better developed in the *Pandanus-Utricularia* sublittoral region (Station II), and mainly show peak densities in the south-

Table 106 *Closterium dianae:* Mean, minimum and maximum population density of *Closterium dianae* at various stations (I-V) in Tasek Bera (1971–1972). (R. P. Lim and J. I. Furtado)

Stations		Population density (No. of cells · litre^{-1})		
		Mean	Minimum	Maximum
Open water region	Station I	0.654	0 (March 1971, Sept.-Nov. 1971, Jan. 1972)	4.5 (April 1971)
	Station II	1.660	0 (Jan. 1971, April 1971, July-Aug. 1971, Nov. 1971, Dec. 1971)	16.0 (Oct. 1971)
Littoral region	Station III	1.080	0 (Feb. 1971, April 1971, Oct. 1971)	4.9 (June 1971)
	Station IV	1.100	0 (April-May 1971)	4.9 (June 1971)
Swamp-forest region	Station V	0.500	0 (March-May 1971, Oct. 1971-Jan. 1972)	2.5 (July 1971)
Overall		1.000	0	16.0 (Oct. 1971)

Table 107 *Closterium gracile:* Mean, minimum and maximum population density of *Closterium gracile* at various stations (I-V) in Tasek Bera (1971–1972). (R. P. Lim and J. I. Furtado)

Stations		Population density (No. of cells · litre^{-1})		
		Mean	Minimum	Maximum
Open water region	Station I	2.08	0 (July 1971, Nov. 1971)	13.5 (April 1971)
	Station II	11.48	0 (April 1971, Aug. 1971)	131.0 (Oct. 1971)
Littoral region	Station III	2.45	0 (Feb. 1971, April 1971)	7.4 (July 1971)
	Station IV	4.20	0 (Jan. 1971, April 1971)	6.8 (March 1971, May 1971)
Swamp-forest region	Station V	1.50	0 (Jan. 1971, April-May 1971, Jan. 1972)	6.5 (June 1971)
Overall		4.30	0 (Oct. 1971)	131.0

Table 108 *Closterium libellula:* Mean, minimum and maximum population density of *Closterium libellula* at various stations (I-V) in Tasek Bera (1971–1972). (R. P. Lim and J. I. Furtado)

Stations	Population density (No. of cells · litre^{-1})		
	Mean	Minimum	Maximum
Open water region			
Station I	0.75	0 (March, 1971, Aug. 1971, Oct. 1971, Dec. 1971)	6.5 (April 1971)
Station II	2.69	0 (Jan. 1971, April 1971, Aug. 1971)	24.5 (Oct. 1971)
Littoral region			
Station III	0.78	0 (Jan. 1971, Feb. 1971, June 1971, Oct. 1971)	2.9 (Aug. 1971)
Station IV	2.30	0 (Jan. 1971, April 1971, June 1971, Oct. 1971, Nov. 1971, Jan. 1972)	12.7 (March 1971)
Swamp-forest region			
Station V	1.10	0 (Jan. 1971, April-May 1971)	2.4 (June 1971)
Overall	1.52	0	24.5 (Oct. 1971)

east monsoon (April-May) (Table 113). *Bulbochaeta* spp. populations (Oedogoniaceae) show a similar spatial and seasonal pattern. Thus, these less common and rarer species contribute to Chlorophyta dominance at Tasek Bera in species diversity and in population fluctuations with peaks related to the northeast monsoons.

5.2.1.7 *Chlorophyll content of phytoplankton* (T. Okino). For the determination of the amounts of chlorophyll, the water sample was filtered through a glass fibre filter (Millipore Co. Ltd.). The amount of chlorophyll was determined following the method of Strickland and Parsons (1968) in which the pigment was extracted with 90% acetone.

The average chlorophyll content was 1.32 mg chl.-a · m^{-3} (maximum 1.94 mg chl.-a · m^{-3} and minimum 0.70 mg chl.-a · m^{-3}) in the modified limnetic and

Table 109 Hyalotheca disiliens: Mean, minimum and maximum population density of *Hyalotheca disiliens* at various stations (I-V) in Tasek Bera (1971–1972). (R. P. Lim and J. I. Furtado)

Stations		Population density (No. of cells · litre⁻¹)		
		Mean	Minimum	Maximum
Open water region	Station I	2.20	0 (Jan.-March 1971, June 1971, July 1971, Jan. 1972)	9.4 (Oct. 1971)
	Station II	13.85	0 (Jan.-June 1971, Aug. 1971, Jan. 1972)	164.0 (Oct. 1971)
Littoral region	Station III	5.86	0 (Feb.-June 1971, Dec. 1971, Jan. 1972)	59.0 (Oct. 1971)
	Station IV	0.80	0 (Feb.-June 1971, Jan. 1972)	7.4 (Oct. 1971)
Swamp-forest region	Station V	1.60	0 (Feb.-May 1971, Nov. 1971, Jan. 1972)	10.6 (Oct. 1971)
Overall		5.17	0	164.0 (Oct. 1971)

Table 110 Hyalotheca undulata: Mean, minimum and maximum population density of *Hyalotheca undulata* at various stations (I-V) in Tasek Bera (1970–1972). (R. P. Lim and J. I. Furtado)

Stations		Population density (No. of cells · litre⁻¹)		
		Mean	Minimum	Maximum
Open water region	Station I	15.59	0 (Feb. 1971)	59.3 (Oct. 1971)
	Station II	149.87	0 (Feb.-April 1971, Aug. 1971)	1783.0 (Oct. 1971)
Littoral region	Station III	126.0	0 (Feb. 1971, April 1971, Jan. 1972)	1299.0 (Oct. 1971)
	Station IV	27.3	0 (Feb.-April 1971)	146.0 (June 1971)
Swamp-forest region	Station V	15.8	0 (Feb. 1971, April-May 1971)	64.2 (June 1971, (Oct. 1971)
Overall		66.9	0	1783.0 (Oct. 1971)

Table 111 Micrasterias foliacea: Mean, minimum and maximum population density of *Micrasterias foliacea* at various stations (I-V) in Tasek Bera (1970–1972). (R. P. Lim and J. I. Furtado)

Stations		Population density (No. of cells · litre^{-1})		
		Mean	Minimum	Maximum
Open water region	Station I	12.58	0 (March 1971, April 1971)	154.2 (Nov. 1971)
	Station II	0.48	0 (Jan.-March 1971, Aug.-Oct. 1971, Jan. 1972)	2.5 (April 1971)
Littoral region	Station III	0.78	0 (Feb. 1971, March 1971, Aug. 1971, Oct.-Dec. 1971, Jan. 1972)	2.5 (June 1971)
	Station IV	0.40	0 (Feb.-April 1971, Nov. 1971, Dec. 1971)	1.6 (June 1971)
Swamp-forest region	Station V	0.50	0 (Feb.-May 1971, Oct. 1971, Jan. 1972)	2.9 (July 1971)
Overall		2.95	0	154.2 (Nov. 1971)

Table 112 Filamentous green algae: Mean, minimum and maximum population density of filamentous green algae at various stations (I-V) in Tasek Bera (1970–1972). (R. P. Lim and J. I. Furtado)

Stations		Population density (No. of cells · litre^{-1})		
		Mean	Minimum	Maximum
Open water region	Station I	71.46	0 (Feb.-July 1971, Jan. 1972)	658.0 (Sept. 1971)
	Station II	110.36	0 (Feb.-Aug. 1971)	597.0 (Oct. 1971)
Littoral region	Station III	38.10	0 (Feb.-July 1971)	286.3 (Dec. 1971)
	Station IV	12.40	0 (Feb.-June 1971)	46.6 (Dec. 1971)
Swamp-forest region	Station V	5.30	0 (Feb.-June 1971)	33.9 (Oct. 1971)
Overall		47.54	0	658.0 (Sept. 1971)

Table 113 Less common Chlorophyta: Maximum abundance of less common Chlorophyta species at designated stations in Tasek Bera (1970–1972). (R. P. Lim and J. I. Furtado)

	Maximum	Station	Date
Desmidiaceae			
Gymnozyga moniliformes	12.7	I	December 1971
	172.0	II	October 1971
	16.4	III	October 1971
	16.8	IV	October 1971
Staurastrum longispinum	2.5	I	April 1971
	10.9	IV	May 1971
	4.9	V	June/July 1971
Staurastrum sp. 3	106.0	II	October 1971
	6.6	III	October 1971
Staurastrum columbetoides	16.0	II	October 1971
	4.0	III	April 1971
	4.1	IV	May 1971
Staurastrum cerastes	16.0	II	October 1971
	3.7	IV	May 1971
Staurastrum sexangulare	3.7	IV	May 1971
Micrasterias lux	32.7	II	October 1971
Euastrum sublobatum	8.0	II	October 1971
Desmidium coarctatum	16.0	II	October 1971
Desmidium aptogonum	2.5	V	January 1971
Closterium setaceum	65.4	II	October 1971
Oedogoniaceae			
Bulbochaeta sp.	3.7	III	December 1971
	16.0	II	October 1971
Staurastrum megacanthum	2.9	I	June 1971
	16.0	II	October 1971
	3.3	III	June 1971
	16.0	IV	May 1971
	6.1	V	June/July 1971

forested limnetic regions (Stations I and V), and was 2.39 mg chl.-a$\cdot$m^{-3} (maximum 3.74 mg chl.-a$\cdot$m^{-3} and minimum 1.25 mg chl.-a$\cdot$m^{-3}) in the littoral *Lepironia* region (Stations XI, XII and XIII) (Table 114). The consistently higher chlorophyll content in the littoral is due to contamination from the attached algae, which were released from the stems and leaves of macrophytes when the water was agitated during sampling.

The chlorophyll content increased when phosphate and other nutrients were added (Table 115). With the addition of nutrients (KH_2PO_4 20 mg$\cdot$l^{-1}, $(NH_4)_2SO_4$ 20 mg$\cdot$l^{-1}, $CaCl_2\cdot 2H_2O$ 10 mg$\cdot$l^{-1}, $MgSO_4$ 10 mg$\cdot$l^{-1}) into the lake

Table 114 Chlorophyll a content of phytoplankton in Tasek Bera (Chlorophyll a $mg \cdot m^{-3}$) (September-October 1973). (T. Okino)

Stations	Chlorophyll a content ($mg \cdot m^{-3}$)		
	Mean	Minimum	Maximum
Modified limnetic and forested limnetic regions			
Stations I and V	1.32	0.70	1.94
Littoral region			
Stations XI, XII and XIII	2.39	1.25	3.74

Table 115 Growth of phytoplankton with nutrient enrichment. Swamp water was enriched with the following reagents: KH_2PO_4 20 ppm, $(NH_4)_2SO_4$ 20 ppm, $CaCl_2 \cdot 2H_2O$ 10 ppm, and $MgSO_4$ 10 ppm (September 1973). (T. Okino)

	Incubation time 33 days Chlorophyll ($mg \cdot m^{-3}$)		
	Chl.-a	Chl.-b	Chl.-c
Control	3.86	5.22	10.7
Enriched	54.5	22.8	72.6

Table 116 Lake water enriched with KH_2PO_4 at the rate of 4.0, 0.4, 0.04 and 0.00 PO_4–$P \cdot mg^{-1} \cdot day^{-1}$, respectively. The phytoplankton was incubated under natural light conditions at 27–31°C (10-24 September 1973). (T. Okino)

		Incubation time (days)			
		0	3	7	14
Control					
(0.00 Pmg)	pH	6.0	7.2	6.9	6.8
	Chl.-a ($mg \cdot m^{-3}$)	2.35	2.66	1.84	–
0.04 $Pmg \cdot l^{-1} \cdot day^{-1}$	pH		7.4	8.0	8.4
	Chl.-a ($mg \cdot m^{-3}$)		3.84	16.3	14.8
0.40 $Pmg \cdot l^{-1} \cdot day^{-1}$	pH		6.8	7.2	7.2
	Chl.-a ($mg \cdot m^{-3}$)		3.57	18.4	17.7
4.0 $Pmg \cdot l^{-1} \cdot day^{-1}$	pH		6.4	6.4	6.2
	Chl.-a ($mg \cdot m^{-3}$)		3.37	20.0	16.6

* The culture in 0.04 $Pmg \cdot l^{-1} \cdot day^{-1}$ increased the filamentous algae *Spirogyra* sp., *Desmidium bongarium*, *Hyalotheca disilius;* and *H. undulata*.

water of Tasek Bera, the chlorophyll amounts of the seston reached values of 54.5 mg chl.-a·m^{-3} after 33 days incubation, while the control had values of 3.86 mg chl.-a·m^{-3}. When only phosphate was added to lake water, the chlorophyll content increased above 16 mg chl.-a·m^{-3} after 7 days incubation (Table 116). Thus, addition of nutrients resulted in a remarkable change in algae composition when compared with the control. Chlorophyta, particularly, tended to increase.

5.2.2 *Epiphytic and benthic algae*
(T. Watanabe, S. Kumano and T. Mizuno)

Although the phytoplankton standing crop is relatively low, the epiphytic algal community is very rich especially those attached to *Utricularia* and *Lepironia* plants. The benthic algal community is also quite rich. In this section the standing crop of the aufwuchs community on *Utricularia* and *Lepironia* was measured assuming the algal community to be dominant. Experiments on growth and colonisation processes of algae were carried out.

Quantitative samples were collected from the *Utricularia* stand by removing the macrophytes from a 50 × 50 cm^2 area, and by filling a 500 ml bottle with branches of *Utricularia*. Dry weights of the attached algae were then determined.

Epiphytes on *Lepironia* were collected by carefully pulling four stems of the reed and washing off the epiphytes. The dry weights were subsequently determined.

The standing crop of epiphytes on *Utricularia* decreased during the north-east monsoonal period (December-January) with increases in the dryer periods (June-July and September-October (Fig. 86)). The main peak occurred in June with a value of about 750 g dry wt per kg dry wt of *Utricularia*. The decrease in the monsoonal period is due to the effect of rain drops impinging directly on the plants and increased water velocity which tended to detach the aufwuchs community. Similarly the standing crop of the epiphytic community on *Lepironia* showed an increase at the dryer periods (September-October and February-March) and lower values at the peak of the monsoon. These variations can be attributed to the factors stated above for the epiphytic community on *Utricularia*. During the extreme dry period (July-August) no readings were available as the littoral dried up completely.

5.2.2.1 *Colonisation of epiphytes on artifical substrates* (T. Mizuno). Experiments to determine the growth rates and succession of epiphytes were conducted using 40 plastic plates of 10 × 10 cm^2. These were suspended at 20

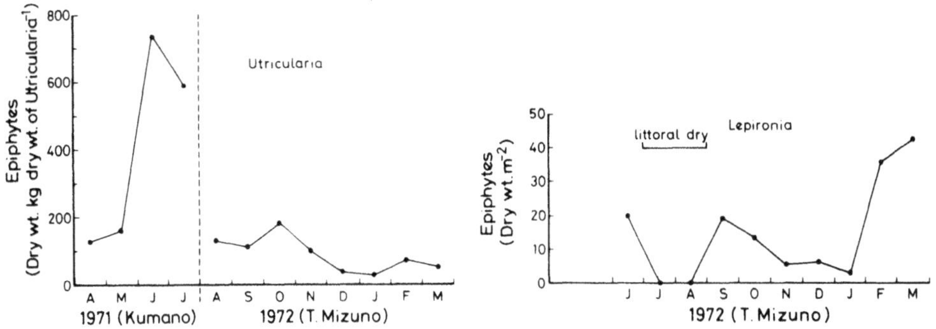

Fig. 86 Seasonal fluctuations in epiphytes on *Utricularia* and *Lepironia*. (T. Mizuno and S. Kumano)

cm depth in the limnetic region (Station I), the forested stream channel (Station V) and the open water at the outlet of the swamp and forest. Another set was placed on the bottom in the littoral (Station XI) among the *Lepironia* reeds. A plate from each station was carefully removed after a threeday interval. Subsequent plates were removed at longer intervals of one, two and four weeks. The aufwuchs community was washed from the plates and preserved in 5% formalin. Dry weights were determined after filtration through millipore filters. Species composition was also determined.

The mean increase biomass of the colonising aufwuchs community is given in Fig. 87. At Station V, the readings were incomplete due to the fact that the plates were washed away by increasing water velocity at the onset of the monsoon.

At Stations I and V initial growth was slow but increased rapidly after the third week despite the onset of the monsoonal season. It reached a maximum at Station I after two months. Colonisation of the plates at the second open water (i.e. at the entrance of the swamp-forest channel) was slower than Station I or V. This could be due to abrasion of the plates by aquatic macrophytes especially *Utricularia* growing around the immidiate vicinity. Growth, however, increased markedly after the monsoon presumably because most of the *Utricularia* had been washed away. The biomass of algae on plates placed at the bottom of the littoral (Station XI) was 4–5 times greater than the plates placed at the other localities (Fig. 87). This indicates that a significant epiphytic mat does dorm on the bottom and contributes significantly to primary production.

The initial colonisers of the plates were diatoms like *Frustulia*, *Eunotia* and filamentous green algae like *Hyalotheca*, *Desmidium*, *Mougeotia* and *Oedogonium*. The diversity in species increased with time with some of the

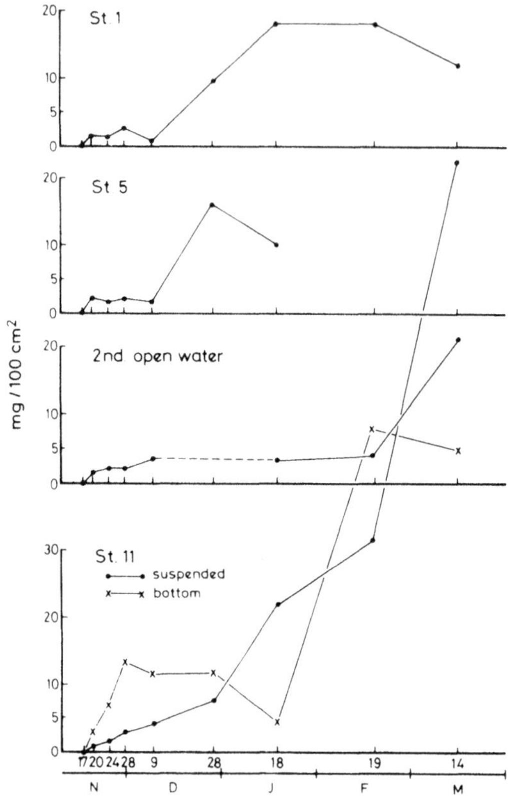

Fig. 87 Experiments showing growth rates of epiphytic algae on artificial substrates at various stations at Tasek Bera. (T. Mizuno)

initial colonisers being replaced by other species of algae (Figs. 88a and b). Although *Batrachospermum* was the dominant macroalgae at Station V, it did not colonise the plates even after two months. There was a greater diversity of algae colonising the plates placed within the littoral (Fig. 88b). In general fungi increased markedly during the latter part of the experiment.

5.2.2.2 *Vertical distribution of standing crop of attached forms on the bamboo pole* (T. Watanabe). A bamboo pole four meters long was planted in Station I in Tasek Bera on December 8th, 1971. After one month, six samples of algae and fungi attached to this bamboo pole were quantitatively collected at intervals of 50 centimeters. The volume of attached forms per unit area (25 cm^2) is shown in Table 116a and the common organisms in each samples are shown in Table 116b. From the results, it is considered that the attached algae grow abundantly, between 1 m and 2 m deep in the open water of this lake.

Table 116a Volume of attached forms on the bamboo pole. (T. Watanabe)

Depth (cm)	Volume (ml) of attached forms per 25 cm	Depth (cm)	Volume (ml) of attached forms per 25 cm
0– 10	2.6	150–160	10.5
50– 60	6.4	200–210	11.0
100–110	10.9	320–330	5.0

Tabel 116b Common organisms occurring in attached forms on the bamboo pole. (T. Watanabe)

Species	Depth (cm)					
	0–10	50–60	100–110	150–160	200–210	320–330
Blue-green algae						
Lyngbya sp.		++	++			
not identified species					++	++
Diatom						
Eunotia curvata		+++	++			
E. serra var. *serra*		+++	++	+		
Eunotia spp.		++	+	+		
Frustulia rhomboides		++++	++	++		
Green algae						
Ankistrodesmus sp.		+	+	++	++	+
Bulbochaete praereticulata	+	+				
Fungi	+++	+++	++++	+++++	+++++	+++++

+++++ = very rich; ++++ = rich; +++ = common; ++ = rare; + = very rare.

5.2.3 *Submerged macrophytes*
(I. Ikusima and S. Kumano)

5.2.3.1 *Utricularia flexuosa* (I. Ikusima and S. Kumano). *Utricularia flexuosa,* which is well adapted for life in slow flowing waters, occurs in abundance along the edges of the open waters (limnetic region) and channels which are not shaded. They form dense mats with stems that reach lengths of up to 2 metres. The standing crop of *Utricularia flexuosa* fluctuates with the water level of the swamp. During the dry season the areas occupied by *U. flexuosa* tend to dry up especially within the littoral region. This results in exposure and subsequent death of most of the plants. Thus the standing crop varies markedly with the seasons, being low in the dry season and increasing as the water level rises. However, the standing crop also decreases at the height of the northeast monsoon as the current becomes strong enough to

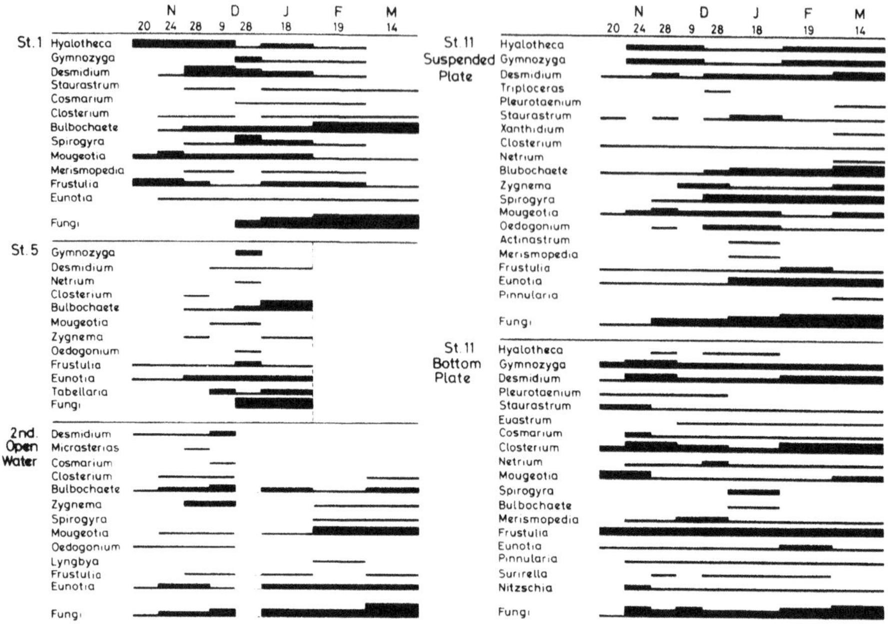

Fig. 88 Seasonal succession of epiphytic communities on plastic plates incubated between November 1972 and March 1973. (T. Mizuno)

Table 117 Seasonal changes of standing crop of living parts in *Utricularia flexuosa* at Station XIII. (S. Kumano)

Date	Standing crop ($g\ dry\ wt \cdot m^{-2}$)	Date	Standing crop ($g\ dry\ wt \cdot m^{-2}$)
17 April 1971	156	12 July 1971	0
1 May 1971	0	1 September 1971	0
15 May 1971	64	5 September 1971	12
13 June 1971	134	5 November 1971	152
9 July 1971	170	7 January 1972	15

wash away portions of the mats (Table 117). But within limits of the velocity of the water and depth it has been found that the standing crop increases with velocity and depth (Ikusima 1973).

The chlorophyll content of *U. flexuosa* was 5.0 $mg \cdot g^{-1}$ dry wt, and the chlorophyll content was estimated 1.9 $g \cdot m^{-2}$ which are comparable to the values on the upper scale obtained for terrestrial communities or with algae in hypereutrophic conditions (Aruga and Monsi 1963).

5.2.3.2 *Cryptocoryne griffithii* (I. Ikusima). *Cryptocoryne griffithii* occurs only along the stream channels and in the open water regions. The standing crop was estimated by placing 10 quadrats of 30 × 30 cm randomly in pure stands in August 1972 when the water level was at its lowest. Standing crop, leaf area index, growth of the plants and development of periphyton on the plant surface were determined. The mean total biomass of *C. griffithii* was greater in Station V which was partially shaded than at the open water which was exposed to direct sunlight (Table 118). Stems and roots comprised the

Table 118 Standing crop (dry weight) and total leaf area ($m^2 \cdot m^{-2}$) of a *Cryptocoryne griffithii* stand, and the standing crop of periphyton growing on the plant surface at Station V and the open water near the forest, in August 1972. (I. Ikusima)

		Dry weight ($g \cdot m^{-2}$)				Total leaf area ($m^2 \cdot m^{-2}$)	Standing crop of periphyton ($g \cdot m^{-2}$)
		Leaf	Stem	Root	Total		
Station V	Mean	98	131	339	568	4.2	19.9
	Max.	116	152	416	646	5.0	
	Min.	70	99	256	453	3.0	
Open water	Mean	75	99	190	364	4.3	0.9
	Max.	122	131	263	501	6.1	
	Min.	45	75	149	291	2.6	

main portion of the total biomass. Chlorophyll content ranged between 0.6 to 0.7 $g \cdot m^{-2}$, being lower than that of the *U. flexuosa* and other macrophyte communities (see later). Specific leaf area was 4300 $cm^2 \cdot g^{-1}$ for *C. griffithii* growing in Station V and 5700 $cm^2 \cdot g^{-1}$ for plants growing in the open water. These values are nearly four times as large as corresponding values of common aquatic macrophytes such as *Potamogeton, Vallisneria, Hydrilla* etc. Periphyton on *C. griffithii* was also better developed in the forested stream channel (Station V) than in the exposed open water.

5.2.4 *Emergent macrophytes*
(I. Ikusima, S. Kumano and T. Mizuno)

5.2.4.1 *Lepironia articulata* (I. Ikusima and S. Kumano). The standing crop of *Lepironia articulata* was determined monthly from April 1971 to May 1973. To determine the biomass, four small quadrats of 50 × 50 cm^2 were randomly placed in pure stands of *L. articulata*. Plants within each quadrat were clipped just above the bottom and dried for 2-3 days in an air-forced

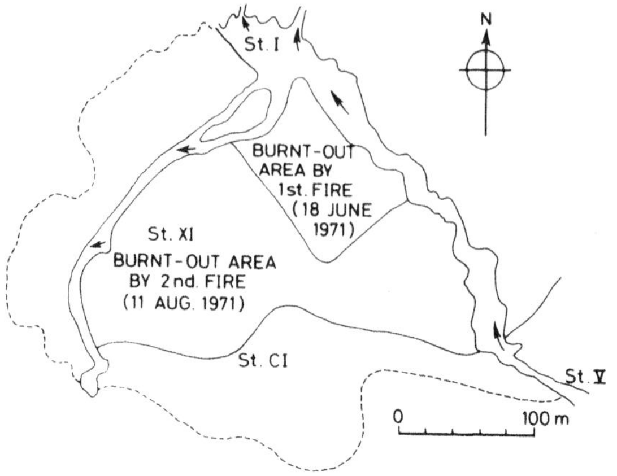

Fig. 89 Map of the *Lepironia* swamp area showing the areas burnt. Dotted line shows the boundary of the study site. (S. Kumano)

oven at 85-90°C. Dead and living culms were separated and their biomasses were determined. The rhizome systems of the plants within each quadrat were removed, dried and their biomass was determined.

During the course of this study a large area was burnt twice by the inhabitants (Fig. 89). The first fire on 18 June 1971 did not affect the study site but the second fire on 11 August 1971 completely destroyed all plants in the study site (Station XI).

The mean standing crop of living and dead portions of culms of *L. articulata* at Station XI is about 910 $g \cdot m^{-2}$ (minimum: 200 $g \cdot m^{-2}$, maximum: 1400 $g \cdot m^{-2}$), and 350 $g \cdot m^{-2}$ (minimum: 190 $g \cdot m^{-2}$, maximum: 720 $g \cdot m^{-2}$) respectively. These results represent the mean from April to August 1971, before the fire. The total standing crop for the period between April and August amounted to 1260 $g \cdot m^{-2}$.

A new station (Station XIII) was established in an unburnt part of the *Lepironia* reed stand in September 1971. Seasonal changes in standing crop of the aerial portions of the stalks and the percentage of flowering culms against total living culms were determined (Table 119). The mean standing crop was 696 $g \cdot m^{-2}$ (maximum: 1427 $g \cdot m^{-2}$, minimum: 306 $g \cdot m^{-2}$) of which 477 $g \cdot m^{-2}$ were live culms while the rest were dead culms.

The standing crop of the subterranean portions of the plants (rhizomes, roots and culms) comprised 40% of the standing crop of aerial portions of the plants. The total standing crop (i.e. aerial and subterranean) in terms of dry weight and chlorophyll content* was about 2000 $g \cdot m^{-2}$ and 1.6 $g \cdot m^{-2}$ respectively. This value is slightly higher than the mean standing crop of terrestrial grasslands in temperate regions.

Table 119 Seasonal changes of standing crop of the culms above ground (mean values of four quadrat samples of 50 × 50 cm²) and the percentage of flowering culms of *Lepironia articulata* at Station XIII. (I. Ikusima, S. Kumano, N. Mizuno, T. Watanabe, O. Sato and T. Mizuno)

Date		Standing crop in dry weight (g · m⁻²)			% of flowering culms
		Living parts	Dead parts	Total	
1971	4 October	375	115	490	12
	21 October	518	136	654	9
	5 November	414	139	553	20
	26 November	284	145	429	19
	9 December	243	63	306	18
1972	5 January	463	108	571	13
	7 February	439	108	547	30
	8 March	322	70	392	19
	28 April	431	124	555	24
	22 May	362	94	456	15
	23 June	363	106	469	3
	28 June	562	113	675	–
	27 July	503	437	940	7
	21 August	322	702	1024	–
	1 September	370	297	667	7
	24 September	565	255	820	–
	16 September	520	298	818	1
	17 October	663	166	829	–
	28 October	526	490	1016	1
	28 November	978	449	1427	11
	31 December	448	139	587	5
1973	19 January	615	204	819	7
	19 February	646	227	873	8
	10 March	514	264	778	9

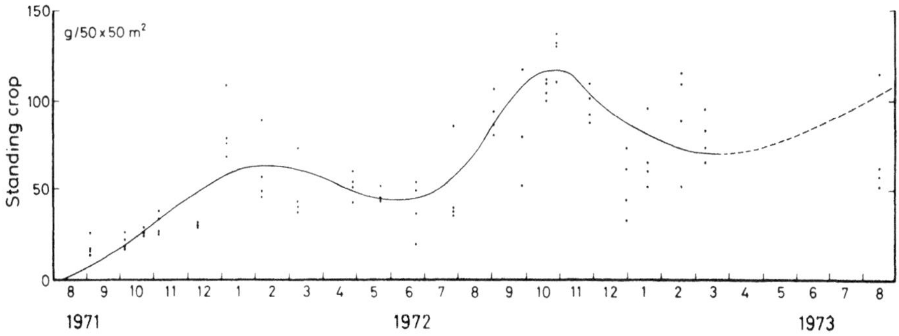

Fig. 90 Recovery of the standing crop (air-dried weight per quadrat 50 × 50 m²) of *Lepironia* stand at Station XI after the second fire in 1971. (S. Kumano, I. Ikusima, N. Mizuno, T. Watanabe, O. Sato and T. Mizuno)

Recovery of L. articulata after the fire. New culms emerged from apices of rhizomes within a month of the fire followed by a gradual increase in standing crop of the aerial portions (Fig. 90). There is a cyclic oscillation in standing crop of the reeds. These oscillations may be due to (1) initiation of regeneration after the fire, (2) a time interval of emergence of the culms and (3) the longevity of the culms which will be discussed later. The standing crop of the *Lepironia* reed stand reaches 330 $g \cdot m^{-2}$, one third of the maximum value of the pre-fire situation, one year after the fire. Full recovery was not achieved two and a half years after burning.

Culms began flowering three months after their initial emergence. There is no clear relationship between the percentage of flowering culms and seasons.

5.2.4.2 *Eleocharis ochrostachys* (I. Ikusima). A narrow zone of *Eleocharis varigata* fringes the border between the reed swamp and the stream channel of the open water. The culms are dense and measure up to 1 m in height. It ranks second in aerial size among the reed swamp communities. The plants die if they are completely submerged for few days at high water level, so that the culms seem to renew after every rise in water level. The standing crop in September is 150 $g.m^{-2}$ for living culms, 30 $g \cdot m^{-2}$ for living roots and 165 $g \cdot m^{-2}$ for dead culms.

5.2.4.3 *Pandanus helicopus* (T. Mizuno). *Pandanus helicopus* is the second most dominant species outside the swamp forest (Plate 3). It grows mainly along stream channels perhaps due to its requirement for a continuous supply of dissolved oxygen. The growth rate and dissolution of organic matter into water from fallen leaves were determined.

Twenty plants were selected, numbered and their growth rates and the amount of fallen leaves were observed at intervals of about half a year. The lengths of the stems were measured from the level of the water which was marked on the stem to the lower most leaf. All subsequent measurements were made from the initial water level mark. The growth rate and amount of litter was obtained from the difference between the previous measurement and that after half a year. The number of fallen leaves during the half year interval was easily determined from traces of excision scars which remained quite clearly and regularly on the stem.

Observations on the growth of the stems which were arbitrarily divided into upper, middle and lower regions showed that growth occurred only in the upper region. This region included the part where leaves had fallen in the last half year.

* The chlorophyll content of the stalks was 1.1 mg chlorophyll $\cdot g^{-1}$ dry weight.

Plate 3 Community of ***Pandanus helicopus***. Tags show the plants measured. Plants were measured for stem elongation and the number of fallen leaves at intervals of about half a year from September 1972 to August 1974. (T. Mizuno)

Table 120 Growth rates and fallen leaves of *Pandanus* plants. (T. Mizuno)

Height of plants (m) at beginning of the observations	0-0.4	0.5-0.9	1.0-1.4	2.5-2.9	3.0-3.4	3.5-3.9	4.0
Growth in length of stems (cm) per half-year	22.4	17.9	17.4	18.0	16.5	14.0	8.0
Number of dehised leaves per half-year	19.8	18.5	19.4	19.0	16.0	13.0	7.0

Table 121 Relationship between length and dry weights of the allen leaves of *Pandanus* plants. (T. Mizuno and O. Sato)

	Mean	Minimum	Maximum
Length (cm)	157.3	117.0	205.0
Dry weight (g)	15.4	8.6	21.6

During the course of this study some of the plants were destroyed by the natives, or were cut down to measure the lengths of their stems. Some of the plants did not grow normally due to heavy infestations of ants. However, plants that grew normally were divided into seven classes according to the height of stems. The growth rate tends to decrease with time (Table 120). Among the growing young plants sixteen to twenty-five leaves fell over a half-year interval. During a half-year interval the plants grew about 20 cm and on the average 19 leaves dehisced. There was a relationship between leaf length and dry weight (Table 121). It is estimated that there are about 40 stems per 10 m^2, and that *Pandanus* stands occupy 7000 m^2 of the study area. Based on the estimate that 38 leaves dehisce per stem per year the annual litter fall of *Pandanus* is 16.4 tons. This recycles about 140 kg Ca per year (based on 8.55 mg $Ca.g^{-1}$ leaf (see later).

5.3 Structure of Lepironia stands (I. Ikusima)

5.3.1 *Vertical profile* (I. Ikusima and S. Kumano)

The structural profile of various photosynthetically active components or productive structures (Monsi and Saeki 1953) of the reed *Lepironia articulata* was studied (Ikusima 1978). The plant was arbitrarily divided vertically into

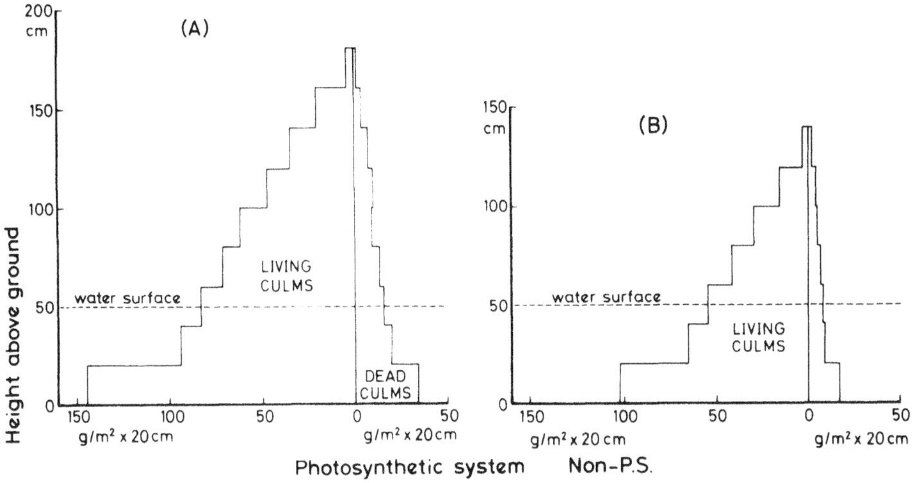

Fig. 91 Profile structure of *Lepironia articulata* stand in June 1972. (A) Station XIII (unburnt), (B) Station XI (burnt in August 1971). The density of living culms is 225 m^{-2} at Station XIII and 212 m^{-2} at Station XI. The standing crop above ground is 562 $g \cdot m^{-2}$ at Station XIII and 309 $g \cdot m^{-2}$ at Station XI. (I. Ikusima)

strata of the green parts (i.e. the culms, roots, rhizomes and other dead components). The stratified clipping technique was used.

One structural profile of the reed community at Stations XI and XIII is shown in Fig. 91. The profile is similar to the 'grass type' (Monsi and Saeki 1953) with maximum photosynthetic tissues at the lower stratum of the community. Detached and floating reed stalks formed about half of the total standing crop of dead culms. Although there was a fair amount of dead floating reeds the attenuation of light in the water was not seriously affected. The well-developed rhizomes composed about 20% of the total biomass of the community. Their average lengths were measured up to 7 $m \cdot m^{-2}$ and 3 $m \cdot m^{-2}$ at Stations XIII and XI respectively.

The patterns of light flux within the reed stand are shown in Fig. 92. These curves were based on measurements of the sideway, upward and downward distribution of light intensity and expressed in relative terms. Sideway light intensity was 25% (canopy surface) –6.5% (canopy bottom). Upward light intensity was 0.5–3.7% of the downward light intensity above the reed stand. The downward relative light intensity on the water surface and on the mud-water interface was 52-70% and 12-15% respectively. The amount of light above and below the surface was sufficient for growth of *Lepironia* and other plants in the water (Ikusima 1973).

5.3.2 *Spatial distribution*
(I. Ikusima)

The spatial distribution of *Lepironia* reeds was conducted when the water

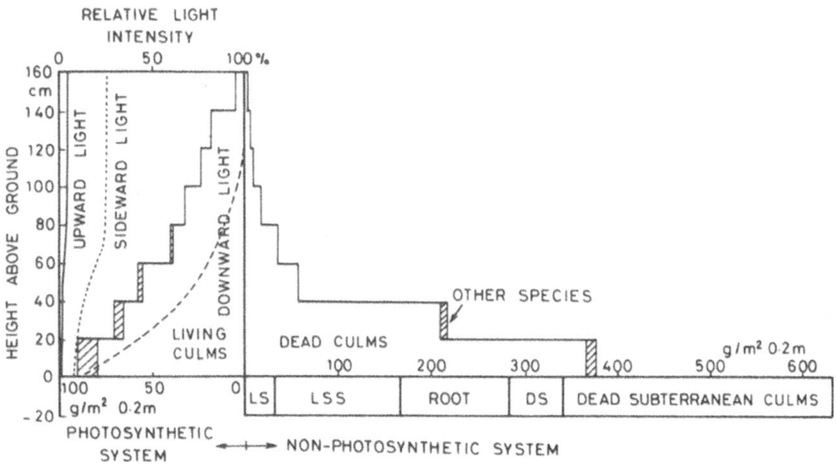

Fig. 92 Profile structure of *Lepironia articulata* stand, with the light profile in terms of the vertical distribution of downward, upward and sideways light flux at Station XIII in August 1972. LS, Living culms in the mud; LSS, Living subterranean culms, DS, Dead culms in the mud. (I. Ikusima 1978)

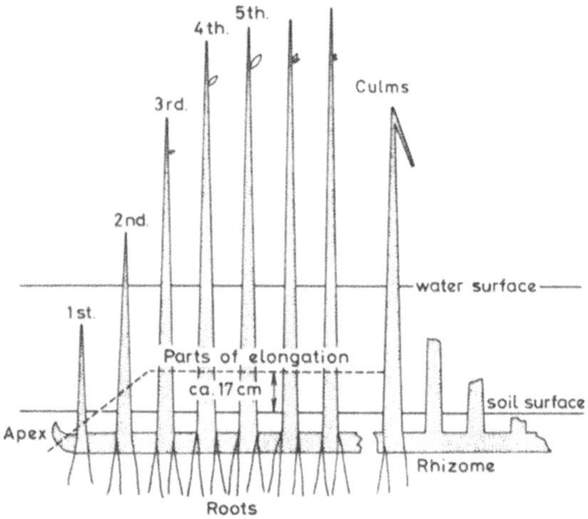

Fig. 93 General structure of a clone of *Lepironia articulata*. (I. Ikusima 1978)

level was at its lowest (Ikusima 1978). A dispersion map was constructed showing the positions of living and dead culms, and rhizomes. The reeds reach a maximum of about 2 m in height. Etiolation of the reed stalk, except in the very young stalks, is restricted to a portion about 17 cm above the mud surface (Fig. 93).

The spatial dispersion map (Figs. 94 A and B) shows that the plant mainly propagates vegetatively. Only three seedlings out of 80 clones (3.8%) and four seedlings out of 52 clones (7.6%) were found in a 1.0 × 2.5 m belt within the unburnt (Station XIII) and burnt (Station XI) stands respectively.

A two-dimensional distribution of the growing apices of rhizomes or the youngest culm was analysed by Morisita's I_δ method (Morisita 1959). Various distributional patterns such as random, uniform, contagious, etc. can be distinguished by the value of the index of distribution (I_δ) and by the trend of changes of I_δ with changing quadrat size. When q quadrats are taken at random from a given area, and n_i (i = 1, 2, 3, . . . q) individuals are found in each of the quadrats, the index I_δ is given by

$$I_\delta = q\delta$$

$$N = \Sigma_i^q = 1^{n_i}$$

$$\delta = \Sigma_i^q = 1^{n_i} (n_i - 1)/N (N - 1).$$

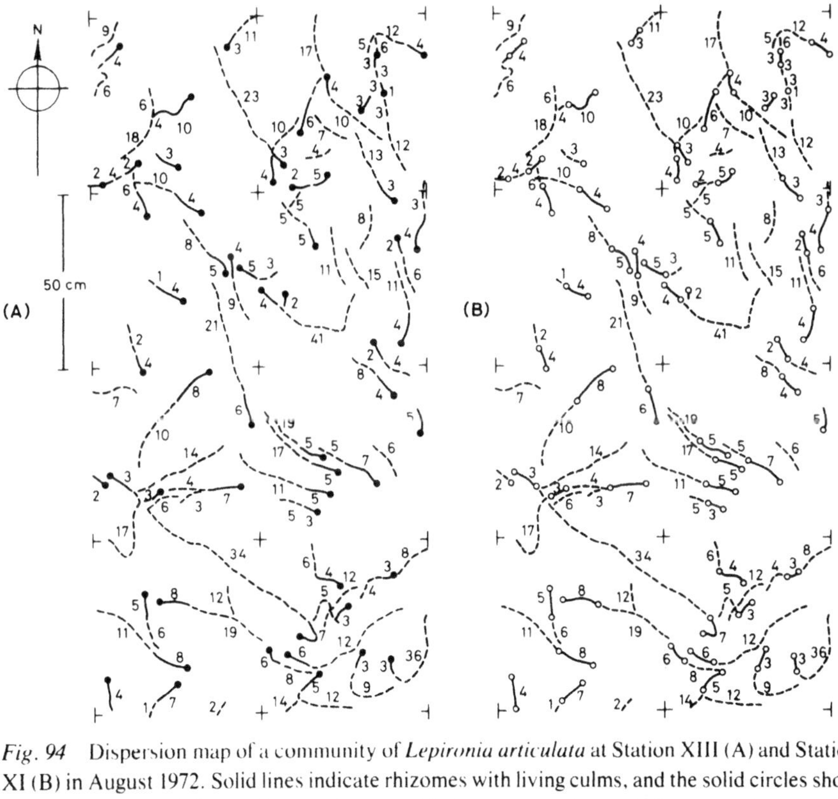

Fig. 94 Dispersion map of a community of *Lepironia articulata* at Station XIII (A) and Station XI (B) in August 1972. Solid lines indicate rhizomes with living culms, and the solid circles show their apices. Dotted lines indicate rhizomes with dead culms. Figures indicate number of culms belonging to a rhizome. (I. Ikusima 1978)

Then

$I_\delta = 1$ for random distribution
$O < I_\delta < 1$ for uniform distribution
$I_\delta > 1$ for contagious distribution

The dispersion of the apices of rhizomes in the two stands (Fig. 95) tends to approach a random pattern, and any departure from randomness is statistically not significant ($p > 0.05$). This probably means that edaphic conditions within the whole *Lepironia* stand are fairly uniform. New culms emerge from the apices of rhizomes sequentially and the number of living culms growing from a rhizome does not exceed 10 (Fig. 93). However, traces of old dead culms can sometimes be found as far back as the 40th node of a rhizome which offers important information for plant growth analysis. The rhizome thus, may be aptly called 'a living picture'.

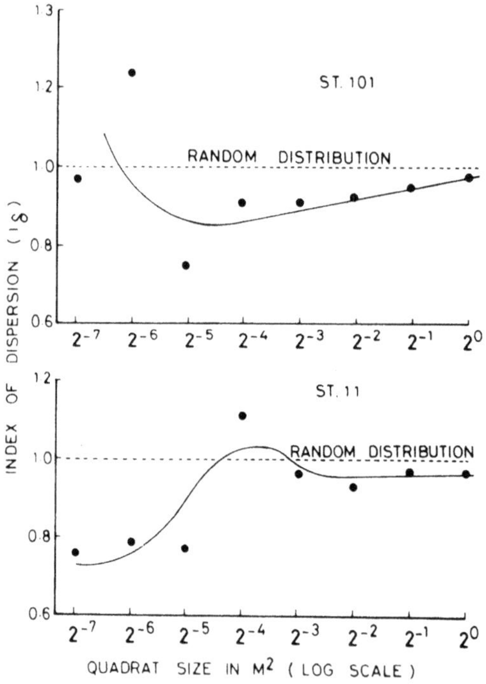

Fig. 95 Trends of index of dispersion-quadrat size curve for the spatial distribution of apices in *Lepironia articulata*. (I. Ikusima 1978)

5.4 Growth of *Lepironia* (I. Ikusima)

5.4.1 *Growht curve*

To obtain the growth curve of the culms, 701 culms belonging to 123 clones were labelled and their heights were measured twice weekly in early August, when the bottom of *Lepironia* reed swamp was exposed during the dry season (Ikusima 1978). At the end of the study all labelled culms were harvested for height and dry weight determinations.

A log-log linear relationship is found between the height (H cm) and the dry weight (w g) of a culm (Figs. 96 A and B). The gradient of the regression line differs between Stations XIII and XI. The relationship can be approximated by the following empirical equations:

$$w = 0.01H^{1.138} \quad \text{for Station XIII,}$$
$$w = 0.01\,H^{1.104} \quad \text{for Station XI.}$$

The dry weight of a culm or the total biomass of culms in a stand can be estimated from these equations, if the heights of the stems and their distribution are known. An attempt was made to formulate the growth curve of the culms assuming the following logistic equation for the basic growth process:

$$\frac{1}{w}\frac{dw}{dt} = \lambda\,(1 - \frac{w}{W}) \quad \text{or} \quad w = \frac{W}{1 + k\,\exp\,(-\,\lambda t)},$$

where w is the dry weight of the culm, λ is the coefficient of growth, W is the upper limit of w, and k indicates the integration constant determined by the initial condition. The values of the two parameters λ and W are determined by a modified difference diagram method (Yoshihara 1951) which change with the stage of growth, or $\lambda = \lambda\,(t)$ and $W = W\,(t)$. The growth curve obtained is therefore Shinozaki's general logistic equation (Shinozaki 1953). The time trends of characteristic values of the logistic growth curve are summarised in Table 122. No appreciable difference is found in the value of λ and W between Stations XIII and XI with respect to the early growth stage. λ reaches a maximum value of $0.102 \cdot \text{day}^{-1}$ at the youngest stage, i.e. the culm can double its dry weight every 7 days. The value of λ falls gradually as stems grow taller and this is accompanied by an increase of W. Growth curves of the mean dry weight of the culms are illustrated in Fig. 97. It takes about 7 months for the culms to mature, and the mean maximum dry weight of a stem reaches 3.40 g (170 cm in height) at Station XIII and 2.25 g (135 cm in height) at Station XI. These differences are probably due to the effect of the fire.

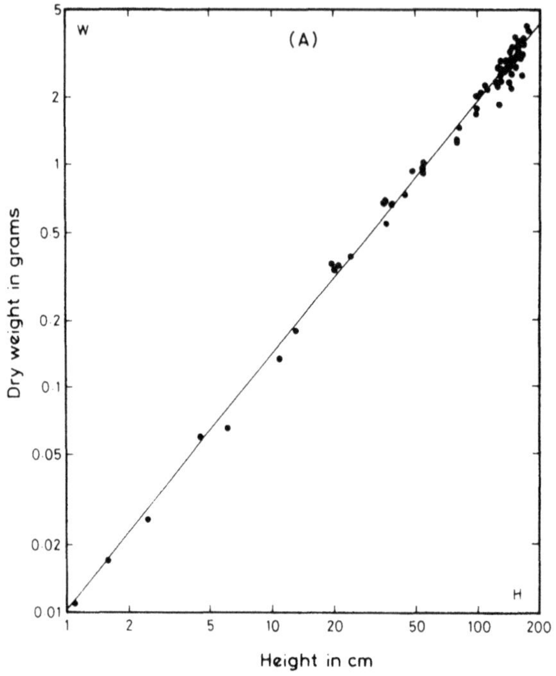

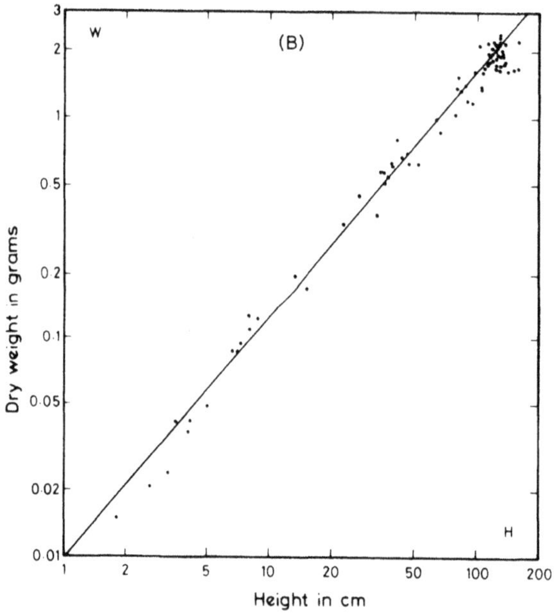

Fig. 96 Dry weight of a culm (w) as the allometric function of height of the culm (H) in *Lepironia articulata* at Station XIII (A) and Station XI (B). Linear regression between the logarithms of w and H can cover a fairly wide range of culm height. (I. Ikusima 1978)

Table 122 Characteristic values in the logistic growth of *Lepironia articulata* at two stations in Tasek Bera. (I. Ikusima)

Station XIII			Station XI		
t (day)	(day^{-1})	W (g)	t (day)	(day^{-1})	W (g)
0			0		
	0.102	0.742		0.101	0.600
33.5			30.9		
	0.0556	2.370		0.0617	1.302
78.7			67.7		
	0.0315	3.398		0.0266	2.250
∞			∞		

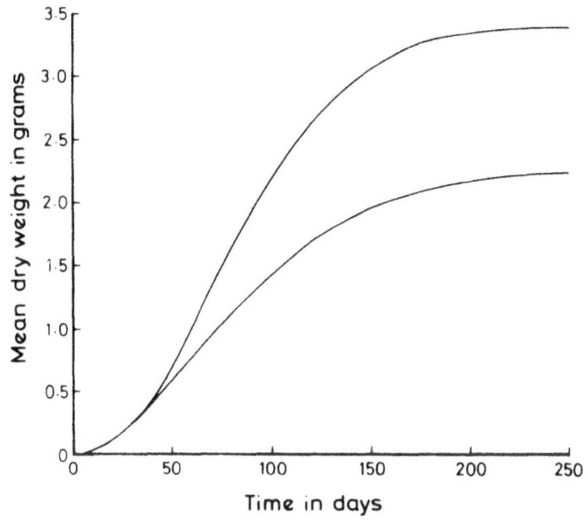

Fig. 97 Growth curves of mean dry weight per culm at Station XIII (upper curve) and Station XI (lower curve). (I. Ikusima 1978)

5.4.2 *Life table*

The life table for *Lepironia* culms at both stations was constructed by applying the following information obtained through field studies and extrapolations from the mean growth curve (Ikusima 1978): (1) the frequency distribution of the number of culms rising from a creeping rhizome (Fig. 98) and (2) the time interval of emergence of new culms and differences in age of the 1st, 2nd and 3rd culms on the same rhizomes calculated from the growth curve (Fig. 97). New culms are estimated to emerge every 51-55 days at Station XIII and 62-77 days at Station XI ($p < 0.01$).

Survivorship curves, death curves and death rate curves are given in Fig.

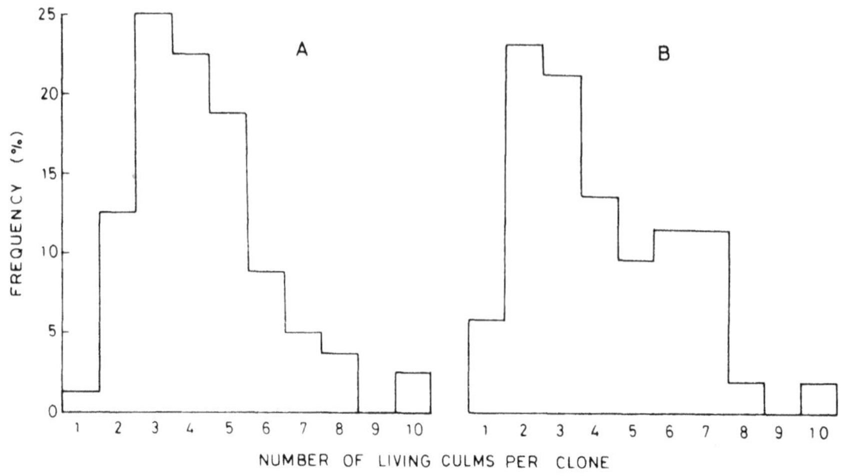

Fig. 98 Frequency distribution histograms of the number of *Lepironia* culms rising from a rhizome in August 1972. A: Station XI (20.8 clones $\cdot$ m^{-2}, 83.6 culms $\cdot$ m^{-2}), B: Station XIII (32.0 clones $\cdot$ m^{-2}, 136.8 culms $\cdot$ m^{-2}). (I. Ikusima)

99. The relative time scale on the horizontal axis is expressed as a percentage deviation from the mean duration of the life-span. The mean length of the life-span of the culms at Stations XIII and XI is statistically estimated at 227 days and 309 days respectively. The survivorship curve and the death curve show clearly that death rate is slow initially, reaches a maximum and finally slows down again. The mortality of young culms is not high as in the seedlings of many terrestrial plants (Kira et al. 1956), probably because the well-developed rhizome contributes to the growth of new culms by translocation.

5.5 Photosynthetic production

(I. Ikusima, R. P. Lim and T. Okino)

5.5.1 *Phytoplankton*

(T. Okino and R. P. Lim)

Photosynthetic production of phytoplankton was measured *in situ* and in tanks using the ^{14}C method. The Winkler method was unsuitable because of interference by humic substances which can be photochemically oxidised (Hutchinson 1967). This reduces the increase in oxygen content of the light bottles and often results in negative production.

Gross photosynthetic production ranges from 0.07–0.25 mg C $\cdot$ l^{-1} $\cdot$ day^{-1} in the limnetic region and 0.19 mg C $\cdot$ l^{-1} $\cdot$ day^{-1} in the stream channels (Table 123). Net production was however greatly reduced ranging from 0.02–0.07 mg

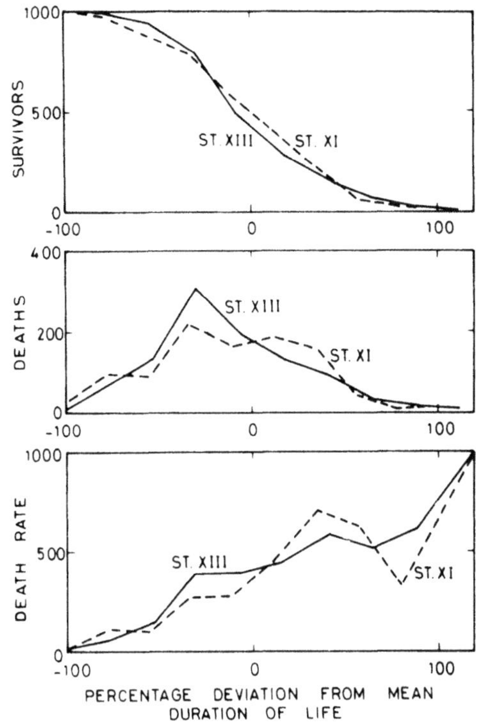

Fig. 99 Survivorship, death, and death rate curves for *Lepironia articulata* at Station XIII and XI. Survivors, deaths, and death rate are plotted against age expressed on a relative scale. (I. Ikusima 1978)

Table 123 Standing crop and photosynthetic production of phytoplankton, community respiration, and environmental factors in Tasek Bera. (T. Okino)

	Still station	Flowing station
Chlorophyll-a (mg/m^3)	1.25– 3.50	0.70– 1.85
Gross production (C mg/l/day)	0.07– 0.25	0.19
Net production (C mg/l/day)	0.02– 0.07	0.01
Community respiration (O_2 mg/l/day)	0.08– 0.75	0.12– 0.64
Water temperature (°C)	26.5 –33.0	24.6 –29.0
Dissolved oxygen (O_2 mg/l)	2.63– 8.41	2.03– 4.83
Total CO_2-C (C mg/l)	2.8	4.3
Transparency (mm)	–	1.9 – 2.4
Light extinction coefficient	0.9 – 1.8	0.5 – 1.8

$C \cdot l^{-1} \cdot day^{-1}$ in the limnetic region and 0.01 mg $C \cdot l^{-1} \cdot day^{-1}$ in the stream channel, which is between 5 and 35% of the daily gross production. Net photosynthetic production in the stream channel was 1/2-1/7th that of the limnetic region.

The photosynthesis-light curve of phytoplankton in Tasek Bera suggests that photosynthesis is inhibited by strong light of over 22,000 lux (Fig. 100). The production rate of a mixed algal population in Tasek Bera was 1.3–1.4 mg $C \cdot h^{-1} \cdot$ mg Chl.-a^{-1} at optimum light intensity which is low when compared to other aquatic situations (Table 124).

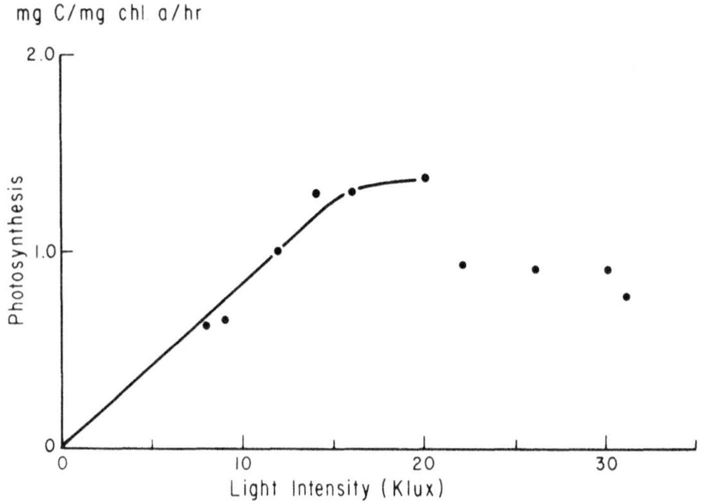

Fig. 100 Photosynthesis – light curve of phytoplankton in Tasek Bera. (T. Okino)

Table 124 Comparison of assimilation rates at optimum light for various situations. (T. Okino)

	Approximate mg $C \cdot h^{-1}$ synthesised per mg chlorophyll at optimum light	References
Natural lake populations	2–6	Gessner 1943, 1949 Manning and Juday 1941
Natural marine populations	1–6	Holmes 1957, Riley 1941 Ryther and Yentsch 1957
Algal cultures	1–10	Clendenning et al. 1956 Fleischer 1935, Ryther 1956 Ryther and Yentsch 1957
Tasek Bera	1.3–1.4	Present study

5.5.2 *Epiphytic algae*
(T. Watanabe, S. Kumano and I. Ikusima)

5.5.2.1 *Photosynthesis of epiphytic algae on Utricularia and Lepilonia* (T. Watanabe). *Utricularia* growing in 50 × 50 cm quadrats were removed and washed to separate the epiphytic algae. The epiphytic algae were then incubated in light and dark bottles at various light intensities and temperatures to estimate primary productivity and respiration. Similarly *Lepironia* growing in 50 × 50 cm quadrats were removed and the attached algae washed from the culms. The algae were also incubated in light and dark bottles at various light intensities.

Photosynthetic activity decreased with decreasing light intensities (Figs. 101 and 102). The rates of decrease in photosynthetic activity also varied with different temperature ranges. The compensation point of the attached algae on *Utricularia* is about 3000 lux while that of the *Lepironia* reeds is about 150 lux. This is due to differences in the algal community between the two stands. The algal community of *Utricularia* is dominated by diatoms such as *Frustulia rhomboides* var. *saxonica*, *Eunotia naegelii* var. *naegelii* and the green algae *Bulbochaeta praereticulata*. Several Conjugales such as *Closterium*, *Cosmarium*, *Xanthidium*, *Staurastrum* and *Desmidium* were also present (Table 125). The algal community on the *Lepironia* reeds is dominated by blue-green algae such as *Stigonema panniforme* and *Hapaloshiphon stuhlmannii* and the green algae. *Bulbochaeta praereticulata* (Table 126).

5.5.2.2 *Production of sediments on the surface of bottom mud* (T. Watanabe). The bottom mud samples were collected by an Ekman-Birge's bottom sampler along the profile of the floor of the swamp of different depths (cf. Fig. 102a), after which 4 ml of the surface sediments of each sample were extracted by a syringe. The surface sediment was put in a bottle containing aerated lake water and then the bottle was sealed tightly. The data of photosynthesis and respiration of each sediment sample was obtained by using the dark and light bottle method under the same light intensity in each depth by hanging the bottles at same places where the samples were obtained. The results are shown in Figure 102a.

In this experiment, the primary net production shows the photosynthesis of epipelic algae on the bottom sediments and respiration shown in decreased oxygen show the oxygen consumption by the respiration of epipelic alae and the decomposition of organisms in the bottom mud. So the value of primary net production was lower than that of the respiration in each point.

From the results, it will be assumed that primary net production of epipelic

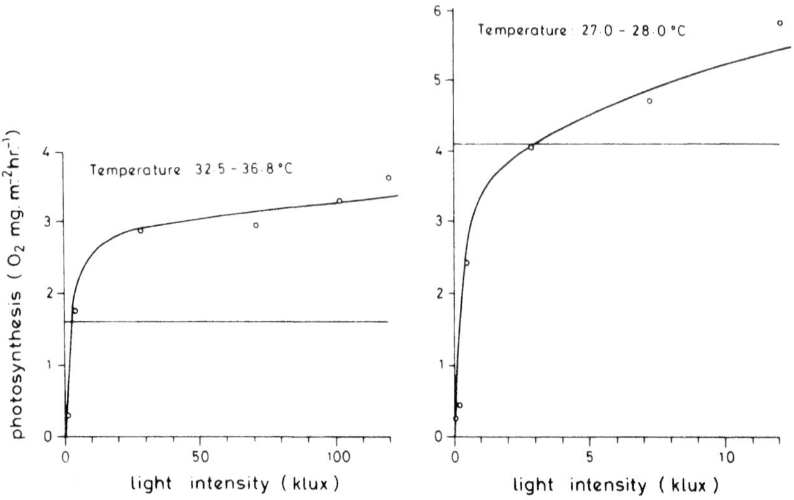

Fig. 101 Photosynthesis – light curve of attached algae of *Utricularia*. (T. Watanabe)

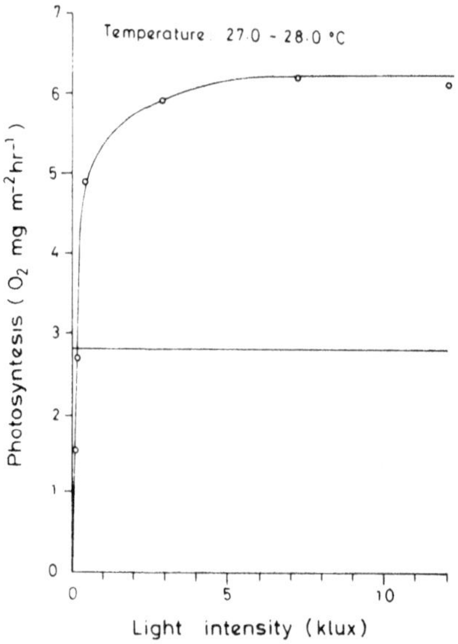

Fig. 102 Photosynthesis – light curve of attached algae on *Lepironia* (27.0–28.0°C).

algae on the bottom mud is higher in the shallow area and oxygen consumption is larger in open water area.

5.5.2.3 *The diurnal change of the efficiency of photosynthesis of sediments*

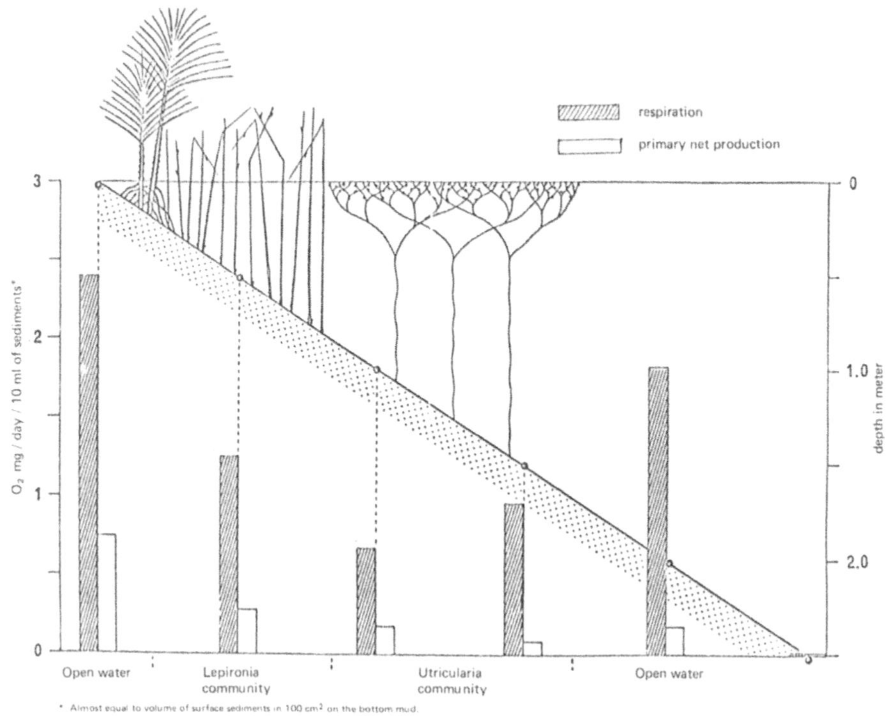

Fig. 102a Photosynthesis and respiration of surface sediments on the bottom mud. (T. Watanabe)

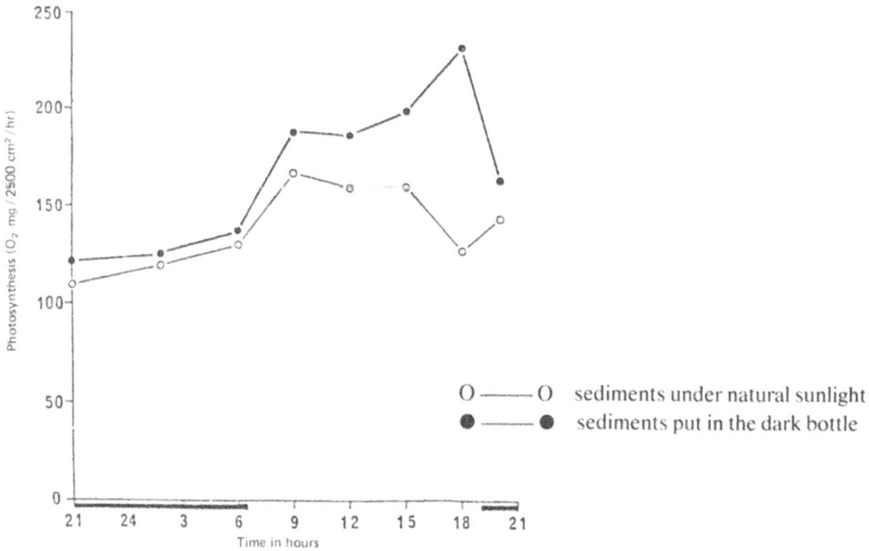

Fig. 102b Diurnal change of the efficiency of photosynthesis of sediments attached to *Utricularia*. (T. Watanabe)

attached to utricularia (T. Watanabe). The study of the diurnal change of the efficiency of photosynthesis has recently become an interesting subject of discussion among researchers. So in the tasek Bera, the diurnal change of the efficiency of photosynthesis of sediments attached to *Utricularia* was measured by an oxygen analyzer – Beckman-Toshiba Model 777 – between 21:00 on the 7th to 21:00 on the 8th of February at intervals of three hours under 11 000 lux of light intensity.

The results are shown in Figure 102b. An outline of the discussion of this experiment is as follows.

1) In sunlight, the efficiency of photosynthesis of sediments attached to *Utricularia* decreased in the course of time from sunrise and a minimum value was obtained at sunset.
2) The efficiency of photosynthesis of the sediments put in the dark bottle increased in the course of time and a maximum value was obtained 22 hours after putting the sediments in the dark bottle.

From this experiment, it will be assumed that the maximum value of the efficiency of photosynthesis of epiphytes to *Utricularia* was obtained within the first short period of 3 hours from sunrise, and decreased during the course of photosynthesis.

5.5.3 *Submerged macrophytes*
(I. Ikusima and S. Kumano)

Photosynthetic and respiration rates of detached shoots and leaves of submerged macrophytes were determined by the light and dark bottles technique. Except for *Batrachospermum* and *Nitella* photosynthesis - light curves were obtained with the tank method in the laboratory using photoreflector lamps (Toshiba, Day-light type, 500 W) as the light source. Light intensity to the bottles was reduced by covering the experimental bottles with black screens of different meshes. Water temperature in the experimental bottles was maintained using a thermoregulated water bath. Photosynthetic production of *Batrachospermum* and *Nitella* was measured using the light and dark bottle technique but *in situ*. Excised pieces of these two species were placed in bottles and incubated for half an hour around noon on a clear sunny day. Light intensity was controlled by wrapping the bottles in a series of envelopes.

The maximum net photosynthetic rate for young apical green shoots of *Utricularia* reached 48.5 mg $0_2 \cdot g^{-1} \cdot h^{-1}$ Fig. 103). This value was nearly twice that of the rates for the more mature portions of the plants, as well as the maximum value for submerged macrophytes in Japan. The light saturated

Table 125 Algae attached to *Utricularia*. (T. Watanabe)

Cyanophyta	
Microcystis incerta	+++
Coelosphaerium kuetzingianum	+
Merismopedia glauca	++
Hapalosiphon flexuosus	++
Bacillariophyta	
Eunotia curvata var. *capitata*	+
E. hexaglyphis var. *hexaglyphis*	+
E. curvata	+++
E. monodon var. *tropica*	+++
**E. naegelii* var. *naegelii*	+++++
E. pectinales var. *minor*	++
E. serra var. *serra*	+
Frustulia rhomboides	+++
**F. rhomboides* var. *saxsonica*	+++++
Anomoeneis serians var. *saxsonica*	+
A. serians var. *brachysira*	++
Pinnularia sp.	+
Surirella engleri f. *angustior*	+
S. spiralis	+
Surirella sp.	+
Chlorophyta	
Gloeocystis gigas	+
Palmella mucosa	+
Ankistrodesmus falcatus	++
**Bulbochaeta praereticulata*	++++
Desmids	
Gonatozygon sudanense	+
Netrium digitus var. *naegelii*	+
Closterium diana var. *minus*	+
C. gracile	+
C. gracile var. *striolatum*	+
C. intermedium	+
C. kuetzingii var. *vittatum*	+
C. porrectum	+
C. ralfsii var. *hybridum*	+
Pleurotaenium ehrenbergii var. *undulatum*	+
Triploceras gracile var. *undulatum*	+
Tetmemogus brebissonii var. *tenissimus*	+
Cosmarium ceylanicum f. *minus*	+
Cosmarium sp.	+
Xanthidium kalimantanum?	+
X. superbum var. *centricornis*	+
Micrasterias mahabuleshwarensis var. *surculifera*	+
Staurastrum freemanii var. *triquetrum*	+
S. menggalense?	+
S. subsaltans var. *indonesianum*	+
S. tetracerum	+
Desmidium aptogonum var. *tetragonum*	+
D. grevillei	+
D. swartzii var. *bicrestatum*	+
Protozoa	
Dinobryon sertularia	+
Centropyxis aculeata	+
Lesquereusia spiralis	+
Euglypha alveolata	+

* Dominant species.
+++++ = very rich; ++++ = rich; +++ = common; ++ = rare; + = very rare.

rate of net photosynthesis and the rate of respiration in the dark of mature leaves of *Cryptocoryne* were 7.4 mg $0_2 \cdot g^{-1} \cdot h^{-1}$ (e.g. 2.4 mg $CO_2 \cdot dm^{-2} \cdot h^{-1}$) and 0.3 mg $0_2 \cdot g^{-1} \cdot h^{-1}$ (0.1 mg $CO_2 \cdot dm^{-2} \cdot h^{-1}$) respectively. These are among the lowest values recorded for submerged macrophytes and terrestrial plants. The light photosynthesis curve shows a pattern characteristic of shade loving plants. There was no significant difference in photosynthetic rates between

Table 126 Algae attached to *Lepironia*. (T. Watanabe)

Cyanophyta		**Desmids**	
Merismopedia glauca	++	*Gonatozygon sudanense*	+
**Stigonema panniforme*	+++++	*Triploceras gracile*	
**Hapalosiphon stuhlmannii*	+++++	var. *undulatum*	+
Oscukkatirua borneti	+	*Closterium biclavatum*	+
O. splendida	+	*C. intermedium*	+
Lyngbya limnetica	++	*Micrasterias foliacea*	+
Microcoleus cathonoplastes	+	*Cosmarium armatum*	
		var. *verrucosum*	+
Bacillariophyta		*Xanthidium superbum*	
Eunotia flexuosa		var. *centricornis*	+
var. *eurycephala*	+	*Staurastrum cerastes* var.	
Frustulia rhomboides	++	*coronatum* fo. *inflatum*	+
Anomoeneis serians var. *acuta*	+	*S. orbiculare* var. *depressum*	+
A. serians var. *brachysira*	+	*Gymnozyga moniliformis*	
Pinnularia abaujensis		var. *gracelescens*	+
var. *abaujensis*	++	*Desmidium grevillei*	+
		Groenbladia neglecta	+
Chlorophyta			
**Bulbochaeta praereticulata*	++++		

* Dominant species.
+++++ = very rich; ++++ = rich; +++ = common; ++ = rare; + = very rare.

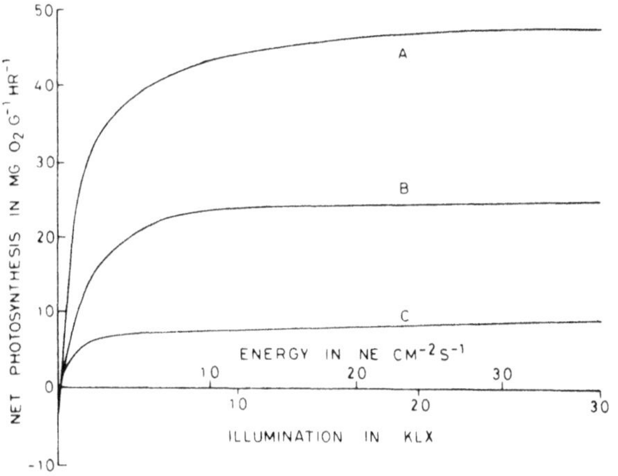

Fig. 103 Light (energy)-response curves of net photosynthesis of submerged macrophytes grown in a sunny open water. A: *Utricularia flexuosa*, 5 cm of apical parts of green young shoot, at 28.0°C. B: *U. flexuosa*, 5 cm of apical parts of yellowish brown mature shoot, at 27.5°C. C: *Cryptocoryne griffithii*, mature leaf without attachment of algae, at 28.4°C (September 1972). (I. Ikusima)

Table 127 Net photosynthesis and dark respiration of *Utricularia flexuosa* measured with *in situ* method. (I. Ikusima)

A. On a cloudy day, 21 September 1972, 26.2~27°C. 10 cm of apical part of mature shoot.

Relative light intensity (%)	Net photosynthesis dark respiration (mg O_2/g)	Time of exposure to the sun
100*	201	0735-1720
31	129	0738-1726
11	78.5	0742-1732
4.4	23.7	0748-1737
1.6	– 7.4	0755-1742
Dark	–21.9	0759-1747

* Ca. 139 cal · cm^{-2} (horizontal radiation during the experimental period).

B. On a fine day, 22 September 1972, 25.5~29.2°C. 10 cm of apical shoot. Numbers in parentheses show the time of exposure.

Relative light intensity (%)	Net photosynthesis and dark respiration (mg O_2 per g dry wt)	
	Young shoot (green)	Mature shoot (brown)
100*	292 (0718-1222)	131 (0715-1219)
31	194 (0727-1228)	70 (0722-1225)
11	210 (0738-1239)	84 (0730-1231)
4.4	96 (0733-1233)	46 (0735-1236)
1.6	40 (0745-1245)	14 (0742-1242)
Dark	–19.9 (0752-1251)	– 8.3 (0747-1248)

* Ca. 189 cal · cm^{-2} (horizontal radiation during the experimental period).

C. On a fine day, 17 October 1972, 25.2-28.0°C. 10 cm of apical shoot. Numbers in parentheses show the time of exposure to the sun.

Depth from water surface (m)	Net photosynthesis and dark respiration (mg O_2 per g dry wt)	
	Apical parts	Basal parts
0.05*	138 (0748-1309)	44 (0748-1313)
0.2	141 (0736-1236)	115 (0736-1238)
0.4	97 (0732-1256)	62 (0732-1258)
0.8	83 (0725-1252)	54 (0725-1248)
1.6	65 (0720-1245)	33 (0720-1243)
Dark	–12.7 (0755-1306)	– 3.7 (0755-1302)

* Ca. 260 cal · cm^{-2} (horizontal radiation at the water surface during the experimental period).

the young and mature leaves.

Experiments of photosynthetic and respiration rates of *Utricularia* and *Cryptocoryne* under varying weather conditions were carried out using the *in situ* method of Ikusima (1965). (Results are summarised in Tables 127 and 128). The estimated daily net production of a pure stand of *Utricularia* is almost 12 g dry wt·m^{-2}·day^{-1} on a clear sunny day. (A conversion factor of 0.85 g dry wt·g·0_2^{-1} was used). It can be extrapolated that the maximum intrinsic growth rate or growth coefficient reaches 0.28 day^{-1}, that is the plant can double its dry weight every 2½ days. The daily net production of *Cryptocoryne* growing in shallow water is about 5 g dry wt on a clear day and has a

Table 128 Net photosynthesis and dark respiration of *Cryptocoryne griffithii* measured with *in situ* method. (I. Ikusima)

A. On a fine and cloudy day, 18 September 1972. At 25.8-28.4°C. Numbers in parentheses show the time of exposure to the sun.

Relative light intensity (%)	Net photosynthesis and dark respiration ($mg\ O_2$ per g dry wt)	
	Young leaf	Mature leaf
100*	3.1 (1104-1355)	3.7 (1106-1352)
31	2.9 (1109-1357)	4.3 (1112-1401)
11	2.2 (1120-1403)	2.8 (1123-1405)
4.4	1.9 (1125-1407)	3.1 (1126-1410)
1.6	1.3 (1135-1412)	2.5 (1138-1415)
Dark	−0.37 (1140-1418)	−0.42 (1142-1420)

* Ca. 122 cal·cm^{-2} (horizontal radiation during the experimental period).

B. On a cloudy day, 20 September 1972. At 25.8-28.7°C. Numbers in parentheses show the time of exposure to the sun.

Relative light intensity (%)		Net photosynthesis and dark respiration ($mg\ O_2$ per g dry wt)	
		Young leaf	Mature leaf
100*		54.8 (0724-1220)	50.5 (0722-1216)
11		48.7 (0728-1225)	32.7 (0726-1223)
Dark		−3.14 (0754-1243)	−1.57 (0752-1244)
Light pass** through colour cellophanes	Red	56.6 (0748-1236)	49.9 (0750-1239)
	Green	50.9 (0732-1232)	34.1 (0740-1234)
	Blue	48.1 (0738-1241)	29.6 (0740-1234)

* Ca. 144 cal·cm^{-2} (horizontal radiation during the experimental period).
** Relative light intensity (RLI) and relative radiation (RR) of light passed through cellophanes with

	RLI	RR
Red cellophane	13 (%)	60 (%)
Green cellophane	23	58
Blue cellophane	29	57

standing crop of 360 mg dry wt. Despite their luxurient growth observed, the maximum daily net production is only 50% or less than the production rates recorded for tempeate terrestrial plants during their growing season.

Photosynthetic-light curves for *Nitella* sp. and two species of *Batrachospermum* are given in Fig. 104. The maximum rates of net photosynthesis and dark respiration for *Nitella* sp. are 19 mg $0_2 \cdot g^{-1} \cdot h^{-1}$ and 0.7 mg. $0_2 \cdot g^{-1} \cdot h^{-1}$ respectively. The rate of net photosynthesis under light saturation conditions for *Batrachospermum cylindrocellulare* and *Batrachospermum beraense* are 23.6 mg $0_2 \cdot g^{-1} \cdot h^{-1}$ and 54.3 mg $0_2 \cdot g^{-1} \cdot h^{-1}$ respectively. Both the *Batrachospermum* species show very high photosynthetic activity and are 1.3 to 4.8 times higher than those of *B. moniliforme* and *B. testale* that occur in Japan. *B. beraense* is more tolerant to sunlight than *B. cylindrocellulare*.

5.5.4 *Emergent macrophytes*
(I. Ikusima)

An attempt was made to estimate the amount of photosynthetic production of *Lepironia articulata* which is one of the dominant macrophytes in Tasek Bera. The estimation of primary production of a swamp reed community in

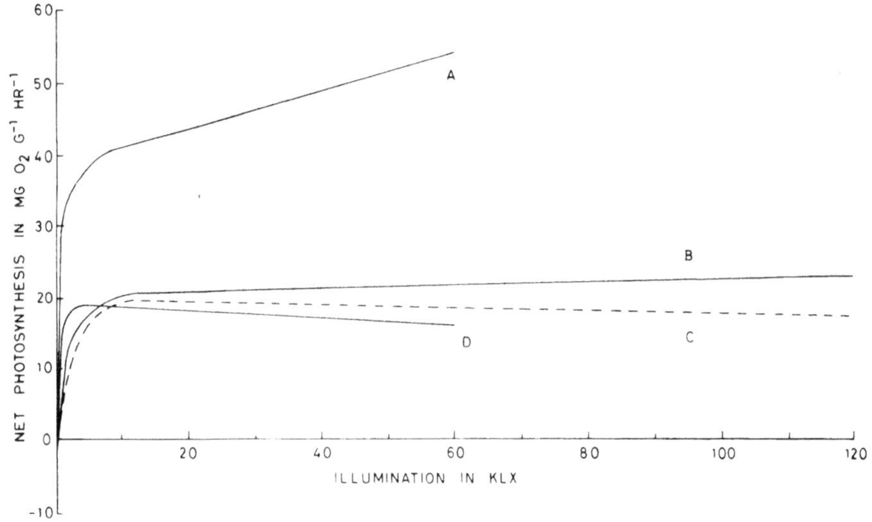

Fig. 104 Light-response curves of net photosynthesis. A: Batrachospermum beraense, at 25.4°C. B: *B. cylindrocellulare*, at 25.5°C. C: *Utricularia flexuosa*, at 30.0°C. D: *Nitella* sp., at 26.5°C. (S. Kumano)

humid tropical regions is more difficult than in temperate regions because the establishment, growth and death processes occur concurrently throughout the year, so that there is not very much change in their standing crop.

The production of a pure *Lepironia* stand was obtained by the harvest method or more properly by the summation method (Ikusima 1978). The method attempts to estimate the amount of photosynthetic assimilation by the growth curve, life table and the interval of emergents in living part, the losses due to death of culms, and the consumption of plants through respiration.

In this method, the gross production rate ($\triangle Pg$) is estimated by the summation (Kira et al. 1967).

$$\triangle P_g = \triangle Y + \triangle L + \triangle G + \triangle R$$

where $\triangle Y$, $\triangle L$, $\triangle G$ and $\triangle R$ refer to the rates of biomass increase, loss due to death of plants, consumption by herbivorous insects, and consumption through respiration in a given period respectively. By definition, the net production rate ($\triangle P_n$) is given by

$$\triangle P_n = \triangle P_g - \triangle R.$$

Each item was calculated using the data obtained twice at weekly intervals in early August. Assuming the standing crop to be in a dynamic equilibrium state over a short period, in which the increments in living culms, newborn culms, rhizomes and roots are compensated by losses through death of these parts, the biomass increments may be zero.

The production of the *Lepironia* stand can be estimated using their growth curves, life table, the interval of emergence of new stems, respiration rate, etc. (Figs. 97, 98, 99, 103).

The dark respiration rate of *Lepironia* culms was measured in an air-tight box with KOH solution as the absorbent (Yoda 1967). The correction for the decline in the absorbing capacity of the alkaline solution with time was taken into consideration (Kirita 1971). The daily rate of respiration was obtained by summing successive 2-3 hours' respiration over a 24-hour period. The dry matter/carbon dioxide conversion factor, was taken to be 0.614, assuming that the substrate for respiration was mainly carbohydrates $(C_6H_{10}O_5)n$.

The respiration rate ranges from 0.8 to 1.9 mg $CO_2 \cdot h^{-1} \cdot g^{-1}$ dry weight (at 27-34°C) for the apex of rhizomes and the first culm, and from 0.3 to 1.4 mg $CO_2 \cdot h^{-1} \cdot g^{-1}$ dry wt (at 27-34°C) for the 2nd, 3rd, 4th culms, etc. The rates vary greatly depending on the age of culms. The mean rate for the first stem is 1.9 times larger than those of the older culms.

Assuming a steady state condition in the *Lepironia* stand over a short period, in which the increments in living culms, newly initiated culms, rhizomes and roots are compensated by the losses through death of corresponding portions, the mean daily production can be calculated (Table 129). Grazing by grasshoppers, *Oxya velox* was observed in the dry season of 1972, but the total amount was negligible (see later).

Tropical aquatic macrophytes are often thought to be exceptionally productive. Some macrophytes such as *Cyperus papyrus, Eichhornia crassipes, Salvinia molesta, Utricularia flexuosus* etc. have high values in biomass and vegetative reproduction (Thompson 1973; Westlake 1963; Mitchell 1974; Ikusima 1973). In contrast, *Lepironia articulata,* has a lower productivity because of the low rate of development of new culms, low growth rate as well as a small standing crop of the aerial parts. The daily gross production at Stations XIII and XI is estimated at 4.89 $g \cdot m^{-2} \cdot day^{-1}$ (17.8 $ton \cdot ha^{-1} \cdot y^{-1}$) and 2.11 $g \cdot m^{-2} \cdot day^{-1}$ (7.7 $ton \cdot ha^{-1} \cdot y^{-1}$) (Table 129). In contrast, another tropical aquatic sedge, *Cyperus papyrus*, has a net production of 17-25 $g \cdot m^{-2} \cdot day^{-1}$ (Thompson 1973). The mean value of daily solar radiation at Pos Iskandar, Tasek Bera in August is 401 $cal \cdot cm^{-2} \cdot day^{-1}$. Therefore, the efficiency of solar energy conversion to gross production is 0.5% at Station XIII and 0.2% at Station XI in August. This efficiency is 50% or less than that of other aquatic macrophyte stands (Ikusima 1972). Perhaps the low rate of production in *Lepironia* is due to the low emergence rate of new culms. A low growth rate as well as small standing crop of aerial parts could all be due to some overall phenotypic limitation. This sedge may also be a less eficient C_3 plant, unlike the majority of aquatic emergents such as *Cyperus papyrus* L., which are more efficient C_4 plants (Westlake 1975).

The daily ratio of net production to gross productions is 46% at Station XIII and 35% at Station XI. On the whole, about 60% of the gross production is consumed through respiration, leaving 40% as net production, all of which are assumed to compensate for losses due to death of some plants.

A large amount of peat is accumulated within the *Lepironia* stand. The dead organic matter that is thus supplied to the bottom mud amounts to 2.24 $g \cdot m^{-2} \cdot day^{-1}$ (8.2 $ton \cdot ha^{-1} \cdot y^{-1}$) at Station XIII and 0.73 $g \cdot m^{-2} \cdot day^{-1}$ (2.7 $ton \cdot ha^{-1} \cdot y^{-1}$) at Station XI. In spite of the fairly small biomass of *Lepironia,* the annual rate of dead organic matter supply is almost equal to that of well developed forests in Japan (Kira and Shidei 1967).

5.6 Production estimates based on oxygen dynamics in a closed water column (I. Ikusima)

Diurnal changes of oxygen concentration in a water column enclosed in a

Table 129 Summary of biomass ($g \cdot m^{-2}$), daily increment in biomass ($g \cdot m^{-2} \cdot day^{-1}$), and daily production ($g \cdot m^{-2} \cdot day^{-1}$) of *Lepironia articulata* at Station XIII and Station XI. (I. Ikusima 1978)

	Station XIII				Station XI			
	Culm	Root	Rhizome	Total	Culm	Root	Rhizome	Total
Living biomass in August	286	70	102	458	150	22	32	204
Increments in living part	1.97	0.11	0.16	2.24	0.65	0.03	0.05	0.73
Losses due to death	−1.97	−0.11	−0.16	−2.24	−0.65	−0.03	−0.05	−0.73
Biomass increments	0.00	0.00	0.00	0.00	0.00	0.00	0.00	0.00
Turnover as dying part	1.97	0.11	0.16	2.24	0.65	0.03	0.05	0.73
Net production (Total)	1.97	0.11	0.16	2.24	0.65	0.03	0.05	0.73
Respiration	2.65	–	–	2.65	1.38	–	–	1.38
Gross production (Sum total)	4.62	0.11	0.16	4.89	2.03	0.03	0.05	2.11

colourless transparent vinyl sheeting allows the estimation of O_2 production by photosynthesis of various aquatic producers and its consumption through respiration by plants and bacteria and by decomposition processes (Strickland et al. 1969, Kemmerer and Neuhold 1969, Hayashi et al. 1972). This method is particularly useful for evaluating the activities of plankton, periphyton, and microbes attached on complicated surfaces of various substrata in shallow water ecosystems.

The following four types of cylindrical plastic tubes, 796 cm^2 in cross sectional area, were installed in the *Lepironia* swamp at Station XI and near Stations I and XI, where the depth of the water was about 55 cm (Fig. 105).

1) A cylinder without covers at both ends in which only the transfer of O_2 from surrounding water was prevented. The cylinder enclosed no *Lepironia* culms.
2) A cylinder with covers at both top and bottom. The supply and transfer of O_2 to and from the outside environment was completely prevented.
3) A cylinder with a top cover only. O_2 supply from the air was cut off.
4) A cylinder with a cover only at its bottom. The transfer of O_2 between mud and water was excluded.
5) The cylinder similar to (1) but enclosing *Lepironia* culms.

The concentration of dissolved O_2 in the water inside and outside the cylinder was determined at sunrise, noon and sunset for two or four successive days. Care was taken to remove the effect of experimental error due to manipulation on biological activities at the bottom of water by making concurrent observations on the undisturbed bottom at the same place. The time trends of dissolved oxygen in five cylinders in the reed swamp and the border between

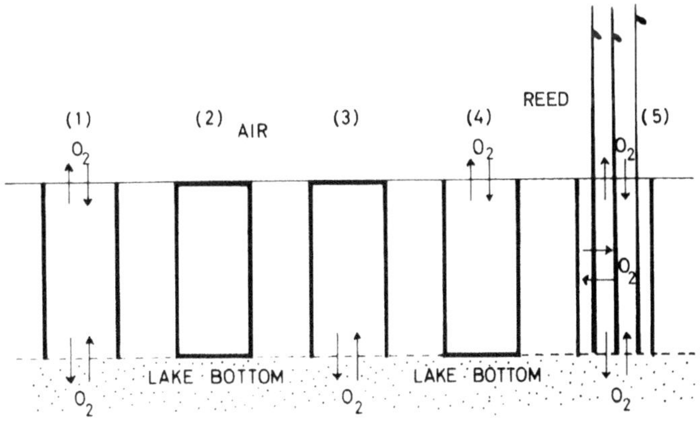

Fig. 105 Five types of plastic cylinders installed in the reed swamp. (I. Ikusima)

the reed swamp and the open water shows a low O_2 concentration for the dark period (Fig. 106).

The oxygen balance in the water column was estimated in terms of the production and consumption at the mud-water interface and of *Lepironia* culms, by seston and dissolved coloured material in the water column as well as by diffusion between the air and water. The production and consumption at the bottom can be calculated from the values of cylinders (1) and (4) and/or cylinders (2) and (3). The oxygen exchange by aufwuchs on *Lepironia* culms is obtained by the difference between the values of cylinders (1) and (5). The oxygen supply and consumption by seston and dissolved matter is calculated from the value of cylinder (2). The supply by diffusion is obtained by combining the results of cylinders (1) and (3) and/or cylinders (2) and (4).

Table 130 summarises the results of a few days' observations at Station XI and near Stations I and XI, from September to November, 1972. O_2 consumption at the mud-water interface as production and consumption by aufwuchs on *Lepironia* culms proved to be most important in controlling the concentration of dissolved oxygen in this case. A large amount of oxygen production by photosynthesis of attached algae in the day time is consumed gradually by a reduction process in the mud.

O_2 concentration decreases to a minimum level just before sunrise. The O_2 consumption in the water is also enhanced under strong light conditions through the oxidation of coloured substances in water, and its amount exceeds that of the oxygen production through photosynthesis. It is possibly caused by some photochemical reaction and may be one of the important characteristics of black waters.

The amount of dissolved oxygen obtained through diffusion from air into water increases remarkably in dark periods since the water is undersaturated with oxygen. The mean rate of oxygen uptake through diffusion changes with the O_2 concentration or % saturation of O_2 in water (Fig. 107).

The daily oxygen production by aufwuchs contributes little to the total oxygen exchange (Table 131). Oxygen production and consumption in water are largely dependent on light intensity and the amount of dissolved and suspended coloured material in the water.

The daily O_2 exchange in water columns with and without *Lepironia* culms after the dry season are estimated at 0.45 g $O_2 \cdot m^{-2} \cdot day^{-1}$ (October) – −1.00 g $O_2 \cdot m^{-2} \cdot day^{-1}$ (November) and 0.39 g $O_2 \cdot m^{-2} \cdot day^{-1}$ (October) – −1.02 g $O_2 \cdot m^{-2} \cdot day^{-1}$ (November) respectively. The supply of O_2 from the air is 0.38 g $O_2 \cdot m^{-2} \cdot day^{-1}$ on the average and plays an important role for the balance of dissolved O_2 of lake water. *Utricularia flexuosa* and a few species of filamentous algae are sporadically found in the *Lepironia* swamp. The well

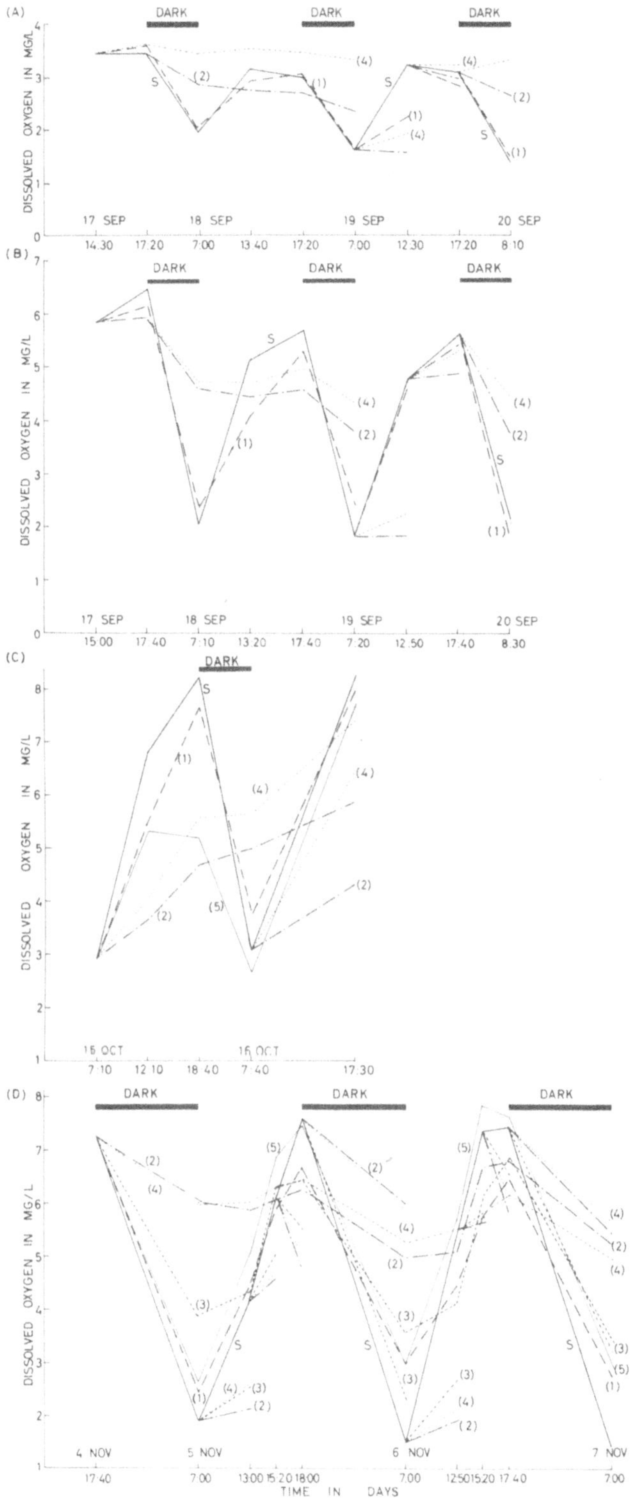

Fig. 106 Daily fluctuation in oxygen concentration in water columns. Five types of plastic cylinders (Cf. Fig. 107) are shown in parentheses. S shows the dissolved oxygen outside of the cylinders. (A) Near Station I in September 1972; (B) Near Station XI in September 1972; (C) At Station XI in October 1972; (D) At Station XI in November 1972. (I. Ikusima)

Table 130 Rates of oxygen production (positive values) and consumption (negative values) and those of oxygen transfer from air to water (positive values indicate oxygen uptake by water) in shallow water of the *Lepironia* reed swamp. Water temperature and percentage saturation of dissolved oxygen outside of the cylinder, and estimates of total solar radiation received at the water surface for the successive observations are summarised. (I. Ikusima)

A. Near Station I, in September 1972

Date and hour		17th 14:30	17:20	18th 6:50	13:40	17:20	19th 7:00	12:30	17:20	20th 8:10
Water temperature °C		27.1	27.1	25.8	27.0	27.0	25.9	27.9	27.7	26.2
% saturation of dissolved oxygen		42	43	24	37	37	20	37	35	16
Solar radiation cal · cm^{-2} per period		151	0	249	137	0	238	222	0	
Mean rate of	At the bottom surface	20	−60	60	30	−50	30	−60	−70	
oxygen exchange	By seston and dissolved matter	0	−20	−10	−10	−10	0	−20	−10	
mg $O_2 \cdot m^{-2} \cdot h^{-1}$	By diffusion from air to water	40	20	10	0	10	40	40	30	
	Total	60	−60	60	20	−50	70	−40	−50	

B. Near Station XI, in September 1972

Date and hour		17th 15:00	17:40	18th 7:10	13:20	17:40	19th 7:20	12:50	17:40	20th 8:30
Water temperature °C		29.3	29.3	25.8	28.7	28.6	25.8	30.0	29.5	26.6
% saturation of dissolved oxygen		57	76	24	57	66	20	62	47	23
Solar radiation cal · cm^{-2} per period		107	0	237	149	0	255	205	0	
Mean rate of	At the bottom surface	30	−90	130	90	−70	200	10	−80	
oxygen exchange	By seston and dissolved matter	0	−40	−10	10	−30	20	20	10	
mg $O_2 \cdot m^{-2} \cdot h^{-1}$	By diffusion from air to water	10	0	10	20	0	40	40	20	
	Total	40	−130	130	120	−100	260	70	−50	

C. At Station XI, in October 1972

		15th						16th		
Date and hour		7:10		12:10		18:40		7:40		17:30
Water temperature °C		27.6		31.3		31.2		27.2		29.0
% saturation of dissolved oxygen		37		52		97		87		70
Solar radiation cal · cm^{-2} per period			177		164		0		238	
Mean rates of oxygen exchange mg $O_2 \cdot m^{-2} \cdot h^{-1}$	At the bottom surface		100		10		–90		80	
	By aufwuchs on *Lepironia* stems		–90		–40		20		20	
	By seston and dissolved matter		40		40		0		30	
	By diffusion from air to water		20		10		0		50	
	Total		70		20		–90		180	

D. At Station XI, in November 1972

		5th						6th						7th
Date and hour		7:00		13:00		18:00		7:00		12:50		17:40		7:00
Water temperature °C		25.5		27.0		29.2		25.9		28.5		30.3		26.0
% saturation of dissolved oxygen		23		53		102		17		70		99		16
Solar radiation cal · cm^{-2} per period			107		155		0		218		177		0	
Mean rate of oxygen exchange mg $O_2 \cdot m^{-2} \cdot h^{-1}$	At the bottom surface		110		210		–130		180		130		–100	
	By aufwuchs on *Lepironia* stems		130		110		–70		110		120		–120	
	By seston and dissolved matter		40		–60		–90		60		–240		–90	
	By diffusion from air to water		50		30		0		20		30		10	
	Total		330		290		–290		370		40		–300	

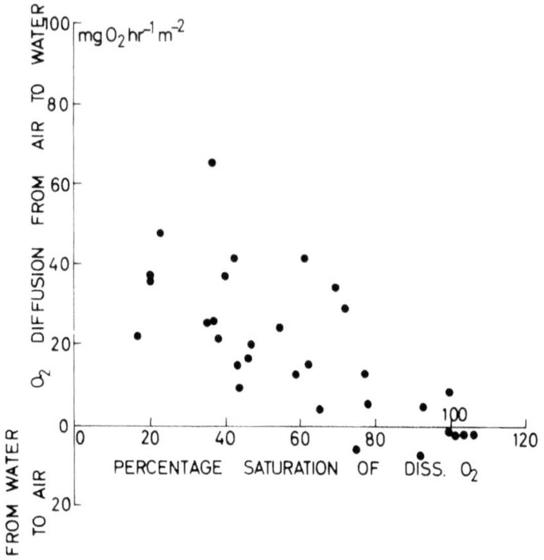

Fig. 107 Relationship between the rate of oxygen diffusion from air to water and water to air, and the dissolved oxygen concentration of the water body expressed as percentage saturation of dissolved oxygen. (I. Ikusima)

Table 131 Breakdown list of mean daily oxygen production and consumption (negative values) in water columns at a typical reed swamp at Pos Iskandar (g $O_2 \cdot m^{-2} \cdot day^{-1}$). (I. Ikusima)

Station	Near St. I	Near St. XI	St. XI	St. XI
Month	September	September	October	November
O_2 production at bottom surface	−0.55	0.08	−0.38	0.18
O_2 production by aufwuchs on *Lepironia* stems	–	–	0.06	0.02
O_2 production in water	−0.30	−0.27	0.40	−1.62
Supply of O_2 from air to water	0.44	0.28	0.37	0.42
Total Site populated by *Lepironia* reeds	–	–	0.45	−1.00
Site without *Lepironia* reeds	−0.41	0.09	0.39	−1.02

developed mat of *Utricularia* reaches about 170 g dry wt $\cdot m^{-2}$. During the day dissolved oxygen is usually supersaturated through intense underwater photosynthesis and the maximum daily net production in the water columns with *Utricularia* is estimated as 30 g $O_2 \cdot m^{-2} \cdot day^{-1}$. In general, O_2 or organic matter production at the *Lepironia* littoral region, is thought to contribute most to the total production of the study site. Net primary production, however, is markedly less than that of other tropical lakes (Lewis 1974).

6. Secondary production

T. Mizuno, R. P. Lim and J. I. Furtado

6.1 Introduction

In this chapter the composition, the standing crop, density, growth rates and productivity of the secondary producers in Tasek Bera are discussed.

The rich fauna of Southeast Asia has attracted the attention of many of the world's biologists. However, the identification and classification of animals is presently very incomplete. During the course of research in Tasek Bera, many aquatic organisms were collected and identified by members of the research team and by specialists in various regions of the world. Ecological studies were also carried out in this swamp.

The aquatic organisms in Tasek Bera can be conveniently divided into the following groups:

1) Zooplankton;
2) Benthic organisms and invertebrates associated with or attached to vegetations;
3) Shrimps;
4) Grasshoppers and spiders.

Each of these groups is discussed with reference to its current state of knowledge in Southeast Asia and Malaysia in particular.

6.2 Zooplankton (T. Mizuno, C. H. Fernando and R. P. Lim)

6.2.1 *History of study* (T. Mizuno)

The zooplankton in Southeast Asia have been fairly well studied since the end of 19th century. Under the auspices of the Deutschen Limnologische Sunda-Expedition carried out between 1928 and 1929 in Sumatra and Java studies on the zooplankton included the Cladocera by Brehm (1933), Copepoda by

Furtado, J. I. and Mori, S., eds., Tasek Bera. ISBN 90 6193 100 2.

Kiefer (1933), Rotifera by Hauer (1938) and several groups by other members of the expedition. Brehm (1938, 1953a, 1953b) reported his findings on zooplankton collected during the Wallace Expedition and the Indian Limnological Survey.

Recent studies were those of Blache (1950, 1951) on the flora and fauna of Tonle Sap in Cambodia, Shirota (1966) in Vietnam, and T. Mizuno and S. Mori (1970) in Thailand, Cambodia and Malaysia, Ueno (1964, 1966) wrote a compendium of aquatic organisms of Southeast Asia. His earlier paper (1964) lists all species of *Diaptomus* collected and their distribution and his later paper (1966) describes six species of Rotifera, eight species of Cladocera and two species of Copepoda which were collected by Prof. R. Yoshii, Kyoto University in 1965. Uneo (1966, 1969) also described several species of the subterranean Crustacea, Bathynellidae which included *Allobathynella malaya (Parabathynella malaya)* found in Batu Cave, Kuala Lumpur, Malaysia. Imamura (1964) described eight species of water mites from Thailand, a group unstudied by limnologists.

Research on the zooplankton in Malaysia is scarce. Johnson (1964) produced a 'Key of Malaysian Cladocera' which included four species of Sididae, one species of Bosminidae, five species of Macrothricidae, 12 species of Daphnidae and 23 species of Chydoridae. A similar key was produced by Lui (1969). Prowse (1958, 1960, 1962) described several species of Eugleninae and Flagellata, and their distribution in Peninsular Malaysia. However, other groups of Protozoa have hardly been studied.

6.2.2 *Classification of zooplankton* (T. Mizuno)

The zooplankton fauna in Tasek Bera is very poor because of the low pH value (T. Mizuno and S. Mori 1970). A total of 64 zooplankton taxa were identified which comprised 13 species of Protozoans, 29 species of Rotifers, 14 species of Cladocera, five species of Copepods and three species of Ostracods (Table 132).

Most of the phyto- and zooplankton in the tropical swamp were similar to those of dystrophic ponds and swamps in temperate zones like those of the northern regions or the high mountains of Japan. There were, however, some typical tropical zooplankton species such as the Cladocera, *Alonella excisa clathratula, Chydorus barroisi, C. faviformis;* the Copepoda, *Ectocyclops phaleratus, Tropocyclops prassinus, Microcyclops varicans;* and the Rotifera, *Lecane papuana, L. ploenensis, Keratella tropica,* which have never been found in Japan. The Rotifera, *Macrochaetus collinsii*, is a rare species.

Table 132 List of zooplankton in Tasek Bera. (collected and identified by T. Mizuno)

Protozoa
- *Arcella vulgaris* Ehrb.
- *Arcella* sp. 1
- *Arcella* sp. 2
- *Centropyxis acureata* Stein
- *Cloen* sp.
- *Difflugia acuminata* Penard
- *D. bacillifera* Penard
- *D. corona* Wallich
- *D. spiralis* Ehrb.
- *Dinobryon sertularia* Ehrb.
- *Euglypha brachiata* Leidy
- *Euglypha* sp.
- *Vorticella campanula* Ehrb.

Cladocera
- *Acroperus harpae* Baird
- *Alona affinis* (Leydig)
- *A. guttata* G. O. Sars
- *Alonella excisa clathratula* Sars
- *Biapertus affinis affinis* (Leydig)
- *Bosminopsis deitersi* (Richard)
- *Ceriodaphnia* sp.
- *Chydorus barroisi* (Richard)
- *C. eurynotus reticulatus* Daday
- *C. faviformis* Birge
- *C. sphaericus* (Müller)
- *Euryalona orientalis* (Daday)
- *Graptoleberis testudinaria* (Daday)
- *Macrothrix spinosa* (King)

Ostracoda
- *Standesia* sp.
- *Eucypris* sp.
- *Pseudocypretta maculata* (Klie)

Copepoda
- *Canthocamptus* sp.
- *Ectocyclops phaleratus* (Koch)
- *Mesocyclops leuckarti* (Claus)
- *Microcyclops varicans* (Lilljeborg)
- *Tropocyclops prassinus* (Fischer)

Rotifera
- *Asplanchna* sp.
- *Brachionus forficula* Wierzyski
- *B. (Platyias) quadricornis* (Ehrb.)
- Bdelloid rotifer
- *Colurella colurus* Ehrb.
- *Conochilus unicornis* Rousselet
- *Dipleuchlanis propatula* (Gosse)
- *Diurella* sp.
- *Euchlanis dilatata* Ehrb.
- *Lecane doryssa* Harring
- *L. leostina* (Turner)
- *L. papuana* (Murray)
- *L. ploenensis* (Voigt)
- *L. tenuiseta* Harring
- *L. (Monostyla) bulla* (Gosse)
- *L. (Monostyla) hamata* (Stokes)
- *L. (Monostyla) quadridentata* (Ehrb.)
- *Lepadella ovalis* (Müller)
- *Keratella cochlearis* (Gosse) var. *tecta* (Gosse)
- *K. tropica* (Apstein)
- *K. valga* (Ehrb.)
- *Macrochaetus collinsii* (Gosse)
- *Monommata* sp.
- *Mytilina ventralis* (Ehrb.)
- *Notommata copens* Ehrb.
- *Testudinella patina* (Hermann)
- *Trichocerca similis* (Wierzejski)
- *Trichotria tetractis caudata* (Lucks)

6.2.3 *Composition and comparison of zooplankton associations* (T. Mizuno, R. P. Lim and J. I. Furtado)

Zooplankton samples were collected with a plankton net (mesh size 10) at Station I to V by R. P. Lim and J. I. Furtado and at Stations I and III by T. Mizuno. Vertical hauls were taken at each station and the depth varied with changes in water level over the year. During the dry season two to three vertical hauls were taken at each station. Samples collected were preserved

Table 133 Checklist of zooplankton species and their presence in the various habitats of Tasek Bera. (R. P. Lim and J. I. Furtado)

	Habitats		
	St. I modified limnetic	St. II, III sub-littoral	St. V forested limnetic
Cladocera			
Bosminopsis deitersi	–	+	–
Chydorus borropsi	–	+	–
Chydorus sphaericus	–	+	–
Macrothrix spinosa	–	+	–
Brachionus forficula	+	–	+
B. (Ilatyias) quadricornis	–	+	–
	+	–	+
Colurella colurus	+	–	–
Euchlanis dilatata	–	+	–
Lecane doryssa	–	+	–
L. leostina	–	+	–
L. papuana	+	–	–
L. (Monostyla) bulla	+	+	–
L. (Monostyla) hamata	+	–	–
L. (Monostyla) quadridentata	+	–	–
Lepadella ovalis	–	+	–
Keratella cochlearis var. *tecta*	+	+	–
K. tropica	–	+	–
Monommata	–	+	–
Mytilina ventralis	–	+	–
Notommata copens	+	+	–
Testudinella patina	–	+	–
Trichocerca similis	–	+	–
Trichotria tetratis caudata	–	+	–
Total	10	17	3

in 5% formalin, concentrated and counted using either an Utermohl's inverted microscope or a compound microscope.

In a particular comparative study of zooplankton assemblages in the open water (Station I), sublittoral *Utricularia* community (Station III) and the limnetic forest stream (Station V) the total diversity of zooplankton was low (36 species). The limnetic region was particularly poor (3-10 species) when compared with the vegetated sublittoral (17 species) (Table 133). This is due to the greater ecological complexity of the vegetated sublittoral as demonstrated by Ang (1971) and Lim (1976) for Canadian lakes. The limnetic is extremely poor in Cladocera (no species), Rotifera (10 species) and Copepoda (no species), while the vegetated sublittoral is richer in Cladocera (4 species) and Rotifera (13 species). *Brachionus forficula* and *Keratella cochlearis* are true planktonic forms, although they also flourish in the littoral. The predominance of the littoral forms of rotifers in the limnetic indicates that they are translocated from the vegetated littoral. This may account for differences in the seasonal fluctuation of zooplankton between habitats in this swamp.

Results of a quantitative study of the structure of zooplankton associations in the limnetic and the vegetated sublittoral are given in Table 134. One metre deep vertical hauls were taken using a plankton net with a coarse wire mesh over its mouth to prevent large vegetation from entering the net.

The number of individuals of Cladocera, Copepoda. Rotifera and Protozoa was significantly greater in the littoral when compared with the limnetic region. This was especially so for cladocerans and protozoans in February 1973 with population densities of 0.8–1.6 individuals $\cdot l^{-1}$ at Station I and 72–169 individuals $\cdot l^{-1}$ at Station III respectively. The total number of individuals $\cdot l^{-1}$ was 60–127 times or a mean of about 100 times greater in the littoral when compared with the limnetic region (Table 134). Therefore the littoral zone can be considered a refuge for zooplankton in the swamp.

6.2.4 *Seasonal fluctuations of zooplankton*
(T. Mizuno and R. P. Lim)

Two separate studies on the seasonal fluctuations of zooplankton in Tasek Bera were conducted. The first study was undertaken by Malaysian participants at five representative stations (Stations I-V) over one year period (April 1971-March 1972). The second study was undertaken by Japanese members at Stations I and V over a year (January-December 1972). The zooplankton density was low with mean of 8.0, a minimum 0.4 at Station IV (October 1971) and maximum of 98.1 individuals $\cdot l^{-1}$ at Station II (October 1971). This popula-

Table 134 Comparison of the number of individuals of zooplankton of the limnetic and the littoral of Tasek Bera. (T. Mizuno)

	January 1973		February 1973		March 1973	
	St. I	St. III	St. I	St. III	St. I	St. III
Cladocera						
Alona	0.1	8.5	0.1	28.7	0	3.2
Chydorus	0	6.4	0	44.6	0	8.5
Macrothrix	0.1	4.2	0	25.5	0	2.1
Copepoda						
nauplius	0.3	2.1	0.1	6.4	0	3.2
adult	0	0	0	6.4	0	0
Rotifera						
Monostyla	0.1	0	0	0	0.1	0
Euchlanis	0	2.1	0	0	0	0
Rotaria	0	0	0	3.2	0	0
Colurella	0.1	2.1	0	3.2	0	3.2
Diploneis	0	0	0	0	0.1	0
Keratella	0	0.1	0	0	0	3.2
Protozoa						
Arcella	0.1	4.2	0	3.2	0	8.5
Euglypha	0.3	17.0	0.1	26.7	0.2	28.7
Centropyxis	0	6.4	0.5	3.2	0	6.4
Difflugia	0.1	19.1	0.8	15.9	0.4	35.0
Total	1.2	72.2	1.6	169.0	0.8	102.0
Ratio	1 :	60.1	1 :	105.6	1 :	127.5

tion was better developed at the sub-littoral *Pandanus-Utricularia* (mean 12.9 individuals·l^{-1}, Station II) and *Utricularia* (mean 12.0 individuals·l^{-1}, Station III) than either the limnetic or the littoral (mean 4.0-4.6 individuals·l^{-1}, Stations I, IV and V), probably because of the greater ecological complexity of these habitats (Table 135).

Although the zooplankton density fluctuated somewhat seasonally, these fluctuations were not consistent between stations (Table 135; Fig. 108). The modified limnetic showed slight fluctuations; the *Pandanus-Utricularia* sub-littoral (Station II) showed marked fluctuations with the peak in April-May and September-October corresponding to the monsoonal season, nutrient influx from the watershed, and to that period after *Utricularia* growth in the dry

Table 135 Mean, minimum and maximum density of zooplankton at Stations I-V in Tasek Bera (April 1971–March 1972). (R. P. Lim and J. I. Furtado)

Stations	Density (No. of individuals · litre^{-1})		
	Mean	Minimum	Maximum
Limnetic, man-modified			
Station I	4.0	1.6	10.5
(Main channel)		(Sept. 1971)	(April 1971)
Sublittoral, man-modified			
Station II	12.9	1.6	98.1
(Pandanus-Utricularia)		(Jan. 1972)	(Oct. 1971)
Station III	12.0	1.2	87.1
(Utricularia)		(Sept. 1971)	(April 1971)
Littoral, man-modified			
Station IV	5.1	0.4	14.7
(Lepironia)		(Oct. 1971)	(April 1971)
Limnetic, swamp-forest			
Station V	6.2	0.8	16.7
(Main channel)		(Sept. 1971)	(May 1971)
Overall	8.0	0.4	98.1
		(Oct. 1971)	(Oct. 1971)

season (see earlier). The *Utricularia* sublittoral (Station III) also showed monsoonal abundance of zooplankton with the April abundance being dominant and corresponding to peak *Utricularia* development (see earlier), and the littoral *Lepironia* (Station IV) and forested limnetic (Station V) showed weak fluctuations with peak zooplankton abundance in May.

In the second study the findings were similar to the preceding study. Zooplankton were very poorly developed at both Stations I (limnetic, main channel) and V (limnetic, swamp-forest). There were 0.1–0.6 individuals · l^{-1} of Cladocera, Copepoda and Rotifera. At both stations peak populations occurred in July and August 1972 (Figs. 114c and i). Protozoans such as *Arcella, Euglypha, Centropyxis* and *Difflugia* which are littoral in habit, were more abundant during the dry period of July and August (Figs. 109b and g).

True zooplankters, especially the Cladocera and naupli of the Copepoda had peak populations during July and August at Station I (Figs. 109h and i). This consistent increase in population of zooplankton in the limnetic region could be due to their migration from the vegetated littoral as it gradually dried up. Also the decrease in water velocity to 0.036 m · sec^{-1} may reduce the flushing rate of zooplankton in relation to its immigration rate.

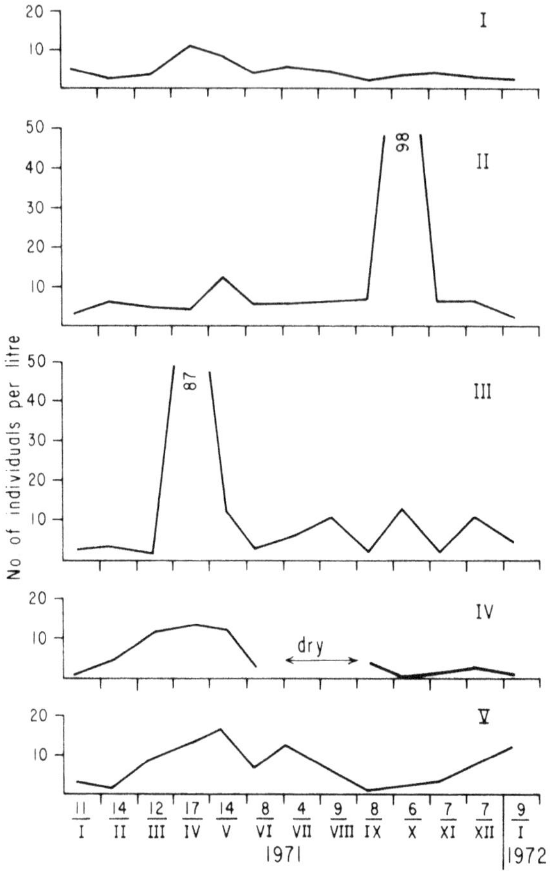

Fig. 108 Seasonal change in population density of zooplankton. (R. P. Lim and J. I. Furtado)

The population density of zooplankton was greater at Station I when compared with Station V. This was due to the greater flushing capacity at Station V which had an average velocity of 0.286 $m \cdot sec^{-1}$ at 0.25 m depth when compared with 0.038 $m \cdot sec^{-1}$ at for the same depth at Station I.

6.3 Benthos and attached animals (T. Mizuno, K. Gose, R. P. Lim and J. I. Furtado)

6.3.1 *History of study* (T. Mizuno)

Identification of benthic organisms has been difficult, but the following publications have been helpful: Tipulidae by Alexander (1931), Chirono-

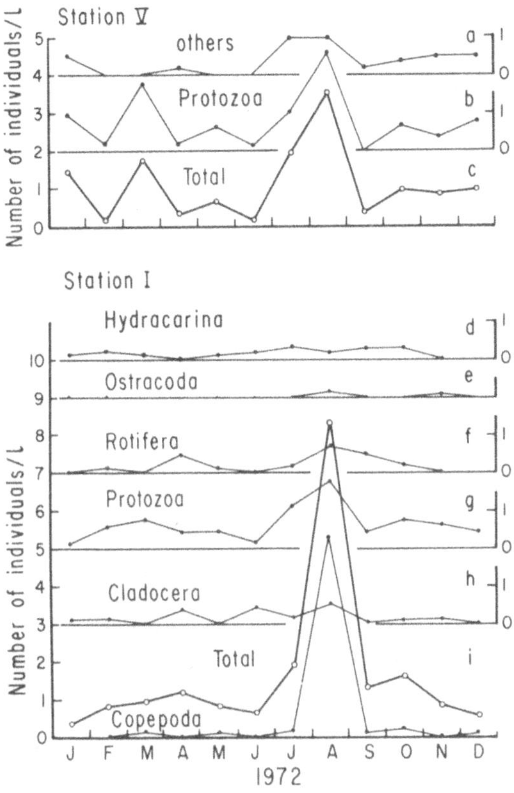

Fig. 109 Seasonal fluctuation of zooplankton at Stations I and V (main channel, limnetic) from January to December 1972. (T. Mizuno)

midae by Thienemann (1944) from the Deutschen Limnologische Sunda-Expedition, Micronesian Chironomidae by Tokunaga and Stone (1964). Other useful monographs for the identification of the Chironomidae are by Johansen (1937a, 1937b), Lenz (1941) and Coe, Freeman and Mattingly (1950), although most of the species described by them are not from Southeast Asia. The present status of Chironomid taxonomy is poor in Malaysia. The most substantial study was carried out by Karunakaran (1969) on the bionomics and taxonomy of chironomids of Singapore.

For the identification of aquatic insects several publications were used, e.g. Ephemeroptera by Vejabhongse (1937), Ueno (1961); Plecoptera by Kawai (1961, 1968), Furuya (1971a) for the Lepidoptera of Thailand. *Nymphula* sp. MNA collected by T. Mizuno from Tasek Bera in 1968 was described by Furuya (1971b). Decapoda and Mollusca will be discussed in Section 6.4.

Table 136 Lists of the aquatic insects and the benthic animals in Tasek Bera.

Diptera (collected by K. Gose and T. Mizuno and adults identified by T. Mizuno)

Ceratopogonidae
- *Antichopogon* sp.

Chironomidae
- *Chironomus* sp. Bera-1
- *Chironomus* sp. Bera-2
- *Chironomus* sp. Bera-3
- *Chironomus* sp. Bera-4
- *C. (Prochironomus) bifascipennis* Tokugawa
- *Chironomus* sp. Bera-5
- *Chironomus* sp. Bera-6
- *Chironomus* sp. Bera-7
- *Polypedium nigribasalis* Tokugawa
- *Tanytarsus* sp.

Tanypodinae
- *Procladius* sp.
- *Pentaneura* sp.
- *Pentaneura monilis* Linn.
- *Clinotanypus* sp. 1
- *Clinotanypus* sp. 2
- *Tanypus punctipennis* Meigen
- *Tanypus* sp.

Culicidae
- *Culex* sp.
- *Chaoborus* sp.
- *Anopheles* sp.

Trichoptera (collected by T. Mizuno and adults identified by G. B. Wiggins)

Hydropsychidae
- Diplectroninae – *Diplectrona* sp.

Polycentropodidae
- Dipseudopsinae – *Dipseudopsis* sp.
- Polycentropodidae – *Polycentropus* sp.

Psychomyiidae
- Econominae – *Economus* spp.

Leptoceridae
- Leptocerinae – *Oecetodella* sp.
- *Oecetinella?* sp.

Hydroptilidae – unidentified

Ephemeroptera (collected and adults identified by K. Gose)

Caenidae, Bera-1 and Bera-2
- *Cleon* sp.
- *Thraulus* sp.

Lepidoptera (collected by T. Mizuno and larvae identified by Y. Furuya)
- *Cataclysta* sp.
- *Chilo* sp.
- *Nymphula* sp. NNA

Odonata (collected and adults identified by J. I. Furtado)

Coenagrionidae
- *Aciagrion hisopa hisopa* (Selys)
- *Agriocnemis femina* (Brauer)
- *Agriocnemis rubescens rubella* Selys
- *Ceriagrion cerinorubellum* (Brauer)
- Ischnura senegalensis Rambur
- *Onychargia atrocyana* Selys
- *Pseudagrion australasiae* Selys
- *P. coomansi* Lieftinck
- *P. microcephalus* (Rambur)

Libellulidae
- *Acisoma panorpoides* Rambur
- *Brachydiplax c. chalybea* Brauer
- *Brachygonis fluviatilis* Lieftinck
- *Chalybeothemis fluviatilis* Lieftinck
- *Diplacodes nebulosa* (Fabricius)
- *D. trivialis* (Rambur)
- *Hydrobasilens crocus* (Brauer)
- *Lathrecista a. asiatica* (Fabricius)
- *Nannophya pygmaea* Rambur
- *Neurothemis fluctuans* (Fabricius)

Table 136 (Continued)

Gomphidae	Libellulidae (continued)
Ictinogomphus acutus (Laidlaw)	*Orchithemis pulcherria* Brauer
I. decoratus melaenops (Selys)	*Orthetrum chrysis* (Selys)
	O. luzonicum (Brauer)
	Rhyothemis obsolescens Kirby
	R. P. phyllis (Sulser)
	R. triangulare Kirby
	Tholymis tillarga (Fabricius)
	Tramea transmarina euryale Selys
	Trithemis aurora (Burmeister)
	T. festiva (Rambur)
	Urothemis abbotti Laidlaw
	U. signata insignata (Selys)

Hemiptera-Heteroptera (collected by L. Cheng, C. H. Fernando and J. I. Furtado and identified by L. Cheng and C. H. Fernando)

Veliidae	Gerridae
Microvelia diluta Distant	*Gerris adelaidis* Dohrn
M. douglasi	*Metrocoris termicornis* Esaki
M. albolineata	*Limnometra insularia* Hung. and Mat.
Mesovelia orientalis	*L. ciliata* Mayr
Naucoridae	*Limnogonus parvulus* (Stal.)
Naucoris acutellaris Stal.	*L. fossarum* (Fabr.)
Hydrometridae	Corixidae
Hydrometra maidli	*Micronecta insa* (Distant)
H. lineata Each.	*M. quadristingata* Bredd.
H. orientalis	*M. ludibunda malayana*
Notonectidae	*M. scutellaris* Stal.
Anisops sp.	
Nepidae	
Laccotrophes robustus	
Rantra sp.	
Cercotmetus sp.	

Coleoptera (collected by T. Mizuno)

Dytiscid nymph

Other attached animals and benthos (collected and identified by D. S. Johnson and T. Mizuno)

Annelida	Decapoda
Oligochaeta	*Caridina thambipillaii* (D. S. Johnson)
Tubificidae	*Macrobrachium trompi* (De Man)
Limnodrilus sp.	
Hirudinae	
Rhynchobdellida	

6.3.2 *Classification of aquatic insects*
(T. Mizuno)

In Tasek Bera at least eight species of *Chironomus*, two species of *Clinotanypus, Pentaneus* and *Tanypus* respectively, and one species each of *Tanytarsus, Atrichopogon, Polypedium, Procladius,* and three species of Culicidae have been recognized. They have not been fully identified.

Seven specimens of Trichoptera collected by T. Mizuno have been classified into six genera by Wiggins, Canada, but have not been identified to species. *Cleon* sp., *Thraulus* sp. and two unidentified species of Ephemeroptera, and three species of Lepidoptera have also been collected from the swamp. Larvae and adults of Odonata have been collected and identified by J. I. Furtado over several years and are listed in Table 136. Most of them are found among vegetation or in the substrate that has vegetational associations. The Hemiptera - Heteroptera were identified by L. Cheng and C. H. Fernando.

6.3.3 *Composition and comparison of aquatic organisms*
(R. P. Lim, J. I. Furtado, T. Mizuno and K. Gose)

6.3.3.1 *Habitats and methods.* This study was aimed at determining the community structure and seasonal changes of the animal community inhabiting the *Utricularia* mats and *Cryptocoryne* stands in the swamp. Eight quantitative samples were taken from *Utricularia* mats between August 1971 and April 1972 by R. P. Lim and J. I. Furtado, using a 25 × 30 cm tray with a fine-mesh bottom. The tray was placed beneath the mat and a portion of the vegetation was lifted. Excess vegetation was trimmed of the edges of the tray. Macro-organisms were flushed out of the vegetation and hand picked to be phyletically sorted, counted and weighed after drying in an oven at 60°C for 18 hours. The remaining debris with micro-organisms were filtered through mesh size 10 bolting cloth and preserved in formalin for further processing.

A similar study at Stations III and XII was conducted by T. Mizuno and K. Gose between April 1971 and March 1973 with a break between July 1971 and June 1972. The method of sampling involved enclosing a portion of the *Utricularia* with a large polyethylene bag and cutting the plants immediately below the bag before removal. The results of both are discussed separately because in the first study the units were expressed as the biomass and number of individuals per unit space i.e. as per m^2. However, the seasonal changes in biomass of the *Utricularia* were also measured. In the second study the population changes of the fauna was expressed as the number of individuals per unit weight of *Utricularia*.

6.3.3.2 *Composition of microfauna in Utricularia mats.* The fluctuations of the population densities for all the major taxa of the microfauna closely corresponded to that of the bladderwort, the microfauna being most abundant during the dry periods. With the exception of the ephemeropterans, there were greater increases in the populations during the February-April dry period than during the August dry period. This could be attributed to a more abundant source of food during the senescent condition of the bladderwort in the February-April dry period. The mean standing crop of each major taxa is given in Table 137.

Dipterans had the largest standing crop, 91,145 individuals $\cdot$ m^{-2} (43.2% of the total standing crop), with chironomids predominating (Fig. 110). They were followed by cladocerans, ostracods and ephemeropterans which constituted 25.3%, 11.2% and 10.4%, respectively, of the total standing crop. The less important groups - the nematodes, oligochaetes, copepods and early-instar odonate larvae - collectively comprised 8.8% of the total standing crop. Some groups showed seasonal fluctuations with peak abundance during the dry seasons, while others exhibited erratic fluctuations due to the small numbers of organisms collected each time. The *Utricularia* mats probably provided a marginal habitat for most of the minor taxa of the microfauna.

Table 137 Standing crop the microfauna in *Utricularia* mats. (R. P. Lim and J. I. Furtado)

	No. of individuals $\cdot$ m^{-2}		No. of individuals $\cdot$ m^{-2}
	Major taxa		Minor taxa
Nematoda	2,228	Hydracarina	375
Oligochaeta	5,020	Hemiptera	541
Cladocera	53,300	Trichoptera	389
Copepoda	10,764	Coleoptera	6
Ostracoda	23,594	Lepidoptera	24
Ephemeroptera	21,957		
Odonata	661		
Diptera			
Chironomidae	82,418		
Culicidae	236		
Ceratopogonidae	8,491		
		Total	210,913

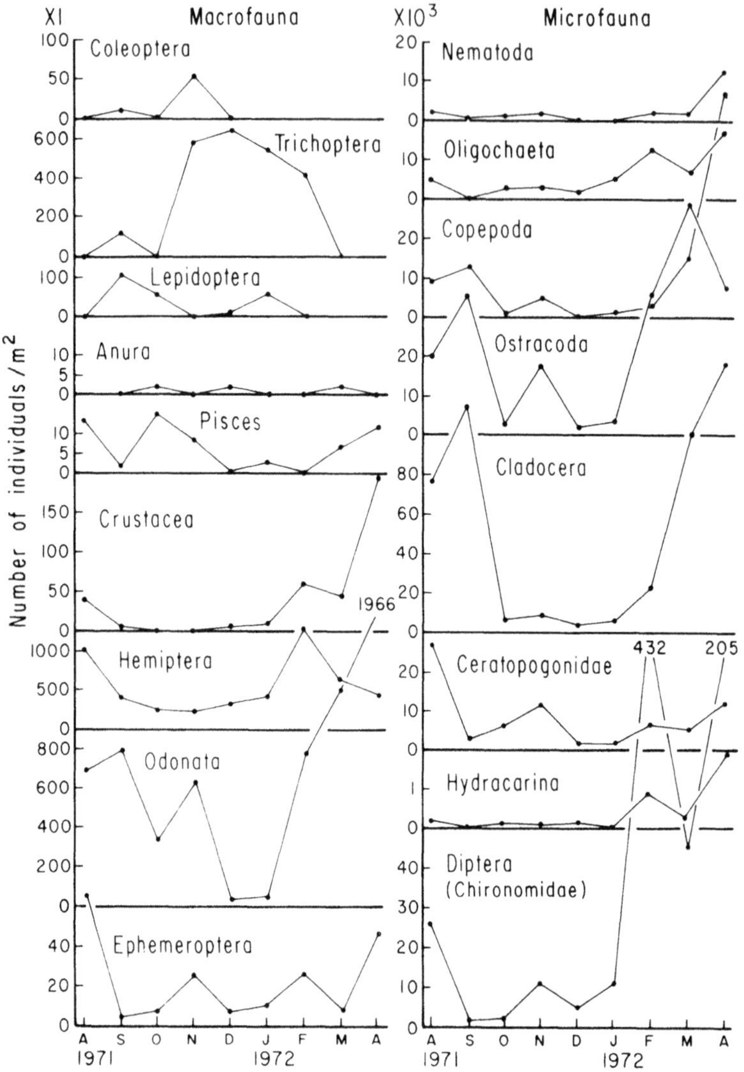

Fig. 110 Seasonal fluctuation of macro- and micro-fauna in *Utricularia* association at Stations II and VI. (R. P. Lim and J. I. Furtado)

6.3.3.3 *Composition of macrofauna in Utricularia mats.* The macrofauna was dominated by four main groups, the Crustacea, Pisces, Odonata and Hemiptera, in order of abundance. The decapod crustaceans, *Macobrachium trompi* (De Man) and *Caridina thambipillai* (D. S. Johnson), comprised the major portion of the macrofauna with a mean standing crop of 2.0 g dry wt $\cdot$ m^{-2} and 40.9 individuals $\cdot$ m^{-2}. *M. trompi* was the more common of the two species. The shrimps were seasonal, disappearing almost completely during the monsoon seasons and reappearing in large numbers during the dry periods, especially in April 1972 when the *Utricularia* mat was in a senescent state (Fig. 115). This was probably due to the abundance of food at this time as they are considered scavangers (Pennak 1953). Immigration to the *Utricularia* mats from other regions of the swamp probably caused the rapid population increase in April. This assumption was based on the sudden increase in large-sized individuals of *M. trompi* and the presence of ovigerous shrimps in April. However, a more detailed study is needed to verify that such migration occurs.

Fish formed the second most important component of the macrofauna with a mean standing crop of 0.4 g dry wt$\cdot$m^{-2} and 6.7 individuals$\cdot$m^{-2}. They constituted 17.4% and 4.0% of the biomass and numerical abundance, respectively. These values are underestimates as the fish are mobile and sensitive to the sampling technique used. Nevertheless, the fish showed a seasonal fluctuation approximating that of *Utricularia*. This was probably due to changes in food supply, shade and protection from predators.

The odonate population comprised several species with libellulids predominating. The larvae formed the third most important component of the macrofauna with a mean standing crop of 0.2 g dry wt$\cdot$m^{-2} and 48.2 individuals$\cdot$m^{-2}, which constituted 8.5% and 41.8% of the biomass and numerical abundance, respectively. Their abundance during the dry season was probably due to their congregation in the favourable *Utricularia* habitat, to their higher reproductive rates during this period and to the abundance of food organisms for the carnivorous larvae (Furtado 1966). Mature instars rarely occurred in the mats, possibly because they migrated closer to the shore and the emergent vegetation such as *Lepironia* and *Pandanus* which they use as emergence supports (Corbet 1962; Furtado 1966, 1969; McLachlan 1969).

Hemipterans common to the *Utricularia* association were *Micronecta* spp. and gerrids. They comprised 2.0% and 8.8% of the biomass and numerical abundance respectively, the mean standing crop being 0.5 g dry wt$\cdot$m^{-2} and 10.2 individuals$\cdot$m^{-2}. Large numbers of hemipterans were present between August and November but none was collected between February and April. It is possible that large numbers migrated to the *Utricularia* mats when the *Lepironia* reed stands dried up in August.

The minor taxa formed the remainder of the macrofauna with a mean standing crop of 0.04 g dry wt$\cdot$m^{-2} and 10.6 individuals$\cdot$m^{-2}, which constituted 1.3% and 8.0% of the biomass and numerical abundance, respectively. They were most abundant between the end of the dry season and the onset of the monsoon season.

The bladderwort *Utricularia flexuosa* provides an excellent habitat for a rich and diversified aquatic fauna in Tasek Bera. In this study, it was found that crustaceans, fish, odonate larvae and dipteran larvae predominated in the biota associated with *Utricularia*. Similar dominant groups (with the exception of the fish) occurred in *Ceratophyllum*, a finely-dissected plant studied by Petr (1968).

The seasonal fluctuations in the *Utricularia* mat and the population of the biota associated with it, were obviously affected by the monsoons which occurred twice during the year. Large portions of the mats were washed away at the beginning of each monsoon and only when the bladderwort re-established itself during the dry season did the periphyton and debris accumulate. They then provided the substrate for a detrivorous and herbivorous microfauna, which in turn provided the substrate for carnivorous macrofauna. This community provided a concentrated food source for larger transient grazers and predators, especially fish. Thus, the rich standing crop of the *Utricularia* community contributed significantly to the Tasek Bera swamp ecosystem.

In the second study by T. Mizuno and K. Gose two stations were used. Station XII was initially selected as a representative *Utricularia* region. However, it dried up completely during the dry season of 1971 and 1972. Thus a new sampling site, Station III, was established although this was not completely satisfactory as part of it was disturbed by canoes passing through it.

Seasonal fluctuations of the dominant taxa such as the Odonata, Ephemeroptera, Diptera and Ostracoda were similar to those of the first study, although their abundance was expressed as per unit weight of *Utricularia*. Pisces and Crustacea were more abundant during the dry season (July-September) (Fig. 111). This could possibly be due to immigration as other regions of the littoral dried up, further substantiating the conclusion drawn from the first study as regards the shrimps.

6.3.3.4 *Composition of attached animal associations in Cryptocoryne mats* (T. Mizuno and K. Gose). Organisms associated with the rooted *Cryptocoryne* plants (Stations V and 0_5) were sampled during the dry period when the plants were more accessible. The data indicate that *Cryptocoryne* is as good a habitat as the *Utricularia* mats for the attached and benthic organisms (Table 138). Although a persistent current was present these organisms were able to maintain their position among the macrophytes.

6.3.4 *Benthic animal associations*
(T. Mizuno, R. P. Lim and J. I. Furtado)

6.3.4.1 *Habitats and methods.* The bottom substrate of the study area is

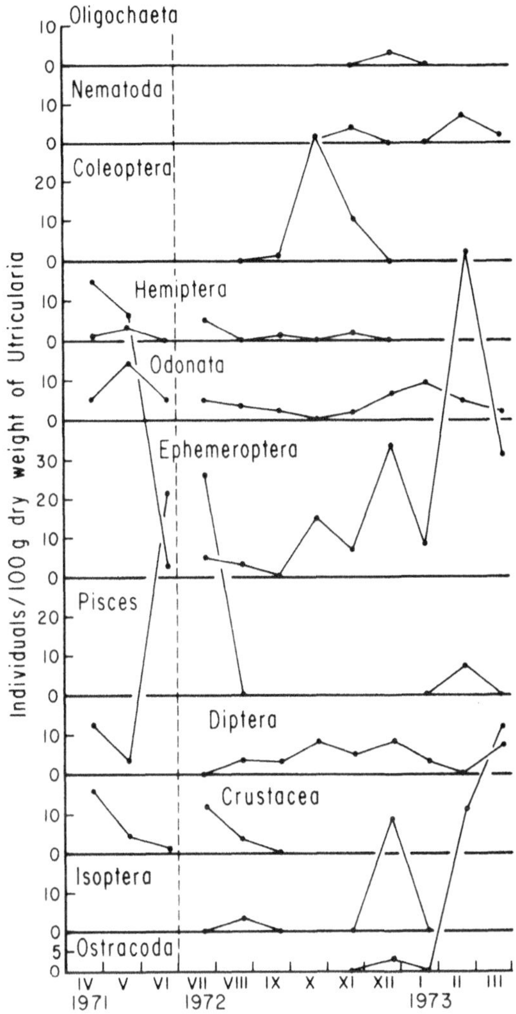

Fig. 111 Seasonal fluctuation of the attached animals among *Utricularia* mats at Stations III and XII. (T. Mizuno and K. Gose)

composed of a thick layer of detritus which is comprised wholly of dead plant material. The bottom substrate at Station I in the main channel is made up of fine detrital material because of the very slow water velocity. It is also devoid of vegetation. At Station V within the swamp-forest the bottom is covered with fallen logs, twigs and a dense mat of *Cryptocoryne* and *Blyxa*. Within the *Lepironia* littoral (Station XI) the bottom substrate is composed of decayed plant material and recently dead reeds, *Utricularia* and *Eleocharis*. Around the *Pandanus* stands are dense matrices of the rhizoids of these plants. The

Table 138 Structure of attached animals associated with the *Cryptocoryne* mats at Stations V and O_5 during the dry season of 1972 (number of individuals · m^{-2}). (T. Miuno)

	27 July		9 August		Mean
	St. O_5	St. V	St. O_5	St. V	
Oligochaeta	1.2	0	4.0	0	1.3
Insecta					
Diptera	9.2	74.0	70.0	402.0	138.8
Ephemeroptera	10.8	6.0	14.0	0	7.7
Odonata	5.2	26.0	14.0	4.0	12.3
Hemiptera	0	2.0	0	0	0.5
Crustacea					
Isopoda	1.2	0	0	2.0	0.8
Decapoda	26.8	12.0	10.0	0	12.2
Pisces	2.8	0	0	0	0.7

Fig. 112 Preliminary survey of the distribution of chironomid larvae individuals 15 × 15 cm^2 in February and March 1971. (T. Mizuno)

benthic animal community was relatively rich in abundance but poor in taxa. They were also contagious in their distribution (Fig. 112).

Sampling for benthic organisms using an Ekman grab (25 × 25 cm) was carried out between August 1971 and February 1972 at Stations I and V which represented two different substrate types (see earlier) by R. P. Lim and J. I. Furtado. Samples were also taken using an Ekman-Birge grab (15 × 15 cm) in the *Lepironia* reed littoral (Station XI) and in the *Utricularia* community (Station XII) by Gose between February and May 1971 and by T. Mizuno between June 1972 and March 1973. Four samples were taken at each station on each occasion. Each sample was washed in the field through a sieve of mesh size 0.5 mm diameter to remove the mud and fine detritus. The organisms were hand-picked and preserved in 5% formalin. Sieved sub-samples were examined through a binocular microscope in the laboratory for organisms. All data were converted to number of individuals per metre^2.

Difficulties were faced in collecting a representative sample and separating organisms from the samples. Since the bottom of the swamp is composed of detritus including pieces of wood, closure of the grab was sometimes incomplete resulting in loss of some of the samples. Often these were discarded and sampling continued until a reasonable sample was obtained. This problem was compounded at Station V where submerged rooted vegetation like *Cryptocoryne* and *Blyxa* grew. The grab could not penetrate the vegetation and had to be pushed in with a pole. Only reasonably good samples were analysed. Further, organisms obtained at Station V composed of those that lived in the mud and those that lived among the plants. In the process of sieving the samples some animals would invariably be lost. However, subsamples of the filtrate were examined in an effort to overcome this problem.

6.3.4.2 *Structure of the benthic community.* The benthic community was composed of insect larvae such as dipterans, ephemeropterans, trichopterans, odonates and young and adults of coleopterans and hemipterans. Crustaceans such as shrimps and crabs, oligochaetes and nematodes also comprised the benthic community. The dominant taxon was the Diptera, principally chironomid larvae while the subdominant taxa were the Ephemeroptera and Odonata. It is realised that problems are faced in sampling for the more mobile organisms like the hemipterans, coleopterans and ephemeropterans and that their population densities may be underestimated. The various taxa were identified by specialists.

Benthic organisms were poorly developed at Station I while they were better developed at Station V both numerically and in species diversity. This is due to the fact that at Station V the benthos has a rich mat of aquatic macrophytes which provides stability to the substrate, a greater variety of microhabitats to live in, shelter from predators, and a supply of oxygen and food. Such a relationship between the benthic community and the development of aquatic macrophytes has been previously demonstrated by McLachlan (1969). Benthic organisms were also abundant and diverse at Station XI and XII, being dominated by chironomid larvae and oligochaetes.

6.3.4.3 *Seasonal fluctuations.* The seasonal changes in population of the benthic organisms were studied over two years, although in some months no samples were taken (July 1971 and between March and May 1972). The population densities differed between stations, due to differences in substrate types, but their seasonal patterns were somewhat similar (Fig. 113). There was an increase during the wet period especially during the northeast monsoon. This increase may be due to migration of organisms to the limnetic region from the vegetated littoral (Fig. 113a). Concurrent with this increase in population was an increase in diversity of taxa. This increase in diversity is unclear and further work is needed.

Similar trends in fluctuation in population of the benthic fauna were found in Stations XI and XII in that there were increases in the population during the wet seasons (Figs. 113c and d). The stations were completely dry for three months between June and August 1972 and this had severe consequences on the benthic fauna but there was a gradual recovery over the following three months. This also affected the emergence of insects as will be discussed later.

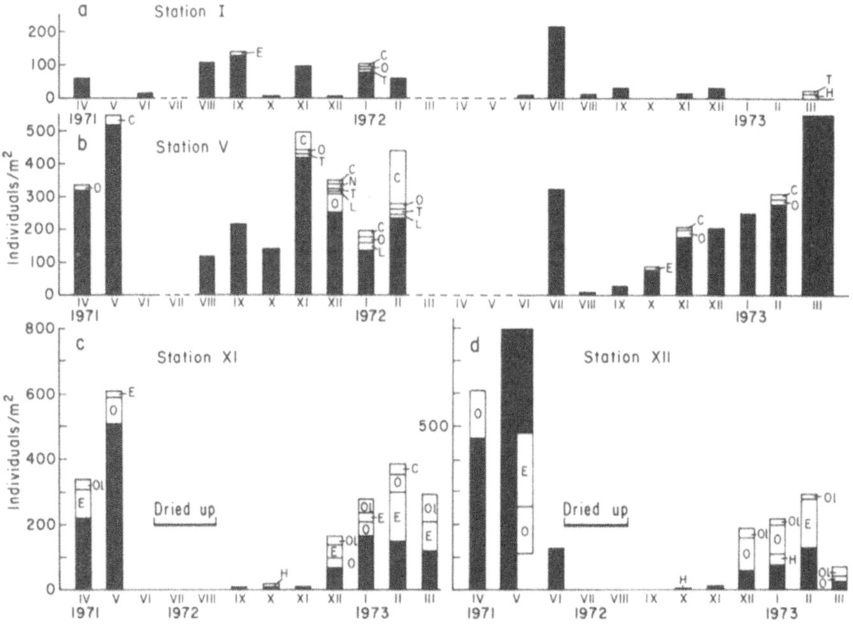

Fig. 113 Density of benthic organisms (individuals/m²) and taxa diversity at Stations I, V, XI and XII in Tasek Bera. (T. Mizuno, K. Gose, R. P. Lim and J. I. Furtado)

C: Crustacea	O: Odonata	OL: Oligochaeta
E: Ephemeroptera	L: Lepidoptera	N : Nematoda
T: Trichoptera	H: Hemiptera	■ : Chironomids, Diptera.

6.3.5 *Emergence and growth of aquatic insects* (K. Gose and T. Mizuno)

6.3.5.1 *Diurnal variation of aquatic insect emergence* (K. Gose). To augment the sampling of benthic fauna, which in general gives only a rough estimate of its true population due to its contagious distribution, emergence traps were set to assess the diurnal and seasonal changes of the benthic fauna.

The emergence traps used were rectangular mosquito nets of 1 m^2 in area. These were placed on the surface of the water at respective stations and secured with the aid of bamboo rafts. Emerged insects which were mainly chironomids, ephemeropterans and trichopterans were collected with a suction trap. Samples were preserved in 70% alcohol.

Diurnal changes over six-hour intervals were observed throughout a day. Insects began their emergence after sunset and stopped emerging abruptly before sunrise (Fig. 114).

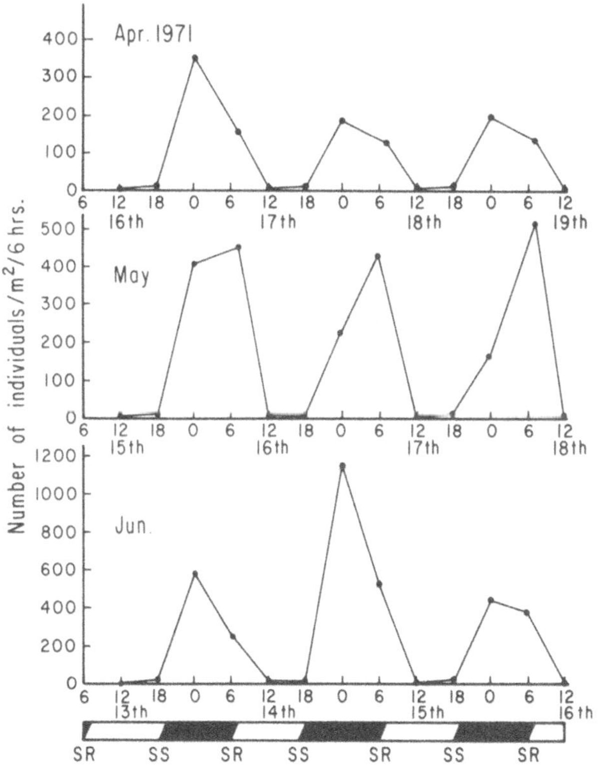

Fig. 114 Diurnal fluctuation of insect emergence at Station XII. (K. Gose)
SR: Sunrise; SS: Sunset

6.3.5.2 *Seasonal changes in aquatic insect emergence* (K. Gose and T. Mizuno). Emergence traps were set at Stations I, V, XI and XII to collect insects for three to six days every month between April and June 1971 and between June 1972 and March 1973. The specimens collected were phyletically sorted and counted. They were identified to orders and differentiated into morpho-species (Table 136).

The samples collected contained several species such as large-sized *Chironomus* (3 species), *Procladius* sp., *Pentaneura* sp. 1, *Tanytarsus* sp. 1 and *Clinotanypus* sp.1 and sp.2, but these were not abundant. Small sized chironomids (< 3 mm) were abundant; the dominant species among them were *Pentaneura monilis*, *Pentaneura* sp.2, Chironomus sp. 1 and 4, *Atrichogon* sp. and occasionally *Anopheles* sp. was collected.

Although the peaks of insect emergence varied between species and between stations, the patterns of fluctuation agreed approximately with that of the benthic fauna except at Station I. This could be due to sampling error as the numbers emerging were few.

During the dry season (August-September 1972) no emergence occurred at Stations XI and XII which were dry and this continued until October 1972. In contrast the emergence rate was high during the dry period at Station V. The seasonal changes of the chironomids at Stations I and V appeared to be constant throughout the study period and little affected by the dry season (Fig. 115). However, at Station XI their emergence was affected by the drying up of the littoral in June. At Station XII emergence continued until the end of July because of the thick mat of *Utricularia* which still provided conditions conducive for their development and emergence (Table 139).

The seasonality in emergence of the Ephemeroptera was similar to that of the Chironomidae (Fig. 115). At Station V the peak emergence rate in June and July may be due to endogenous rhythms inherent in the insects themselves. In comparison the trichopterans had peak emergence rates during the wet season between December and March (Fig. 115).

One hundred and fifty individuals of the large chironomids (> 3 mm body length) weighed 10.7 mg dry wt. and had a calorific value of 4174.77 $cal \cdot g^{-1}$. Also, three hundred individuals of the small chironomids (< 3 mm body length) weighed 8.7 mg dry wt and had a calorific value of 4060.66 $cal \cdot g^{-1}$. Table 140 shows the monthly number of individuals of the large and small chironomids and their calorific values at each station.

In terms of the amount of energy that is transferred from the aquatic ecosystem through insect emergence the estimated amounts at Stations I, V, XI and XII were 2,929.3 $cal \cdot m^{-2} \cdot y^{-1}$, 8,350.6 $cal \cdot m^{-2} \cdot y^{-1}$, 4,350.6 $cal \cdot m^{-2} \cdot y^{-1}$ and 17,688.9 $cal \cdot m^{-2} \cdot y^{-1}$ respectively. Thus a mean of 8,300 $cal \cdot m^{-2} \cdot y^{-1}$

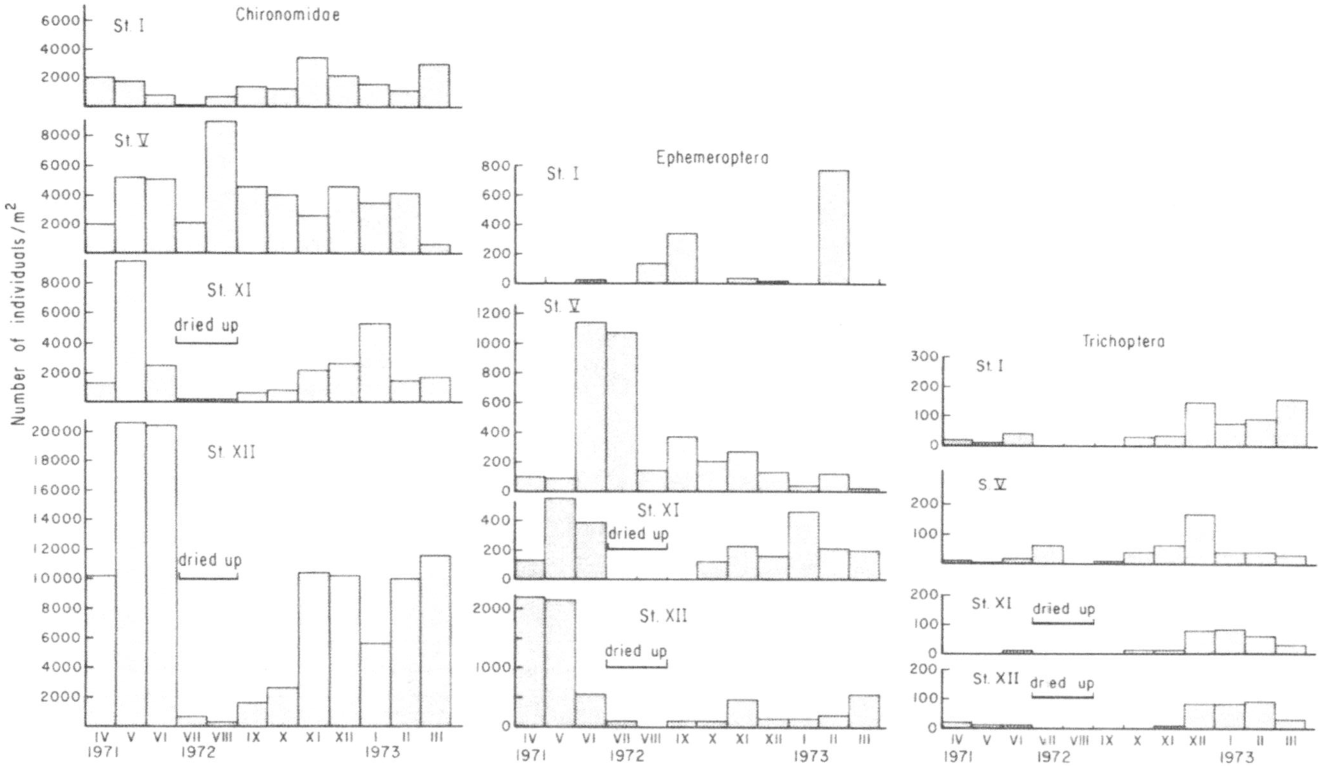

Fig. 115 Seasonal fluctuation of emerging insects in the study area. (T. Mizuno and K. Gose)

Table 139 Summary of the emergence rates of various taxa at four stations (Number of individuals · $m^{-2} \cdot day^{-1}$). (T. Mizuno)

	Station I			Station V			Station XI			Station XII		
	Mean	Max.	Min.	Mean	Max.	Min.	Mean	Max.	Min.	Mean	Max.	Min.
Chironomidae	65.4	150.5 Nov.	21.4 July	179.8	421.9 Aug.	64.3 April	98.2	317.7 May	2.7 July, Aug.	406.2	712.7 Dec.	7.9 Aug.
Ephemeroptera	3.1	25.0 Feb.	0 April, May, July, Oct., March	8.2	32.5 June	0.3 March	6.5	18.3 May	0 July	17.3	74.0 April	0 Aug.
Trichoptera	2.7	12.3 July	0 Aug., Sept.	3.4	9.7 July	0 Sept.	1.1	1.7 March	0 April, May, July, Aug., Sept.	0.5	1.3 Jan.	0 July, Aug., Sept.
Others	8.8	2.7 Oct.	0 April, May, June, July	1.1	3.3 Sept.	0 Feb.	1.1	5.7 July	0.4 June	2.6	7.0 April	0.7 Oct.
Total	71.9	153.3 Nov.	30.7 June	192.4	429.0 Aug.	66.0 April	106.2	336.7 May	3.4 Aug.	426.5	760.0 May	8.7 Aug.

Table 140 Amount of energy transferred out of the aquatic ecosystem through chironomid emergence in Tasek Bera (Number of individuals · m^{-2} · $month^{-1}$; large: > 3 mm body length; small: < 3 mm bocy length). (T. Mizuno and K. Gose)

Station	Size	1971		1971-72							1973				(g)	
		IV	V	VI	VII	VIII	IX	X	XI	XII	I	II	III	Total indiv.	Total biomass	cal · m^{-2} · y^{-1}
I	Large	20	62	90	301	4	8	22	39	84	9	8	31	678	0.05	208.7
	Small	2,060	1,798	790	363	1,290	1,269	2,130	4,476	2,930	1,528	2,428	2,046	23,108	0.67	2,720.6
																2,929.3
V	Large	20	505	140	0	214	1,005	115	285	481	207	48	208	3,228	0.23	960.2
	Small	1,910	4,867	5,410	6,098	12,865	4,557	3,679	3,960	5,968	4,287	4,060	4,982	62,643	1.82	7,390.4
																8,350.6
XI	Large	0	40	230	0	0	0	31	60	6	270	28	9	674	0.05	208.7
	Small	1,280	9,805	2,945	84	84	318	1,944	4,011	4,774	5,992	1,820	2,006	35,063	1.02	4,141.9
																4,350.6
XII	Large	440	1,652	740	0	0	8	9	39	40	71	0	71	3,070	0.22	918.4
	Small	9,760	19,632	20,015	394	245	1,719	5,465	16,746	22,050	11,670	16,059	17,059	142,359	4.13	16,770.5
																17,688.9
															Total (cal)	33,319.2
															Mean (cal)	8,329.8

through chironomids is exported from this swamp (Table 140).

6.3.5.3 *Growth rates of chironomids* (T. Mizuno). Since chironomids are the most dominant taxon in the benthos, it is pertinent to study them in greater detail. The life history of a large species, *Chironomus* sp. Bera-1 was studied under the laboratory conditions.

Culture method. Eggs were obtained by placing bowls of water below a light by the edge of the swamp. Adults attracted to the light laid gelatinous masses of eggs in the bowls. Each egg mass was about 10 mm in length and contained about 500 eggs (Fig. 116). The eggs were transferred to petri dishes where they hatched. For experimental purposes five newly hatched individuals were randomly selected, measured and placed in culture bowls simulating different field conditions. The conditions simulated are as follows:

Bowl no.	Simulated aquatic condition
1	500 ml of swamp water
2	500 ml of swamp water and a suitable volume sand-clay
3	500 ml of swamp water and attached algae scraped off a canoe.
4	500 ml of swamp water and detritus from the bottom of the open water region (Station 1)
5	500 ml of swamp water and detritus from the bottom of a channel

The body lengths of the chironomids were measured daily at noon. Temperature of the water in the culture bowls varied between 24°C and 30°C during the day (Fig. 117).

Results and discussion

Chironomid larvae cultured in Bowl 3 containing algae grew normally and took 13-14 days to complete their life cycle from eggs to emergence. The larvae in Bowl 1 stopped growing when they reached 2.0 mm in length. In Bowl 2 growth terminated at 2.5 mm body length. Most of them died within a week. No larvae were found a few days after the commencement of the experiments in Bowls 4 and 5 (Fig. 117).

In general, artificial conditions can at best be poor simulations of the natural environment due to several constraints. However, the results in growth rates and generation times obtained in Bowl 3 agreed well with field observations where it took two weeks for a generation to complete its cycle. The lack of sufficient food in Bowls 1 and 2 may have resulted in the

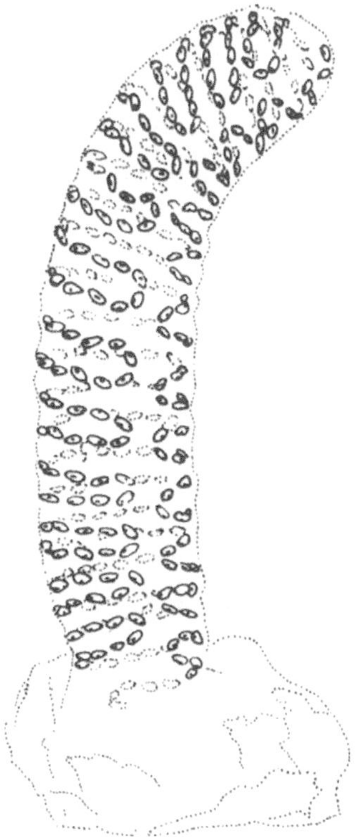

Fig. 116 Egg mass of *Chironomus* sp. Bera-1. (T. Mizuno)

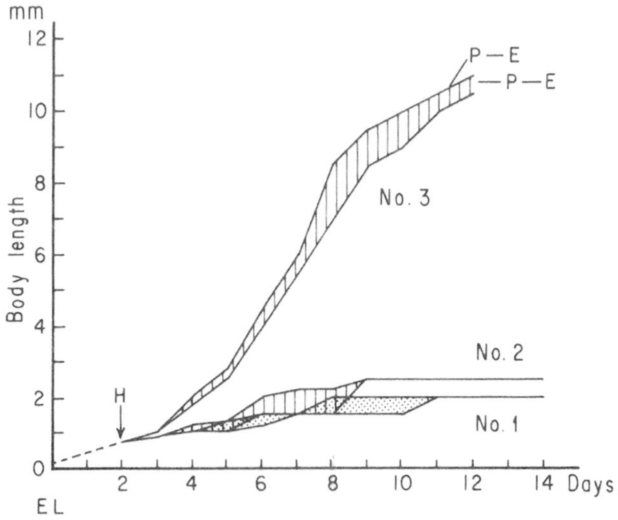

Fig. 117 Growth rates of *Chironomus* sp. Bera-1 under simulated environmental conditions. (T. Mizuno) (EL: egg laying, H: hatching, P: pupa, E: emergence)

termination of their growth. Under natural conditions chironomids are able to survive in sand-clay conditions because of replenishment of food material through wave action. The rapid deaths of chironomids in Bowls 4 and 5 were unexpected and may have been due to an extreme reduction in dissolved oxygen. Under natural conditions oxygen at the bottom is replenished through water movement.

In general, several studies have shown that the generation time for temperate species of chironomids is longer than that of tropical species. Nose (1962) showed that *Chironomus plumosus* took 44 days to reach instar IV and Shirota (1968) reported that it took 23–25 days for a generation to complete its life cycle under optimum conditions (25°–27°C). *Chironomus dorsaris* required 30–33 days in June, 28–30 days in July, 24–28 days in August, 28–30 days in September and 32–48 days in October to complete its life cycle. From the foregoing it is seen that chironomids in temperate regions take about 24 days to complete their life cycles under optimum conditions. In contrast, *Chironomus* sp. Bera-1 in Tasek Bera has about half generation time (Fig. 118).

It has also been observed that chironomids in Tasek Bera laid eggs continuously over the period of study from September 1972 to March 1973. Thus it can be inferred that chironomid production in a tropical swamp must theoretically be greater than that of temperate zone lakes.

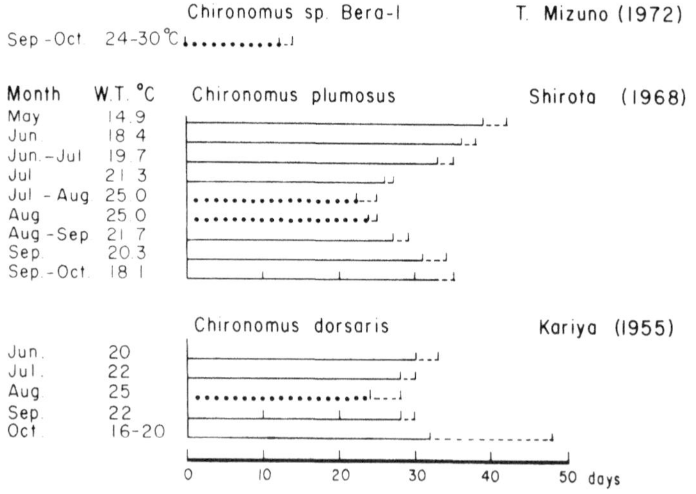

Fig. 118 Comparison of the length of a generation of *Chironomus* in Malaysia and Japan.

6.4 Shrimps, crabs and molluscs
(T. Mizuno)

6.4.1 *History of study*

The taxonomy of shrimps and crabs of Southeast Asia has been studied by several biologists since the end of 19th century. Noted publications are those of J. G. De Man, S. Kemp, and J. Roux.

The freshwater prawns and shrimps of Peninsular Malaysia and Singapore were extensively studied by D. S. Johnson (1958, 1961, 1963, 1966, 1969). In his first paper Johnson (1958) discussed the subspecific and infraspecific variations in some *Macrobrachium* species. His second paper (1961) dealt with Caridinae and Stenopodidae of Singapore with notes on their distribution and a key to the genus *Caridina* occurring in Malaysian waters, which comprised 11 species. The third paper (1963) pertained to the taxonomy of non-marine Caridinae in Peninsular Malaysia extracted from Lanchester's papers (1901, 1906). This list included 12 species of *Macrobrachium* and 8 species of *Caridina*. The rest of his papers discuss the factors affecting the distribution of freshwater prawns and shrimps in Malaysia.

Crabs were also studied by some specialists, such as Roux (1930, 1934, 1936, Smedly (1931), Balss (1937), Tweedie (1938) and more recently by Fernando and Lim (1964). These publications have been used in the identification of the crabs collected in Tasek Bera.

Only two species of shrimps occur in Tasek Bera. These are *Macrobrachium trompi* (De Man) and *Caridina thambipillai* (D. S. Johnson). One species of freshwater crab, *Potamon johorense*, has been collected.

In general, molluscs are normally associated with the benthos and aquatic macrophytes. Molluscs in Southeast Asian inland waters were reported by Van Jutling (1953), Habe (1964), Wenz and Hass (1968), Brandt (1968) and others. Recently T. Mizuno and S. Mori (1970) listed several freshwater molluscs collected from Cambodia, Thailand and Malaysia. However, none was found by them in Tasek Bera. This has been attributed to the soft water of the swamp with calcium ion deficiency.

6.4.2 *Distribution and standing crop of shrimps*

Two dominant species of shrimps occur in Tasek Bera, viz. *Macrobrachium trompi* and *Caridina thambipillai*. They mainly inhabit the *Utricularia* mats and show a diurnal pattern in vertical distribution, migrating to the lower layers of the mat during the day and moving to the upper layers at night. They seem to congregate in the stream channels during the dry season and disperse when the water level rises.

In a preliminary survey in March 1971 shrimps were quantitatively collected from vegetation at various points in the swamp using a 0.15 m^2 area hand net. Shrimps did not occur among the *Lepironia* plants except at R_2. They were present most abundantly among the *Utricularia* mats especially in the channels and Station XII (Table 141).

A more detailed study of the standing crop of shrimps was carried out in August 1974. On five occasions samples were quantitatively collected from *Utricularia* mats at 30 locations with concurrent measurements of the area of the mats. The shrimps were classed into large individuals (> 40 mm body length), intermediate individuals (20 - 39 mm body length) and small individuals (< 19 mm body length) for dry weight analysis. More shrimps were found among old and senescent *Utricularia* mats which grew in areas where there was running water. This preference for such conditions could be due to the abundance of food through accumulation of detritus and growth of periphyton, and sufficiently good supply of dissolved oxygen. Young and stable *Utricularia* communities did not support as many shrimps (Table 142). The standing crop of shrimps is estimated at 21.75 individuals$\cdot m^{-2}$ or 1.49 g dry wt$\cdot m^{-2}$ of *Utricularia*. The total biomass of shrimps in the study area is estimated at about 2 kg dry wt (Table 142, Fig. 119).

6.4.3 *Growth and breeding patterns of shrimps*

The two shrimps that occur in Tasek Bera varied in size. *Macrobrachium trompi* was more dominant and larger than *Caridina thambipillai*. The former had a body length of 7 mm at its youngest stage and 80 mm when it was mature, and the latter measured only 16 mm even when it was mature. *M. trompi* was, therefore, studied in detail. Growth rates and population structure were analysed from monthly samples, since attempts at rearing the shrimps in cages suspended in water for growth observations were not successful.

In the case of the unmutilated shrimps, mature males are usually distinguished by the difference in size of the left and right chelipedes (the external secondary sexual feature) from adult females, but it is not always possible to identify the sex of the shrimps using the chelipedes because they are often lost through autotomy. Specimens without chelipedes are sexed using another secondary sexual characteristic i.e. the endopodites of the second swimming

Table 141 Abundance (no. of individuals · m^{-2}) and biomass (g) of shrimps in Tasek Bera. (T. Mizuno)

	Main channel						Secondary channel					
	St. V		R_4		B_8		O_7		B_4		R_6	
	No.	WW	No.	WW	No.	WW	No.	WW	No.	WW	No.	WW
M. trompi L.	2	4.2	2	2.7	0	0	0	0	0	0	0	0
I.	5	1.0	5	1.0	4	1.2	7	4.7	1	0.9	12	–
S.	11	0.4	12	0.3	17	0.4	10	0.4	2	0.06	7	0.2
C. thambipillai	89	0.3	70	0.3	29	0.1	12	0.05	0	0	4	0.02
Total	107	5.9	89	4.3	50	1.7	29	5.2	3	1.3	23	0.22

	Utricularia community								*Lepironia* community			
	St. XII-1		St. XII-2		St. XII-3		St. XII-4		R_4		R_3	
	No.	WW	No	WW	No.	WW	No.	WW	No.	WW	No.	WW
M. trompi L.	1	1.3	0	0	0	0	0	0	0	0	0	0
I.	2	0.3	0	0	1	0.1	7	1.3	2	0.5	0	0
S.	10	0.3	3	0.1	19	0.5	15	0.6	1	0.01	0	0
C. thambipillai	7	0.04	10	0.06	12	0.07	12	0.08	8	0.05	0	0
Total	20	1.9	13	0.2	32	0.7	34	2.0	11	0.56	0	0

	Lepironia community													
	St. XI		B_3		BR		$B_{1.2}$		R_0		R_1		$R_{1.2}$	
	No.	WW	No.	WW	No.	WW	No.	WW	No.	WW	No.	WW	No.	WW
M. trompi L.	0	0	0	0	0	0	0	0	0	0	0	0	0	0
I.	0	0	0	0	0	0	1	0.5	0	0	0	0	0	0
S.	0	0	0	0	0	0	0	0	0	0	0	0	0	0
C. thambipillai	0	0	0	0	0	0	0	0	0	0	0	0	0	0
Total	0	0	0	0	0	0	0	0	0	0	0	0	0	0

L: large, I: intermediate, S: small, No.: number of individuals, WW: wet weight (g)

Table 142 Standing crop of freshwater shrimp *Macrobrachium trompi*, in *Utricularia* mats (18-19 August 1974). DW: dry weight, L: large, I: intermediate, S: small. (T. Mizuno)

No. of collection locality		1	2	3	4	5	6	7	8	9	10	11	12	13	14	15	16
No. of individuals	L.	3	2	1	0	0	0	0	0	0	0	4	3	9	3	10	2
	I.	5	19	6	0	1	0	0	0	1	1	12	7	20	5	11	5
	S.	7	25	3	0	0	1	0	0	3	4	38	4	70	33	42	31
g DW · m⁻² *Utricularia*		3.20	2.04	1.52	0	0.05	0.01			0.06		2.60		4.51	2.00	5.24	1.61
Total g DW in *Utr.* com.		6.40	142.1	30.40	0	3.1	4.2			2.4		78.0		180.4	32.0	172.9	104.7

No. of collection locality		17	18	19	20	21	22	23	24	25	26	27	28	29	30	31	Total
No. of individuals	L.	2	3	3	5	1	10	4	1	0	1	1	0	0	0	0	68
	I.	1	7	1	2	3	9	4	0	0	3	4	0	1	0	0	128
	S.	22	107	12	4	8	15	3	1	1	0	2	0	0	1	0	426
g DW · m⁻² *Utricularia*		0.81	2.64	0.88		3.67		2.45	1.03	0.01	0.76	0.60	0	0.04	0.02	0	35.74
Total g DW in *Utr.* com.		13.0	34.3	6.2		917.5		49.0	16.5	0.4	19.8	78.0	0	2.0	0.1	0	1951.0

Average of g DW · m^{-2} *Utricularia* = 1.49, Total g DW in *Utricularia* community.

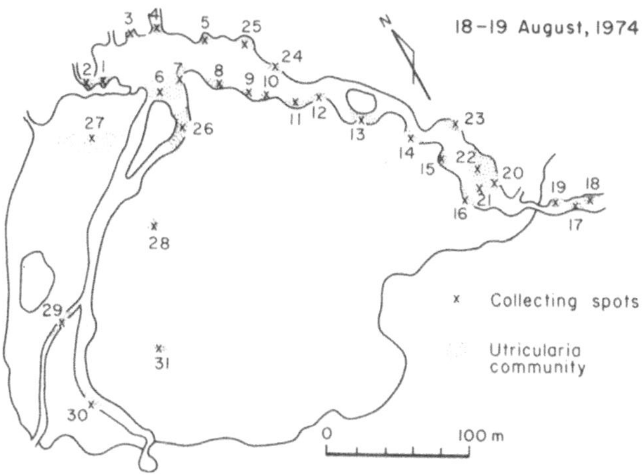

Fig. 119 Study location of standing crop of shrimps. (T. Mizuno)

appendages under the abdomen. As this sexual characteristic is not developed in the juveniles they cannot be sexed. The juveniles were thus equally divided into a 1: 1 ratio of males and females.

Figures 120 and 121 show the population structure in terms of percentages of individuals of each class size and sex collected monthly. Growth curves can be extrapolated from the information (Fig. 122).

Among the females three growth curves were constructed representing possibly three generations (Fig. 122). In the first generation all individuals were mature and reached their maximum size of 60 mm in Janauary 1973. The second generation reached a maximum length of 50 mm also in January 1973. In the third generation juveniles were first caught in November 1972, reaching maturation in March 1973. Only a few individuals of 60 mm body length were found suggesting that most of females die after the breeding season of the second year (Fig. 122).

In contrast five growth curves representing five generations among males can be recognised. But the first and second generations are not very clear, because a few individuals in the size class 75–80 mm body length were caught. Individuals within the size class of 60–70 mm body length were quite abundant in the samples. The growth curves of the third and fourth generations coincided with those of the corresponding growth curves for the females. In the case of the fifth male generation its growth rate lagged behind that of the corresponding growth curve of the females (Fig. 122). From the data it can be inferred that males survived longer than females i.e. for three to four years. If

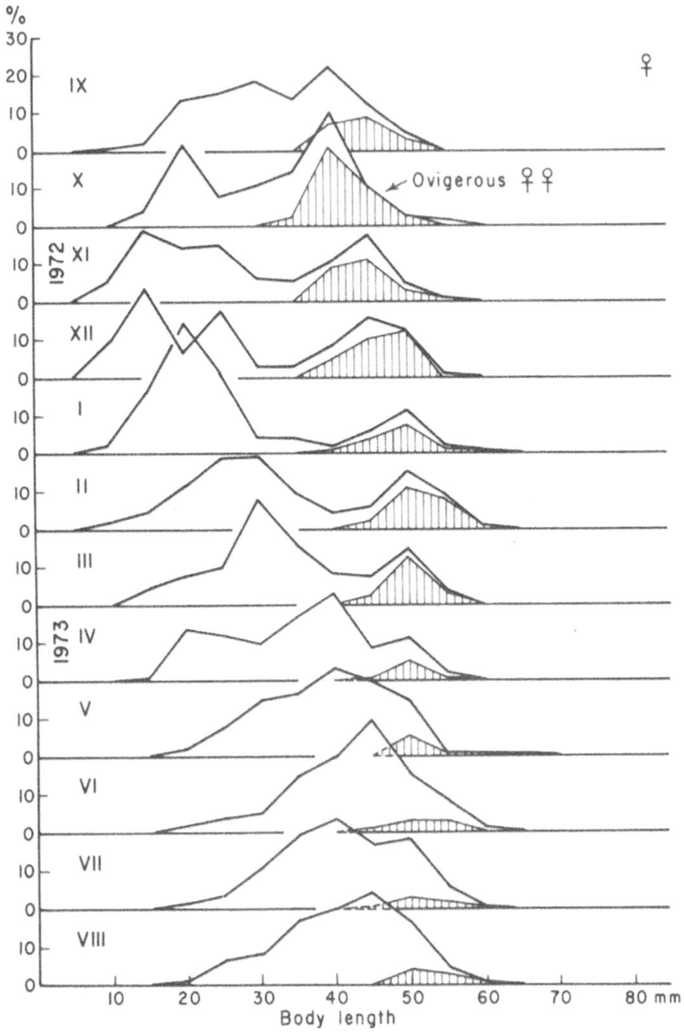

Fig. 120 Size distribution in the population of *Macrobrachium trompi* (female). (T. Mizuno)

M. trompi can be used as an example of a tropical species its period of growth extends over a longer period when compared with the temperate species where growth is restricted to the summer period.

Female shrimps are capable of becoming ovigerous when they reach a body length of 35 mm. Ovigerous females can be found throughout the year, although the ratio between ovigerous and non-ovigerous females fluctuated temporally. Generally the ratio of ovigerous females to the total number of

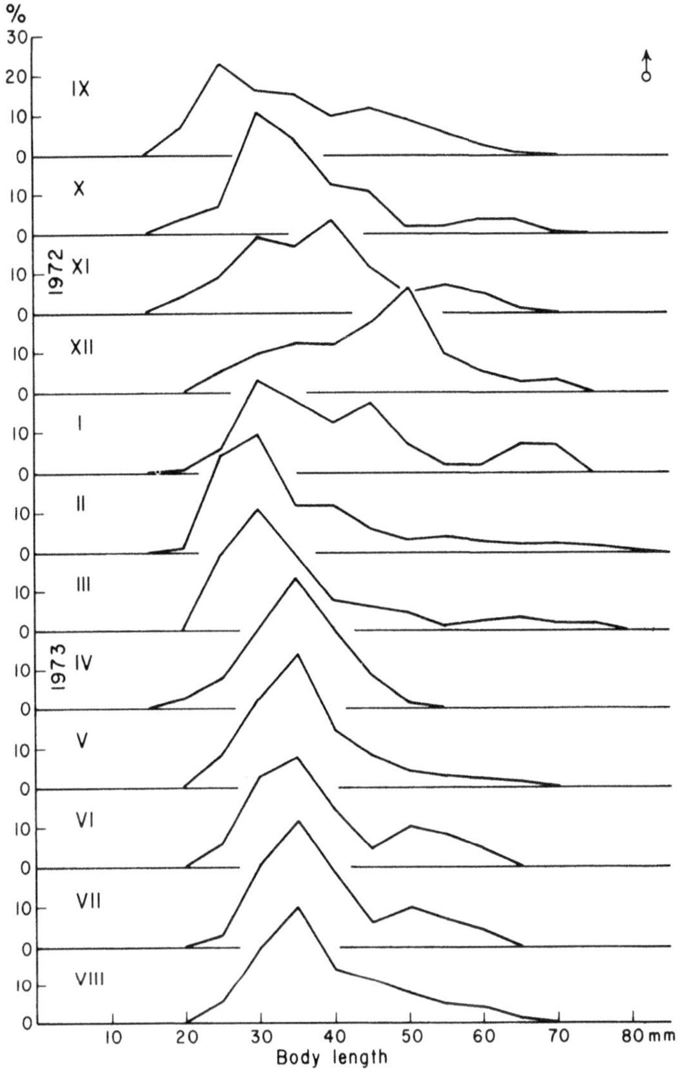

Fig. 121 Size distribution in the population of *Macrobrachium trompi* (male). (T. Mizuno)

females was more than 65% between October and March, and was less than 50% between April and September (Table 143).

In contrast to *M. trompi*, a related Japanese species, *Macrobrachium nipponensis*, which inhabits Lake Biwa, had a shorter breeding season from June to September with peak months in July and August (Fig. 123). Thus annual productivity of shrimps in Tasek Bera is very high when compared to related species in Japan.

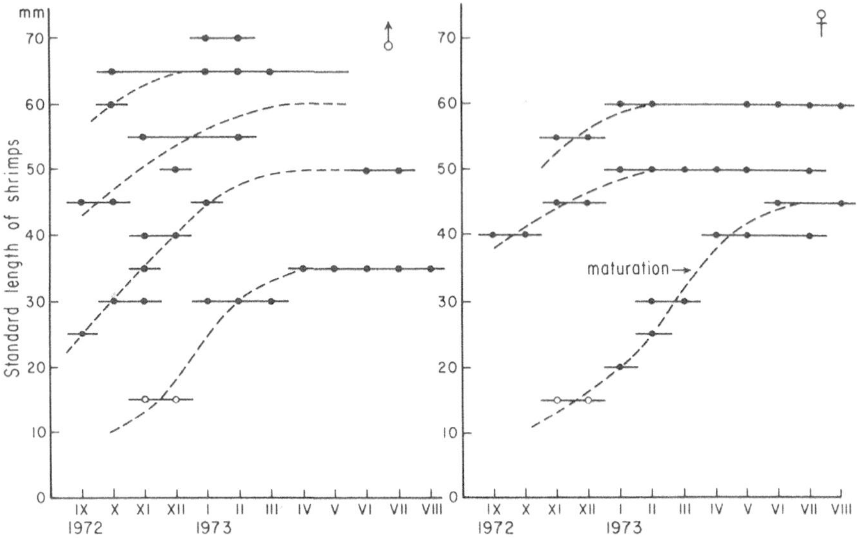

Fig. 122 Growth curves of freshwater shrimps (*Macrobrachium trompi*). (T. Mizuno)

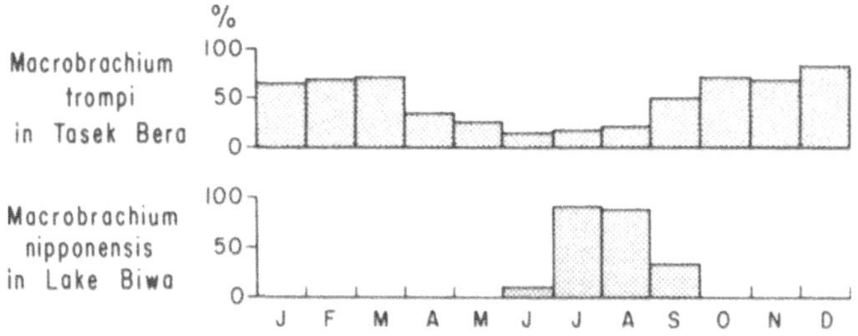

Fig. 123 Comparison of the breeding season and the percentage of ovigerous females in two related shrimp species occurring in Malaysia and Japan. (T. Mizuno)

6.5 Grasshoppers and spiders (T. Mizuno)

6.5.1 *History of study*

The *Lepironia* reeds and *Pandanus* stands harbour a variety of terrestrial insects such as Coreidae, Cicadellidae, Membracidae of the order Hemiptera and *Oxya velox* Fubricius and *Pachytylus (Locusta) migratorious* Linnaeus of the order Orthoptera. These insects have not been well studied in South-east Asia. *Oxya velox* was studied in detail because it is considered an

Table 143 Ratio of ovigerous females to total females of shrimps in each month. (T. Mizuno)

Month	1972					1973							
	VIII	IX	X	XI	XII	I	II	III	IV	V	VI	VII	VIII
Adult females	130	96	39	43	35	62	102	45	29	36	43	58	54
Ovigerous females	19	48	28	30	27	41	67	32	10	10	7	10	12
Ratio (%) of ovigerous females to total females	14.6	50.0	71.8	69.8	77.1	66.1	65.7	71.1	34.5	27.8	16.3	17.2	22.2

important secondary producer, feeding largely on the aerial parts of *Lepironia* reeds.

Documentation of spiders in Southeast Asia is quite extensive. Since 1930, Finnegan (1933), Giltay (1935), Wolf (1937) and recently Okuma (1968) and Okuma and Wongsiri (1973) have studied the spiders in Thailand and Malaysia. In Tasek Bera spiders were collected by T. Mizuno and identified by Prof. T. Yaginuma of Japan. Six species were identified belonging to five families (Table 144).

6.5.2 *Productivity of the grasshopper*

The grasshopper, *Oxya velox*, is an important secondary producer in the *Lepironia* reed swamp, grazing on culms and leaves of a variety of macrophytes like *Lepironia, Eleocharis* and *Pandanus*. In September 1972 an increase in grasshoppers in the *Lepironia* reed swamp was noted, leaving conspicuous traces of grazing on the plants. In this study attempts were made to assess the amount of food consumption, excretion and respiration by an individual per day.

In estimating the rates of food consumption and excretion two different methods were used. In the first method, a square metre of *Lepironia* stand was enclosed with a mosquito net into which several grasshoppers were

Table 144 Terrestrial insects and spider fauna.

Orthoptera (collected and identified by T. Mizuno)
Oxya velox Fubricius
Pachytylus (Locusta) migratorious Linn.
Arachnoidae-Araneina (collected by T. Mizuno and identified by T. Yaginuma)
Heteropodidae
Heteropoda venatoria (Linn.)
Tetragnathidae
Tetragnatha sp.
Salticidae
Harmochirua sp.
Thomisidae
Thomisus sp.
Argiopidae
Nephila maculata (Fabricius)
Nephilengys malabarensis (Walckenaer)

released. After six to seven days grazed portions of *Lepironia* culms were collected. A corresponding number of ungrazed culms, of approximately the same size was collected in the vicinity. These were dried and weighed. The difference between the ungrazed culms and grazed culms gave an estimate of

Table 145 Net production of an individual grasshopper *(Oxya velox)* inhabiting the *Lepironia* reed swamp. (T. Mizuno)

A. Amount of *Lepironia* stems consumed by the grasshoppers and amount of feces excreted

Individual no.	Dry weight of body (mg)	Dry weight of grazed parts (mg)	Dry weight of feces (mg)
1	375	87	110
2	376	118	86
3	252	72	57
4	263	13	15
5	118	80	56
6	113	11	7
7	120	10	15
8	75	18	15
Total	1,692	409	361
Average	211.5	51.1	45.1
$cal \cdot mg^{-1}$	4,526.7	3,818.7	3,429.4
$cal \cdot mg^{-1} \cdot day^{-1}$	–	922.6	731.3

B. Respiration rate of the grasshopper

Indiv. no.	Dry wt of body (mg)	Respiration						
		$\mu l\ O_2 \cdot K \cdot h^{-1}$		$ml\ O_2 \cdot K \cdot mg^{-1} \cdot h^{-1}$			K	$cal \cdot mg^{-1} \cdot day^{-1}$
		min.	max.	min.	max.	mean		
1	336	136.2	211.4	405.4	629.2	517.3	1.474	61.85
2	403	141.4	271.7	350.9	674.2	512.6	1.155	48.03
3	363	271.4	485.6	747.2	1337.7	1042.7	1.513	127.86
Mean								79.28

K = constant, O_2 ml = 3.38 cal

Net production rate = 922.6 – (731.3 + 79.3) = 112.0 $cal \cdot mg^{-1} \cdot day^{-1}$

Table 146 Density of *Tetragnatha* sp. inhabiting *Lepironia* community. (T. Mizuno)

Date		A-section	B-section	C-section		Comments
		10 × 10 m²	10 × 10 m²	5 × 5 m²	100 m²	
1972	Dec. 7	70	—	—	—	
	8	91	62	21	84	
	9	12	37	10	40	heavy rain the night before
		Floods covering the *Lepironia* community				
1973	Jan. 11	0	0	2	8	
	12	6	12	2	8	
	13	33	30	2	8	
	14	15	12	0	0	
	15	14	13	0	0	
	16	25	28	6	24	
	Feb. 8	14	2	2	8	
	9	29	15	3	12	
	10	17	11	2	8	
	Mar. 11	126	89	27	108	
	12	98	129	17	68	

food consumption. The calculated rate of food consumption in two experimental runs was 45.1 mg · indiv.$^{-1}$ · day^{-1} and 49.6 mg · indiv.$^{-1}$ · day^{-1}.

To assess the amount of excretion a sheaf of *Lepironia* culms together with one grasshopper was enclosed in a cloth bag for a fixed period under field and laboratory conditions. The rates of ingestion and excretion were estimated simultaneously. The grazing rate was 51.1 mg · indiv.$^{-1}$ · day^{-1} which is slightly higher than the rate obtained by the first method, and the dry weight of excretion was 45.1 mg · indiv.$^{-1}$ · day^{-1}.

The estimation of net production was based on the results obtained by the second method together with measurements of respiration using a manometer. The rates of food consumption, excretion, respiration and net production were 922.6 cal · mg^{-1} · day^{-1}, 731.3 cal · mg^{-1} · day^{-1}, 79.3 cal · mg^{-1} · day^{-1} and 112.0 cal · mg^{-1} · day^{-1} respectively (Table 145).

6.5.3 *Density of spiders*

Several species of spiders such as *Tetragnatha* sp., *Harmochirus* sp., *Thomisus* sp., *Nephila maculata* (Fabricius), *Heteropoda venatoria* (Linneaus), *Nephilengys malabarensis* (Walckenaer) inhabit the *Lepironia* reed swamp. Studies on the standing crop were conducted on a dominant species, *Tetragnata* sp., Tetragnathidae. This spider has an interesting behaviour. It spins a web between *Lepironia* culms every evening and removes it early the following morning. For this study three areas were selected and plots established. Two plots ($10 \times 10\ m^2$) had a lower density of *Lepironia* while the other plot ($5 \times 5\ m^2$) had a higher density of reeds. The number of spiders was estimated by counting the number of webs before sunrise as they were more visible (Table 146).

The spiders disappeared from *Lepironia* swamp after the floods at the end of December 1972. However, they reappeared in early January and gradually increased to a peak population density of about 100 individuals per $100\ m^2$ in mid-March (Table 146). As tertiary producers they prey upon emerging insects, while at the same time they provide a source of food for fish and birds. They, therefore, form an important link in the food-web of the swamp ecosystem. Studies on their feeding behaviour and mortality rates should be carried out in future to better understand their role in Tasek Bera.

7. Ecological notes on fishes

N. Mizuno and J. I. Furtado

Lowe-McConnell (1969, 1975) has reviewed the ecology of tropical freshwater fishes in relation to their great diversity, mechanisms of speciation and the ability of so many species to co-exist. However, most of the information discussed is from Africa and South America. Most of the studies on freshwater fishes in Asia have been taxonomic with little ecological information. Only two comprehensive ecological studies on Malaysian freshwater fishes have been published since 1962. Inger and Chin (1962) have given a comprehensive account of the ecology of riverine fishes in Sabah, East Malaysia, while Johnson (1967b) conducted an extensive survey of the distribution and community composition of freshwater fishes in Peninsular Malaysia in relation to water chemistry. However, a large area of the central region of Peninsular Malaysia, where Tasek Bera lies, was not studied. Recently Bishop (1973) in an intensive limnological survey of a small Malaysian river, the Sungai Gombak, included a chapter on the ecology of fishes. This included community structure, longitudinal succession, feeding segregation, diversity and annual productivity of the fish community. This section deals with some aspects of the ecology and behaviour of the fishes in Tasek Bera.

The studies were carried out mainly around Pos Iskandar from February 1971 to March 1973. In studying the ecology of these fishes, two methods were used. The first method involved the collection of fishes using gill nets of mixed mesh sizes, cast nets and scoop nets. Each mixed gill net of mesh sizes 1.5, 2.0, 3.0, 4.2 and 5.8 cm diameter was in reality five separate nets connected transversely. Such a net could catch a wide range of sizes of fishes without bias to any particular size class. Cast nets and scoop nets were used at various localities in the study area where reasonably good catches were expected. The second technique was visual observation and identification of fishes through snorkeling. This is a suitable method since most of the smaller Malaysian fishes are distinct and recognisable at a glance (Johnson 1967b). *In*

Furtado, J. I. and Mori, S., eds., Tasek Bera. ISBN 90 6193 100 2.

situ observations were, however, limited to species that were very distinct.

The fishes caught were identified with the aid of several taxonomic papers which include Alfred (1963, 1964b and 1966a), Brittan (1954), Fowler (1938), Herre (1940a and 1940c), Herre and Myers (1938), Hora (1941), Hora and Gupta (1941), Menon (1954), Smith (1945), Sufi (1956), Tweedie (1952 and 1956), Weber and de Beaufort (1916) and others. Doubtful species were confirmed by Mr. E. R. Alfred, National Museum of Singapore, for which the authors gratefully acknowledge.

7.1 Composition of fish communities (N. Mizuno, R. P. Lim, Y. Shiraishi, M. Nagai and J. I. Furtado)

7.1.1 *Faunal composition*

Shiraishi et al. (1972) recorded 70 species of fishes from Tasek Bera. Since then only five additional species were recorded, namely *Luciosoma trinema, Ompok leiacanthus, Prophagorus nieuhoffi, Polyacanthus hasseltii* and *Oxyeleotris marmoratus*. Prior to this study, 17 species of fishes had been recorded from Tasek Bera on various occasions (Alfred 1964a, b and c, Brittan 1954, Menon 1954, Tweedie 1952, 1956 and others). These species were *Scleropages formosus, Rasbora elegans, R. cephalotaenia, R. heteromorpha, R. dorsiocellata, Puntius tetrazona partipentazona, Xenentodon canciloides, Betta pugnax, Polyacanthus hasseltii, Tetraodon palembangensis, Labiobarbus faestiva, Tor clouremis, Helostoma temmincki, Botia hymenophysa, Kryptopterus limpok, Leiocassis micropogon* and *Rasbora myersi*. The latter six species were not recorded during the course of this study. Lim and Furtado (1974) recorded 75 species of fish obtained over a period from 1961 to 1972. Out of these 75 species, 14 species were not recorded in the present survey. These were *Leptobarbus hoevenii, Osteochilus melanopleura, Oxygaster anomalura, Puntius bulu, P. eugrammus, P. tetrazona, Rasbora leptosoma, R. sumatrana, Leiocassis leiacanthus, Mystus nigriceps, Belone canala, Betta splendens; Trichogaster pectoralis* and *Channa maruloides*.

The overall total fish fauna in Tasek Bera from the various studies consists of 95 species (Table 147). The more dominant and exotic species are shown in Plate 4.

Lowe-McConnell (1969) in comparing the ichthyofaunal composition of Thailand (representing tropical Asia), South America and Africa noted the predominance of cyprinids in tropical Asia. In the Malaysian context, no

Table 147 Checklist of fishes in Tasek Bera.
Collected by E. R. Alfred and us from 1961 to 1973; identified by E. R. Alfred, M. Nagai, Y. Shiraishi and N. Mizuno. Species names with asterisks were previously recorded by Alfred (1964), Brittan (1954), Menon (1954), Tweedie (1952) and others

Osteoglossidae
- **Scleropages formosus* (Müller and Schlegel)

Notopteridae
- *Notopterus notopterus* (Pallas)
- *N. chitala* (Hamilton)

Mastacembelidae
- *Macrognathus aculeatus* (Bloch)

Flutidae
- *Fluta alba* (Zuiew)

Cyprinidae
- *Oxygaster oxygastroides* (Bleeker)
- *O. anomalua* Van Hasselt
- *O. hypophthalmus* (Bleeker)
- *Chela* sp.
- *Luciosoma setigerum* (Cuvier and Valenciennes)
- *L. trinema* (Bleeker)
- **Rasbora dorsiocellata* Duncker
- **R. cephalotaenia* (Bleeker)
- *R. sumatrana* (Bleeker)
- **R. elegans* Volz
- *R. caudimaculata* Volz
- *R. dusonensis* (Bleeker)
- *R. trilineata* Steindachner
- **R. heteromorpha* Duncker
- *R. pauciperforata* Weber and De Beaufort
- *R. leptosoma* (Bleeker)
- *R. argyrotaenia* (Bleeker)
- *R. einthovenii* (Bleeker)
- *R. taeniata* Ahl
- *R. retrodorsalis* Smith
- **R. myersi* Brittan

Ophicephalidae
- *Channa striata* (Bloch)
- *Ch. micropeltes* (Bleeker)
- *Ch. lucius* (Cuvier)
- *Ch. maruloides* (Bleeker)

Centropomidae
- *Chanda ranga* Hamilton
- *Ch.* sp.
- *Gynochanda filamentosa* Fraser-Brunner

Nandidae
- *Nandus nebulosus* (Gray)
- *Pristolepis fasciatus* (Bleeker)

Eleotridae
- *Oxyeleotris marmoratus* (Bleeker)

Tetraodontidae
- **Tetraodon palembangensis* (Bleeker)

Bagridae
- *Mystus nemurus* (Valenciennes)
- *M. nigriceps* (Valenciennes)
- *M. cavasius* (Hamilton)
- *Leiocassis poecilopterus* (Cuvier and Valenciennes)
- *L. leiacanthus* Weber and De Beaufort
- **L. micropogon* (Bleeker)

Chacidae
- *Chaca chaca* (Hamilton)

Sisoridae
- *Glyptothorax major* (Boulenger)

Akysidae
- *Parakysis verrucosa* Herre

Belonidae
- *Belone canala* (Hamilton)
- **Xenentodon canciloides* (Bleeker)

Hemiramphidae
- *Hemirhamphodon pogonognathus* (Bleeker)

Synapturidae
- *Synaptura harmandi* (Bleeker)

Anabantidae
- *Anabas testudineus* (Bloch)
- *Osphronemus goramy* Lacepéde
- **Helostoma temminckii* (Cuvier and Valenciennes)
- *Trichopsis vittatus* (Cuvier and Valenciennes)
- **Betta pugnax* (Cantor)
- *B. splendens* Regan
- *Trichogaster leerii* (Bleeker)
- *T. pectoralis* (Regan)
- *T. trichopterus* (Pallas)
- *Sphaerichthys osphronemoides* (Canestrini)

Table 147 (Continued)

**Polyacanthus hasseltii* (Cuvier and Valenciennes)	**Labiobarbus faestiva* (Heckel)
Luciocephalidae	**Tor clouremis*
Luciocephalus pulcher (Cuvier and Valenciennes)	Cobitidae
Leptobarbus hoevenii (Bleeker)	**Botia hymenophysa* (Bleeker)
Hampala macrolepidota Van Hasselt	*Acanthophthalmus kuhli* (Cuvier and Valenciennes)
Cyclocheilichthys apogon (Valenciennes)	*Noemacheilus selangoricus* Duncker
C. heteronema (Bleeker)	*Vailentalla flavofasciata* Tweedie
C. armatus (Cuvier and Valenciennes)	Siluridae
Puntius binotatus (Valenciennes)	*Wallagonia miostoma* (Vaillant)
P. schwanenfeldii (Bleeker)	*Silurichthys hasseltii* (Bleeker)
P. fasciatus (Bleeker)	*Silurodes hypophthalmus* (Bleeker)
P. lateristriga (Valenciennes)	*Ompok leiacanthus* (Bleeker)
P. tetrazona tetrazona (Bleeker)	*Kryptopterus apogon* (Bleeker)
**P. tetrazona partipentazona* (Fowler)	*K. bicirrhis* (Cuvier and Valenciennes)
P. bulu (Bleeker)	**K. limpok* (Bleeker)
P. eugrammus (Silas)	Clariidae
Osteochilus hasseltii (Cuvier and Valenciennes)	*Clarias batrachus* (Linnaeus)
O. spilurus (Bleeker)	*C. meladerma* Bleeker
O. vittatus (Valenciennes)	*Prophagorus niewhofi* (Cuvier and Valenciennes)
O. melanopleura (Bleeker)	

* Dominant species.

comprehensive listing of the freshwater fishes is available. Johnson (1967b) is the only authoritative source. In his paper he lists 37 cyprinid species, 8 catfish species (Siluriformes) with the rest of the taxa comprising 36 species (i.e. 46, 10 and 44% of the fish fauna respectively). Likewise, of the 95 species listed in Tasek Bera, the cyprinids, catfishes and the rest of the taxa comprised 40, 19 and 36 species respectively (i.e. 42, 20 and 38% of the total fish fauna respectively). These rations are similar to those found for the fishes of Thailand (Smith 1945).

Johnson (1967b) proposed two large distribution areas of fishes in Peninsular Malaysia viz. north and south regions. He arbitrarily divided the regions along the southern border of Perak and Kelantan, and along the Sungai Trengganu. The southern crescent, occupying a peripheral portion of his 'south' region and is crescent shaped, was characterised by freshwaters which were acid and poor in nutrients. Only six species which Johnson (1967b) lists as endemic to the 'southern crescent' were found in Tasek Bera. These were *Rasbora cephalotaenia, R. einthoveni, R. pauciperforata, R.*

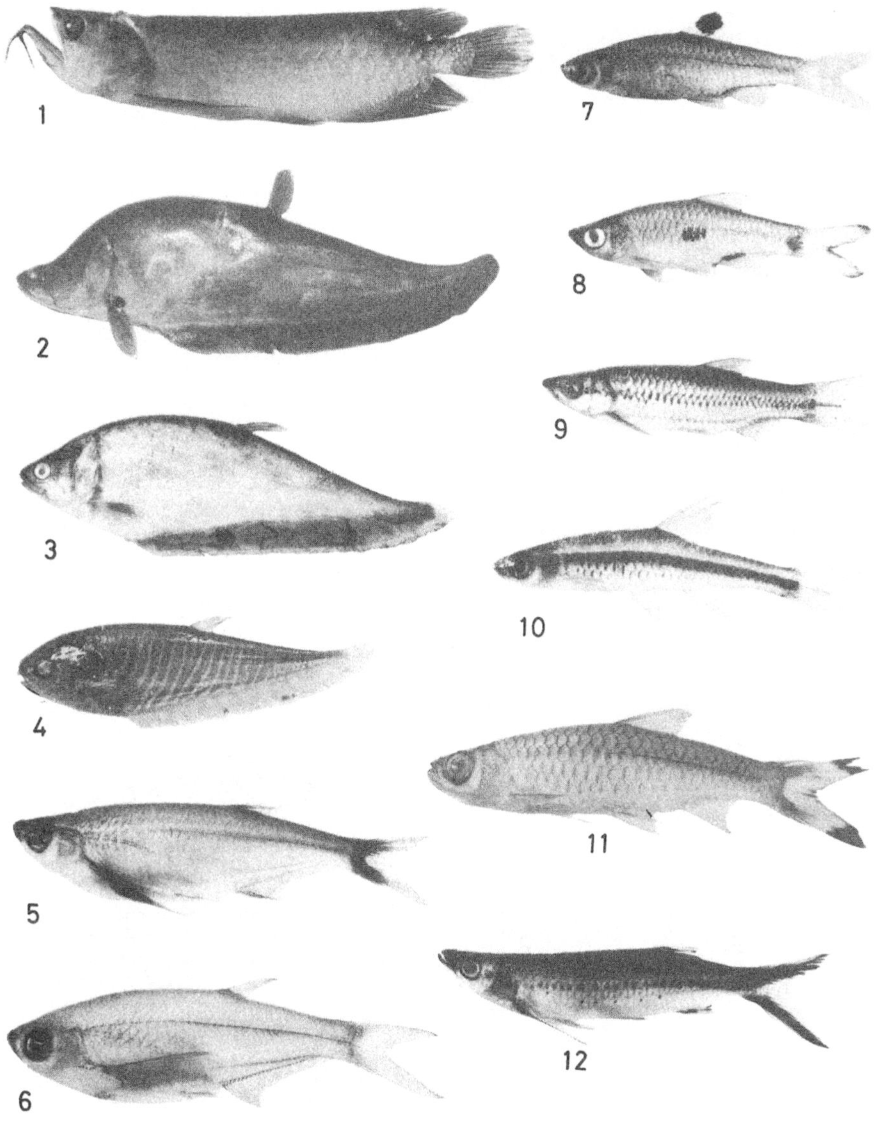

Plate 4A Dominant or exotic fishes in Tasek Bera. Sizes of specimens are standard length. Photos by M. Nagai.

1. *Scleropages formosus*, 45 cm
2. *Notopterus chitala*, 65 cm
3. *Notopterus notopterus*, 14 cm
4. *Notopterus notopterus (young)*, 4.8 cm
5. *Oxygaster oxygastroides*, 8.5 cm
6. *Oxygaster hypophthalmus*, 5.0 cm
7. *Rasbora dorsiocellata*, 3.9 cm
8. *Rasbora elegans*, 3.3 cm
9. *Rasbora cephalotaenia*, 6.2 cm
10. *Rasbora taeniata*, 2.6 cm
11. *Rasbora trilineata*, 5.7 cm
12. *Luciosoma setigerum*, 10 cm

Plate 4B Dominant or exotic fishes in Tasek Bera. Sizes of specimens are standard length. Photos by M. Nagai.

1. *Hampala macrolepidota*, 28 cm
2. *Cyclocheilichthys apogon*, 12 cm
3. *Puntius schwanenfeldii*, 7.5 cm
4. *Puntius lateristriga*, 8.5 cm
5. *Puntius tetrazona partipentazona*, 3.6 cm
6. *Puntius fasciatus*, 6.2 cm
7. *Osteochilus spilurus*, 4.3 cm
8. *Osteochilus hasselti*, 13 cm
9. *Labiobarbus faestiva*, 7.5 cm
10. *Rasbora heteromorpha*, 1.7 cm

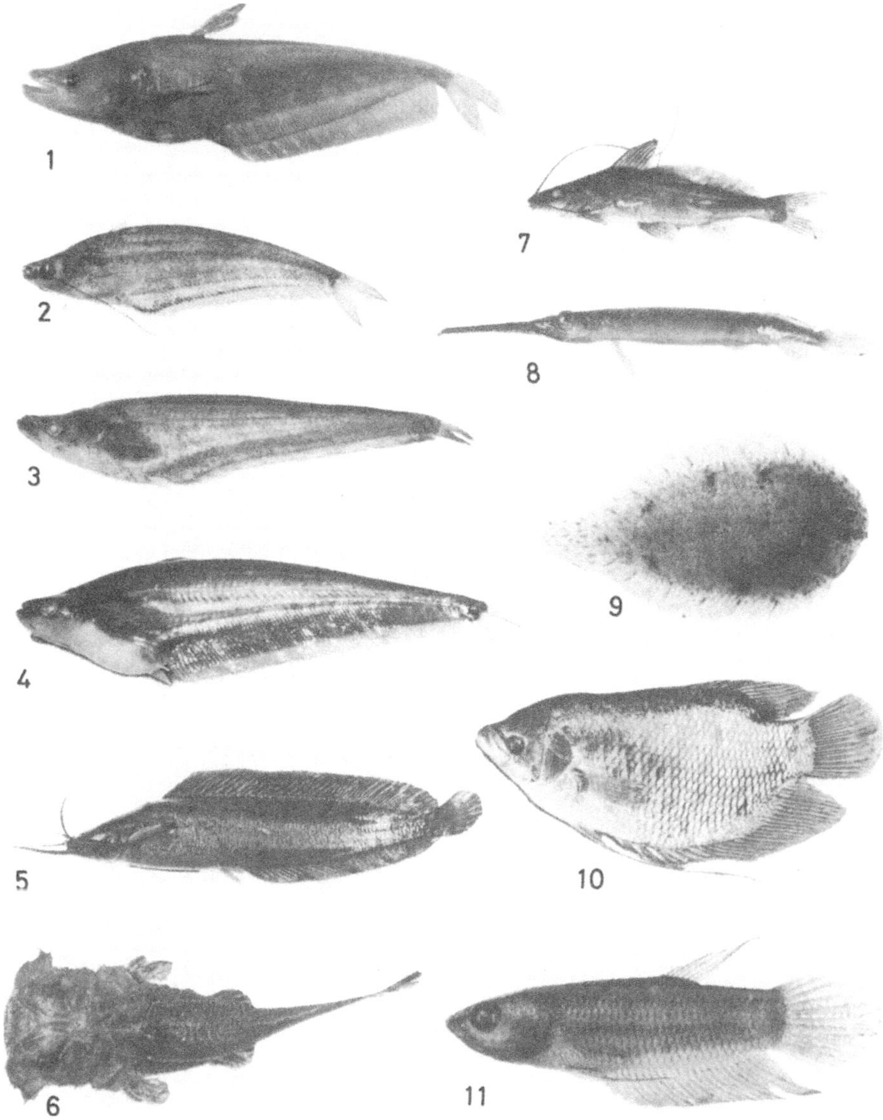

Plate 4C Dominant or exotic fishes in Tasek Bera. Sizes of specimens are standard length. Photos by M. Nagai.

1. *Wallagonia moistoma*, 15 cm
2. *Kryptopterus bicirrhis*, 6.5 cm
3. *Kryptopterus apogon*, 14 cm
4. *Silurodes hypophthalmus*, 14 cm
5. Clarias batrachus, 22 cm
6. *Chaca chaca*, 22 cm
7. *Mystus cavasius*, 12 cm
8. *Hemiramphodon pogonognathus*, 4.9 cm
9. *Synaptura harmandi*, 4.5 cm
10. *Osphronemus goramy*, 26 cm
11. *Betta pugnax*, 6.3 cm

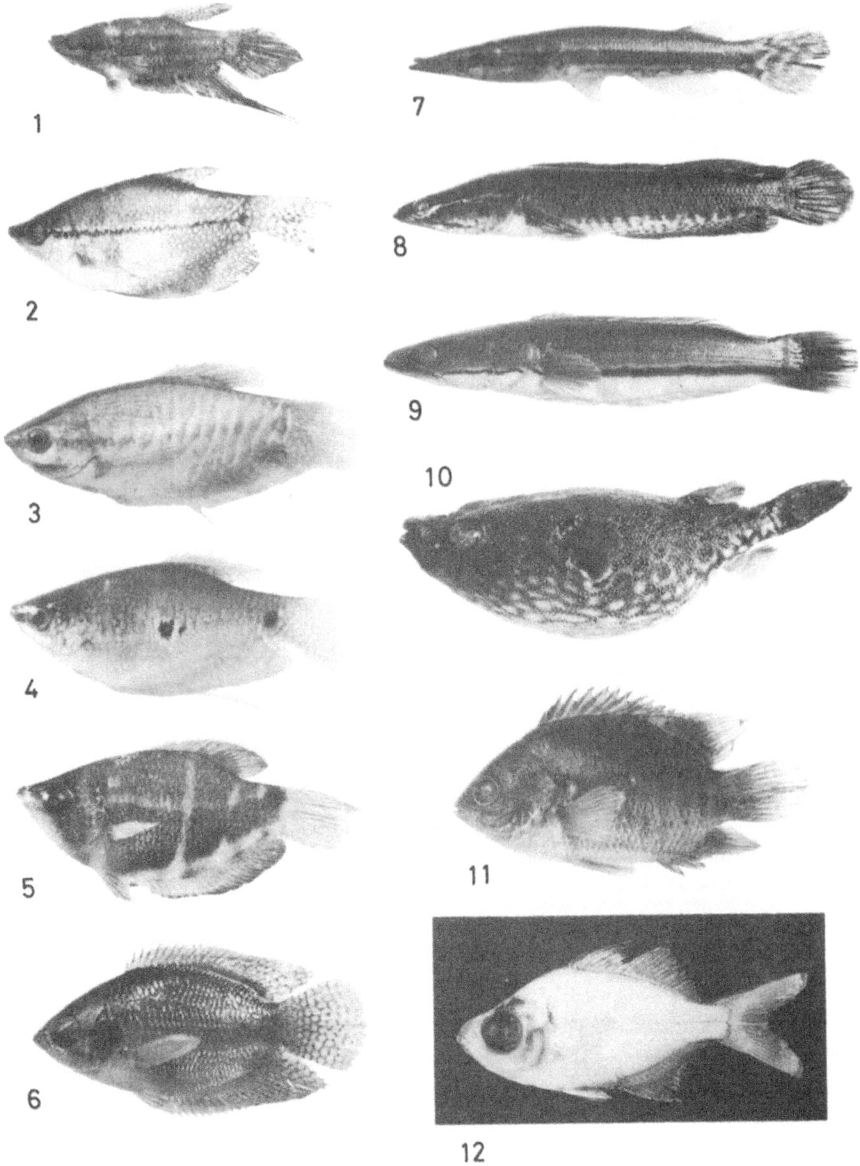

Plate 4D Dominant or exotic fishes in Tasek Bera. Sizes of specimens are standard length. Photos by M. Nagai.

1. *Trichopsis vittatus*, 5.2 cm
2. *Trichogaster leerii*, 8.5 cm
3. *Trichogaster pectoralis*, 8.0 cm
4. *Trichogaster trichopterus*, 9.0 cm
5. *Sphaerichthys osphronemoides*, 3.8 cm
6. *Polyacanthus hasselti*, 6.7 cm
7. *Luciocephalus pulcher*, 6.0 cm
8. *Channa striata*, 38 cm
9. *Channa micropeltes*, 19 cm
10. *Tetraodon palenbangensis*, 18 cm
11. *Pristolepis fasciatus*, 7.5 cm
12. *Chanda ranga*, 3.8 cm

taeniata, *Osteochilus spilurus* and *Luciocephalus pulcher*. In addition 4 species of 7 cosmopolitan species and 13 species of 17 'southern' species were recorded in Tasek Bera. In contrast none of the 14 'northern' species were recorded in Tasek Bera. The ichthyofauna of Tasek Bera clearly has more similarity to Johnson's 'south' region than to his 'north' region.

Almost all the 95 species in Tasek Bera seem to be indigenous to Peninsular Malaysia with the exception of *Trichogaster pectoralis* (Johnson 1967b). Johnson (1967b) however states that *Betta splendens* is probably a recently introduced fish although we do not agree with him.

7.1.2 *Quantitative composition of the ichthyofauna*

The difficulties involved in estimating the abundance of a particular fish species under natural conditions is well known. This difficulty is compounded in Tasek Bera because of several factors: (a) the small size of the fishes, (b) the presence of numerous refuges like the swamp-forest where they become inaccessible, and (c) the low intensity of fish catches by the natives, although they use various techniques. These conditions prevent the use of such standard techniques as the mark-recapture method for assessing the population of the fishes.

Table 148 shows the percentage composition of a total of 10,176 individuals collected from five habitats in the study area using a variety of fishing techniques. Species which had values below 0.05% were omitted from this table.

Since each fishing technique biases a catch it is not surprising that within a given habitat such as the *Utricularia* stand the community composition varies with the fishing technique used. Poisoning the fish with derris (a tuba poison) tends to be biased against the actively swimming cyprinids, which were effectively trapped in gill nets (Table 148). The use of scoop nets among the vegetation also tended to be biased against actively swimming fishes and shows a similar trend to that of derris poisoning. Since gill nets and cast nets had minimum mesh sizes that were larger than the tiny *Rasbora pauciperforata*, none of this species was recorded in these catches. However, they were caught with scoop nets. *Rasbora dorsiocellata*, which was slightly larger than *R. pauciperforata*, was caught in cast nets (Table 148). This illustrates the importance of noting the technique used when interpreting the data obtained. However, it is felt that taking the arithmetic mean of the four techniques used gives one a reasonably realistic estimate of the fish composition in *Utricularia* stands. The same techniques were used in the *Lepironia*

Table 148 Percentage composition of number of fishes collected using several fishing methods from various habitats in Tasek Bera. (Utr.: *Utricularia* stand; Lep.: *Lepironia* stand; Open: Open water area) (N. Mizuno et al.)

Fishing method	Gill net					Cast net	Scoop net	Poison (Tuba)	
Fishing period	Feb.-March			July-March	Oct.-March	Feb.-July	Feb.-July	July	
Habitat type	Utr.	Lep.	Open	Open	Canal	Utr.	Utr.	Utr.	Lep.
No. of individuals caught	1797	650	1635	1043	674	3109	820	396	52
Notopterus notopterus	2.1	7.5	0.6	0.2	0.3	0.1	0.6	45.5	17.2
Oxygaster oxygastroides	10.1	4.9	12.8	26.0	13.5	1.1			
O. hypophthalmus	2.6	1.2	4.5	4.3	1.8	5.4	0.2		
Rasbora dorsiocellata				0.1		11.8	2.0		3.7
R. pausiperforata							5.6	7.4	
R. einthoveni									6.4
Cyclocheilichthys apogon	11.7	11.7	34.6	12.5	7.3	32.3	0.4	0.5	
Puntius tetrazona	25.4	14.8	7.0	15.4	10.5	7.5	3.3		
P. fasciatus	12.8	10.2	13.4	13.2	10.8	21.3			
Labiobarbus lineatus	0.3	0.3	3.6	4.0	1.9	7.6			
Kryptopterus bicirrhis	25.4	34.9	16.5	9.7	24.8	6.9	8.7		
Clarius batrachus							0.1		5.5
Channa micropeltes						1.9	0.2	5.3	
Ch. striatus		0.3				–	1.8	3.7	5.5
Chanda ranga				2.0	0.1	–	0.1	5.3	
Pristolepis fasciatus	3.8	6.0	0.1	3.5	15.3	0.7	14.9		3.7
Nandus nebulosus	0.	3.4	1.2	0.3	1.0	0.3	13.5	1.6	5.5
Betta pugnax							10.7		1.8
Trichogaster trichopterus				0.6	0.4			1.1	11.7
Trichopsis vittatus	1	0.9				0.1	25.6	23.3	20.9
Polyacanthus hasselti		0.6		0.2				0.5	9.8
Luciocephalus pulcher	0.2			1.0	0.3	0.1	1.1	1.6	5.5
Others	4.7	3.3	5.7	7.0	12.0	2.8	11.2	4.2	2.8
Total no. of species collected	20	20	17	24	22	26	31	14	12

reed stands although the scoop net was not used there.

However, in the other two habitats, the open water and the stream channels, fishing methods were more limited. Cast nets, scoop nets and poisoning of the fishes could not be used effectively. The borders around the open water areas and along the stream channels are fringed with *Pandanus* stands. Numerous schools of *Hemiramphodon pogonognathus, Rasbora heteromorpha, R. pauciperforata* and *R. taeniata* were observed within these stands. The former two species were exclusive to the *Pandanus* stands. These two species and 22 other species listed in Table 148 were relatively abundant in our study area, and these 24 species are indicated with double crosses in Table 149.

Johnson (1967b) reported that water chemistry was important in influencing the composition of fish communities in Peninsular Malaysia. He found three distinct habitat types based on water quality. He quantified the fish communities of these habitat types by determining constancy indices based on absence-presence data. He stated that species which consistently occurred in a particular habitat were also numerically more abundant. Table 149 compares the species found in various habitats and their constancy indices given by Johnson (1967b) with that of Tasek Bera. Constancy values above 1.0 in his paper are represented as double crosses and those below 0.9 as a single cross. The ichthyofauna of Tasek Bera is to a greater or lesser degree similar to the three habitat types defined by Johnson (1967b). In considering the cyprinids as a group, Tasek Bera has a fauna most similar to tree country waters. On the other hand, if the air breathing anabantid community is considered, Tasek Bera has a fauna more similar to those inhabiting dilute black waters and riceland waters (Table 149). Thus the fish composition of Tasek Bera represents the sum total of the three habitat types defined by Johnson (1967b). This is due to the fact that Tasek Bera is a complex ecosystem with representative habitats similar to the three defined by Johnson (1967b), i.e. it has lentic, lotic, limnetic, littoral and a variety of vegetated habitats. Such a situation would permit the existance of a diverse ichthyofauna as is found in Tasek Bera.

There are, however, subtle differences in fish composition between Tasek Bera and Johnson's three habitat types. Tasek Bera differs from tree country streams in that it had several species of the Genus *Oxygaster* and a scarcity of true benthic fishes such as cobitid fishes and species belonging to the genera *Parakysis, Glyptothorax* and *Mastacembelus*. The scarcity of benthic fishes can be attributed to two reasons. The low concentration of dissolved oxygen would preclude most non-air breathing fishes and allow only air breathers like *Clarias batrachus, C. meladerma, Macrognathus aculeatus* and *Fluta alba*

Table 149 Comparison of fish communities in the study area in Tasek Bera with the three habitat types as defined by Johnson (1967).

	Tasek Bera	Dilute Black waters	Tree Country waters	Riceland waters
**Notopterus notopterus*	++			++
Oxygaster oxygastoides	++			
O. anomalura	+			+
O. johorensis				++
O. hypophthalmus	++			
Danio tweediei				+
Rasbora dorsiocellata	++			
R. cephalotaenia	+	++	++	
R. pausiperforata	+	++	++	
R. taeniata	++			
R. einthoveni	++	++	++	
R. elegans	+		++	
R. heteromorpha	++		++	
R. sumatrana	+			+
R. bankanensis			++	
R. caudimaculata			++	
R. trilineata			++	+
Hampala macrolepidota	+		++	
Cyclochilichthys apogon	++		++	
Osteochilus hasselti	+		++	++
Labiobarbus lineatus	++			
Puntius tetrazona	++			
P. pentazona		++	++	
P. lineatus		++		
P. fasciatus	++			
P. binotatus	+		++	++
P. lateristriga	+		++	
P. sp.	++			
Noemacheilus selangoricus	+		++	
Acanthophthalmus kuhli	+		++	
A. muraeniformis			++	
Silurichthys hasselti	+		++	
Kryptopterus bicirrhis	++			
**Clarius batrachus*	++			++
Parakysis verrucosa	+		++	
Glyptothorax platypogonoides			++	
Hemiramphodon pogonognathus	++	++	++	
Dermogenys pusillus		+		++
**Channa micropeltes*	++			
**Ch. striatus*	++			++
**Ch. lucius*	+	++	++	
Chanda ranga	++			
Pristolepis fasciatus	++			+
Nandus nebulosus	++		++	
**Sphaerichthys osphronemoides*	+	++	++	
**Betta pugnax*	++	++	++	
**B. splendens*				++
**Trichogaster trichopterus*	++	+	++	++
**T. pectoralis*				++
**Trichopsis vittatus*	++			++
**Anabas testidianus*	+			++
**Polyacanthus hasselti*	++			
Luciocephalus pulcher	++		++	
Dorichthys martensi			++	
Mastacembelus macalatus			++	

* Air breathers.

+ = *constancy value* $\leq$ 0.9

++ = *constancy value* $\geq$ 1.0

to exist (Table 150). However, their population densities are low. The second reason could be due to the fact that the bottom substrate is soft and unstable making it unsuitable for the fishes to live in. This is especially so in the open water and sub-littoral *Utricularia* regions. Although the roots of the *Lepironia* reeds do provide some stability to the substrate, their density and presence of macroalgae do limit the free movement of benthic fishes, thereby limiting their abundance. Most of them were found to inhabit areas in and around the swamp-forest which has a rather stable bottom substrate and on the trunks of fallen trees in the water.

Generally, most fishes caught are rarely larger than 30 cm with the exception of *Chana* and *Clarias* which are air breathers. Again the low concentration of dissolved oxygen may be the cause of this phenomenon. However, larger fish such as *Scleropages formosus, Puntius schwanenfeldi, Osphronemus goramy* and *Oxyeleotris marmoratus* can be found in Tanjung Kuin. Their abundance can be attributed to the fact there are larger open water areas and consequently higher dissolved oxygen content making it more conducive for fish to flourish. Also, this area is subjected to less fishing pressure because it is uninhabited.

7.2 Fish associations in relation to habitat types (N. Mizuno, R. P. Lim, M. Nagai, Y. Shiraishi and J. I. Furtado)

Allopatry and microgeographical isolation between related species are generally recognised as the factors that result in speciation and diversity among fish communities in tropical freshwaters. (Lowe-McConnell 1969, 1975). Among Malaysian freshwater fishes Johnson (1967b) noted that there was spatial and temporal segregation in the dominant species between the three habitat types he defined. Such a phenomenon was also found by Bishop (1973) in the Sungai Gombak, Malaysia.

Although the study area in Tasek Bera can be defined into three distinct habitats, the composition of the fishes species between these habitats are

Table 150 List of benthic fishes obtained in the study area in Tasek Bera. (N. Mizuno and M. Nagai)

Macrognathus aculeatus	*Prophagorus nieuhofi*
Fluta alba	*Leiocassis poecilopterus*
Clarias batrachus	*Chaca chaca*
Clarias meladerma	

quite similar (Table 148). However, the total number of individuals collected for the same collecting effort differed between the habitats (Table 148). The fish density was highest in the *Utricularia* stand, and was lowest in the *Lepironia* stand.

Utricularia stand. From the data obtained it is found that the *Utricularia* stand is a very important habitat for fish for food and shelter. The composition and abundance of fish varied diurnally (Table 151). (The calculation of the index of density per 100 m^2 was based on the number of casts and the area of the cast net). During the daytime, ***Rasbora dorsiocellata*** and ***Puntius tetrazona*** were dominant. They were also spatially separated, the former inhabiting the surface layer and the latter the bottom layer of the water (Fig. 124a). During the night, the total density increased four-fold that during the day. The most dominant species were ***Puntius fasciatus*** and ***Cyclocheilichthys apogon*** with ***Puntius tetrazona***, ***Kryptopterus bicirrhis*** and two species of ***Oxygaster*** as subdominants (Table 151). These species were also spatially separated (Fig. 124a). The two species of ***Oxygaster***, ***Cyclocheilichthys apogon*** and ***Puntius fasciatus*** occupied the upper, middle and bottom layers, respectively. ***Kryptopterus bicirrhis*** and ***Notopterus notopterus*** occupied the upper and lower layers of the fringes of this habitat respectively.

Open water area. The total number and composition of fishes obtained with mixed gill nets from the open water (Station I) was studied intermittently between March 1971 and March 1973 (Table 152; Fig. 124b). During the daytime, ***Puntius tetrazona*** was abundant in the middle layer along the

Table 151 Diurnal changes of fish composition at the *Utricularia* stand. (N. Mizuno and M. Nagai)

Time of day	7-17	18	19-22	6
No. of individuals/100 m	88	432	336	272
	%	%	%	%
Rasbora dorsiocellata	27	6	3	9
Puntius tetrazona	40	26	12	24
Labiobarbus faestiva	8	2	4	4
Cyclocheilichthys apogon	4	11	29	10
Puntius fasciatus	8	54	33	46
Kryptopterus bicirrhis			10	
Oxygaster spp.*		2	6	3
Others	13		3	4

* *Oxygaster oxygastroides* and *O. hypophthalmus*.

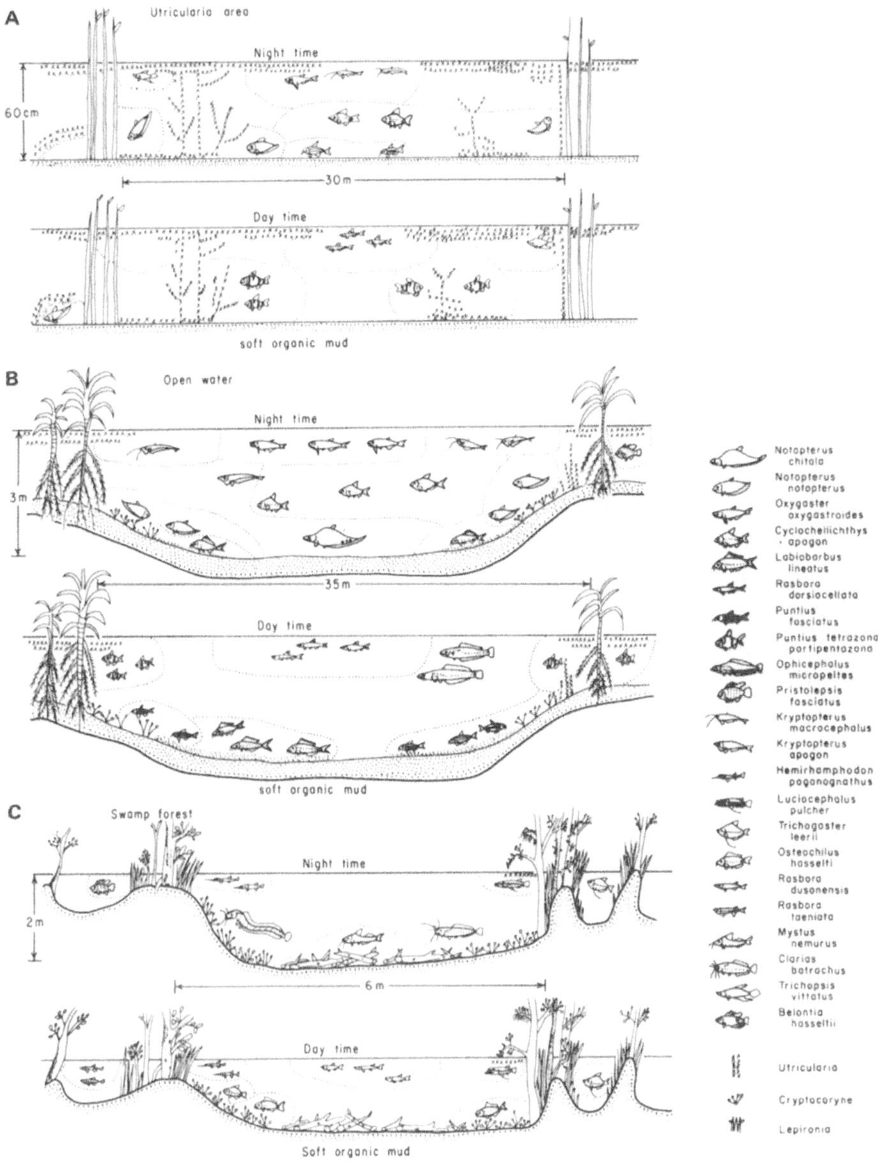

Fig. 124 Diagrammatic representation of spatial and temporal segregation of several dominant fishes in *Utricularia* community (A), at the open water area (B), and in the forest stream (C). (M. Nagai)

margins of the open water. *Rasbora dorsiocellata*, although too small to be caught with the gill nets, was observed to be numerous just below the surface of the water (Fig. 124b). During the night, two species of *Oxygaster* and

Table 152 Diurnal change of fish composition at the open water area. (N. Mizuno and M. Nagai)

No. of individuals	Daytime 932	Nighttime 2152
Puntius tetrazona	38	3
Pristolepis fasciatus	9	1
Puntius fasciatus	22	19
Cyclocheilichthys apogon	7	28
Kryptopterus bicirrhis	6	15
Oxygaster spp.*	7	21
Others	11	13

* *Oxygaster oxygastroides* and *O. hypophthalmus*.

Cyclocheilichthys apogon were found in shoals in the upper and middle layers of the open water. In the marginal waters where *Pandanus* stands occur, *Kryptopterus bicirrhis* and *Notopterus notopterus* were found in the upper and lower layers respectively.

The spatial and temporal segregation of the dominant fishes in the open water (Fig. 124b) was very similar to that of the *Utricularia* stand (Fig. 124b). Several differences, however, are apparent. In the open water area, *Puntius fasciatus* was as abundant during the day as at night. *Pristolepis fasciatus*, which was rare in the *Utricularia* stand, was abundant along the margins of the open water associated with *Pandanus* stands. In the *Utricularia* stand, the smaller *Oxygaster hypophthalmus* was more abundant than the larger *O. oxygastroides*. In contrast, *O. oxygastroides* was more abundant than *O. hypophthalmus* in the open water region. This difference is due to the differences in habitat; the *Utricularia* stand providing shelter and protection to the smaller *O. hypophthalmus*.

In the large open water areas of Tanjung Kuin large fishes such as *Scleropages formosus*, *Puntius schwanenfeldi*, *Osphronemus goramy* and *Oxyeleotris marmoratus* were commonly recorded. These species were rare in the study area. *Labiobarbus faestiva* was observed grazing on benthic algae in large schools during the night with the larger *Notopterus chitala* following them, presumably predating upon them.

Figure 125 shows the change in number of fishes, extrapolated to the number of monthly total catch to the mean number per day each consisting of 6-8 hauls of mixed gill nets throughout the day and night. Fish were more abundant during the April-May and July-August dry season, and were poor during the November-March wet season. This is due to the fact that during the dry season, as the littoral dries up, fishes migrate to the open water which

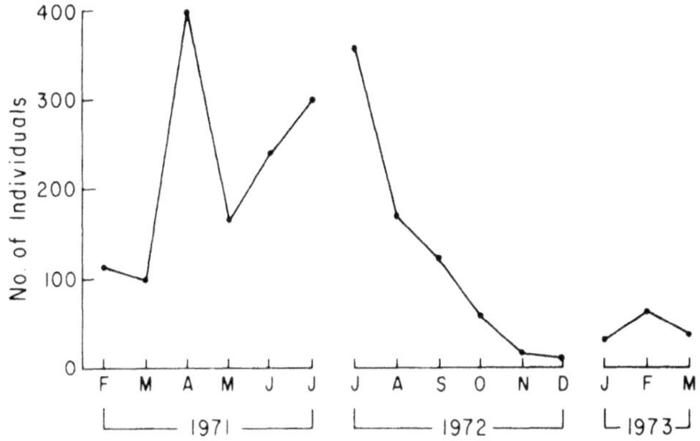

Fig. 125 Monthly change in abundance of fishes from the open water area. (N. Mizuno, M. Nagai and T. Mizuno)

was little affected by the dry season. In fact catches in the *Utricularia* and *Lepironia* stands were poor during the dry season substantiating the hypothesis that migration to the open waters does occur (Table 148). Most of those remaining in these two habitats were air breathers such as *Notopterus notopterus*, *Trichogaster trichopterus*, *Trichopsis vittatus* and *Polyacanthus hasselti*. Thus the open water is important as a refuge during the dry season.

Lepironia stand. The abundance of fish in this area seems to be the poorest amongst the three habitat types in the study area (Table 148). Vere few fishes were observed through snorkeling during the daytime in the wet season. The exceptions were among isolated burnt clumps of *Pandanus* within the *Lepironia* stands where schools of small fish such as *Rasbora*, *Puntius tetrazona*, *P. fasciatus* and *Pristolepis fasciatus* were abundant. The first three occupied the upper, middle and lower layers respectively, while the last species was found among the entangled roots of the *Pandanus* stands.

Table 153 shows the total number of fishes obtained from 15 hauls of a mixed gill net of three mesh sizes for three days, in November (columns 3 and 4). There were fewer fishes in the daytime when compared with the night catches. The majority of them were noctural swimming fishes, which migrate from the forest to the open water via the *Lepironia* reed swamp. They were presumably caught along their migratory route.

With decreasing water levels in the *Lepironia* littoral during the dry season (April-May) the passage ways of the fish become narrower, and this may result in greater catches due to increased efficiency in the fishing gear used (Table 153). Air breahting fishes were found in isolated pools of water as the reed swamp dried further in June and July.

Table 153 Fish composition at the *Lepironia* stand collected with gill nets. (N. Mizuno and M. Nagai)

	Februari March	April May	November	
			Daytime	Nighttime
*Notopterus notopterus**	1%	10%		3
Oxygaster oxygastroides	12	2	16	113
O. hypophthalmus	3	1		3
Rasbora spp.*		—	1	2
Cyclocheilichthys apogon	19	9	8	78
Puntius tetrazona	12	16		
P. fasciatus	10	10	2	10
Osteochilus hasselti	3	—	1	
O. spilurus		—	1	
Labiobarbus faestiva	1			44
Luciosoma setigerum			4	3
Kryptopterus apogon		1		
K. bicirrhis	33	33		3
Mystus cavasius		—		1
M. nemurus		—		1
Silurodes hypophthalmus	1	—		
*Polyacanthus hasselti**		1		3
Pristolepis fasciatus	2	8	7	5
*Trichogaster leerii**				1
*Trichopsis vittatus**		1		
*Channa striata**	1	—		
Nandus neblosus	2	4		
Total number of individuals	177	473	40	270

* Air breathers.
** *Rasbora dorsiocellata*, *R. cephalotaenia*, *R trilineatus* and *R. retrodorsalis*.

Forest streams The swamp-forest which surrounds the study area is traversed by several stream channels. These streams are filled with fallen trees, branches and twigs so that fishing with conventional gear is nearly impossible. Spearing amongst the marginal vegetation was the only effective method. Large noctural, piscivorous fishes such as *Clarias meladerma, Mystus cavasius* and *Chana striata* were caught. They obviously were using the vegetation as shelters during the daytime. *Hemiramphodon pogonognathus* was abundant in the day and night inhabitring the surface waters along the margins of the stream channels. In the open water region of the stream channel, *Rasbora dusonensis* were observed shoaling near the surface while *Osteochilus hasselti, Labiobarbus feastiva* and *Rasbora heteromorpha* were found in the middle and near bottom layers of the stream channel. The first two in the latter three species were also observed to be grazing in submerged macrophytes such as *Utricularia* and *Cryptocoryne* while one of them was observed feeding on aquatic insects associated with the macrophytes.

The spatial distribution is represented diagramatically in Fig. 124c. On several occasions the small benthic fish, *Leiocasis poecilopterus* was caught from the surfaces of *Pandanus* leaves or from holes in fallen trees. A rare benthic fish *Chaca chaca* was also collected from these holes.

Movement of fish between habitat types. On 11 and 12 February 1971, mixed gill nets were set up at the open water region, *Utricularia* stand, *Lepironia* stand and along the borders between the open water and the three habitats viz. *Utricularia*, *Lepironia* and *Pandanus* stands. The nets were examined every two hours over a 24-hour period. The results are shown in Fig. 126.

The highest rate of migration of fishes was between the open water and the *Utricularia* stand. (Fig. 126). A small noctural catfish species, *Kryptopteris bicirrhis*, was caught in abundance only within the *Utricularia* stand and not at the border between this habitat and the open water. This suggests that the fish do not migrate from the open water but from the swamp-forest situated in close proximity to the *Utricularia* stands. The conclusion that the *Lepironia* reed swamp is used as a migratory route by fishes as discussed earlier is substantiated in this study.

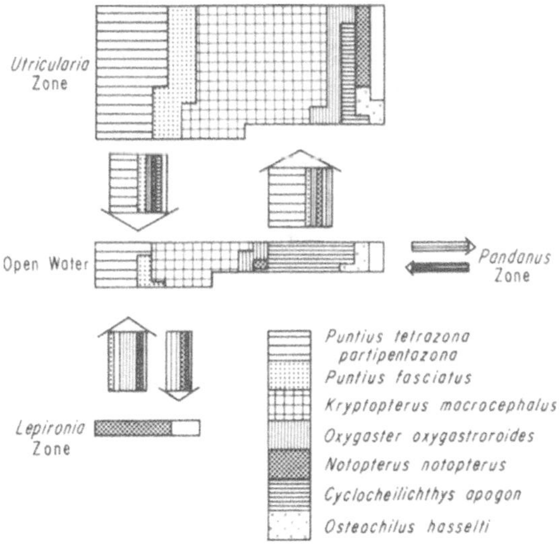

Fig. 126 Diagrammatic representation of fish composition and their diurnal migration between the main habitats in the study area. (Y. Shiraishi)

7.3 Diurnal rhythm of activity
(N. Mizuno, R. P. Lim, M. Nagai, Y. Shiraishi and J. I. Furtado)

The diurnal rhythm of movement of fishes was examined using gill nets. Hourly changes in the abundance of food organisms in the gut of individuals of four species of fish were analysed. Five species of fish, namely *Puntius fasciatus, P. tetrazona partipentazona, Osteochilus spilurus, Pristolepis fasciatus* and *Luciocephalus pulcher*, were more active during the day (Fig. 127). Six species, namely *Notopterus notopterus, Kryptopterus bicirrhis, Nandus nebulosus, Kryptopterus apogon, Oxygaster oxygastroides* and *O. hypophthalmus*, were more active during the night (Fig. 127). The trends in gut content of the four species examined, namely *Puntius fasciatus, P. tetrazona partipentazona, Notopterus notopterus* and *Kryptopterus bicirrhis*, correspond closely to their diurnal activity (Fig. 127). This indicates that the diurnal changes in abundance of fish caught in the gill nets are related to the diurnal activity patterns especially the feeding activity of the fishes.

Fig. 128 shows the seasonal changes in abundance of fish caught in gill nets over five months. The number of fishes caught on each occasion was extrapolated to the square root percentage value of the total number caught in a 24-hour period. The monthly changes in this ratio are presented in Fig. 129.

Puntius tetrazona tended to maintain its monomodal diurnal activity throughout the period of study and can be presumed to maintain such a trend throughout the year (Fig. 128). This had been earlier substantiated by the fact that they were more active in their feeding activity in the *Utricularia* stand and in their movements between habitats during the day.

Pristolepis fasciatus showed a stable diurnal activity throughout the five months (Fig. 129). They showed greater activity at dusk and this is due to their feeding behaviour. As they feed largely on aquatic insects such as chironomids and mayflies which emerge at dusk (see earlier), their activity would correspond to the emergence activity of these insects.

Of the three remaining species *Osteochilus spilurus* and *Puntius fasciatus* showed varied trends in diurnal activity (Fig. 129). During the dry period their diurnal activity patterns were not as marked as in the wet seasons (Fig. 129). The diurnal rhythms of *Osteochilus spilurus* could not be analysed in detail as the samples were too small.

The diurnal activity pattern of *Cyclocheilichthys apogon* was more marked. They showed a bimodal diurnal activity pattern with a higher peak just after dawn in November and a clear monomodal activity pattern in February. In April and May they showed marked nocturnal activity (Fig. 133b).

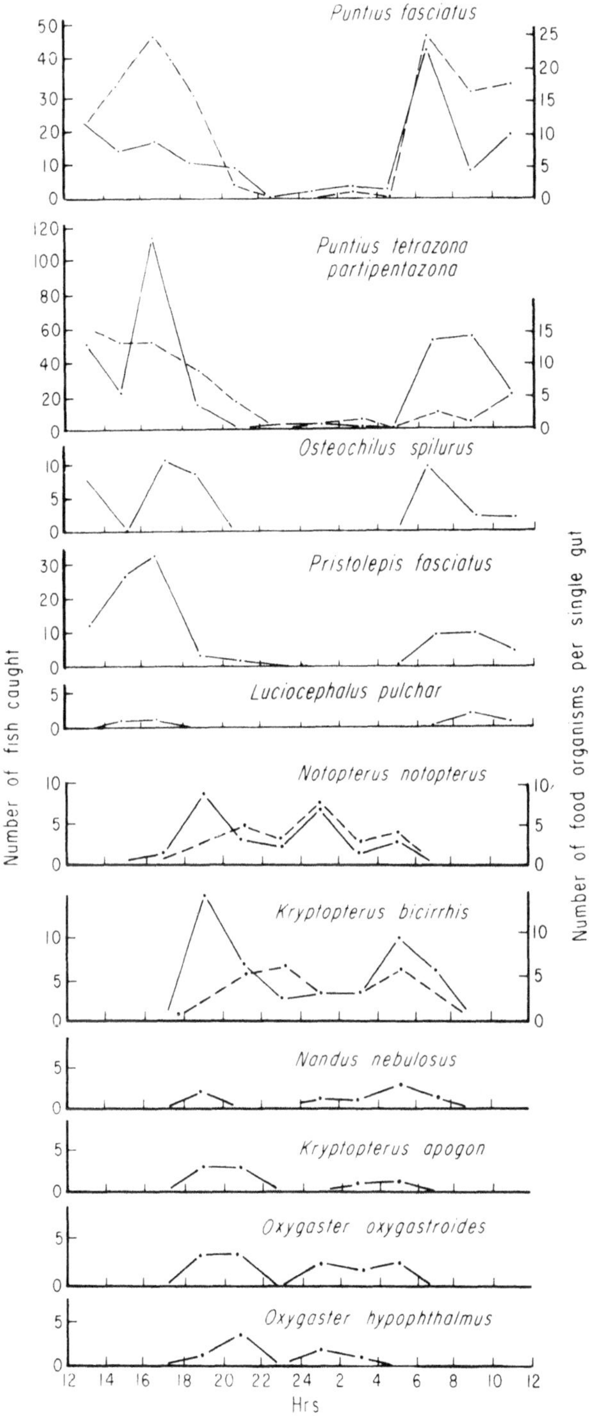

Fig. 127 Diel changes of 11 species of fishes caught in gill nets in February 1971. For four species of fishes the hourly fluctuations of the number of food organisms (. — — — .) in their guts are depicted. (Shiraishi et al. 1972)

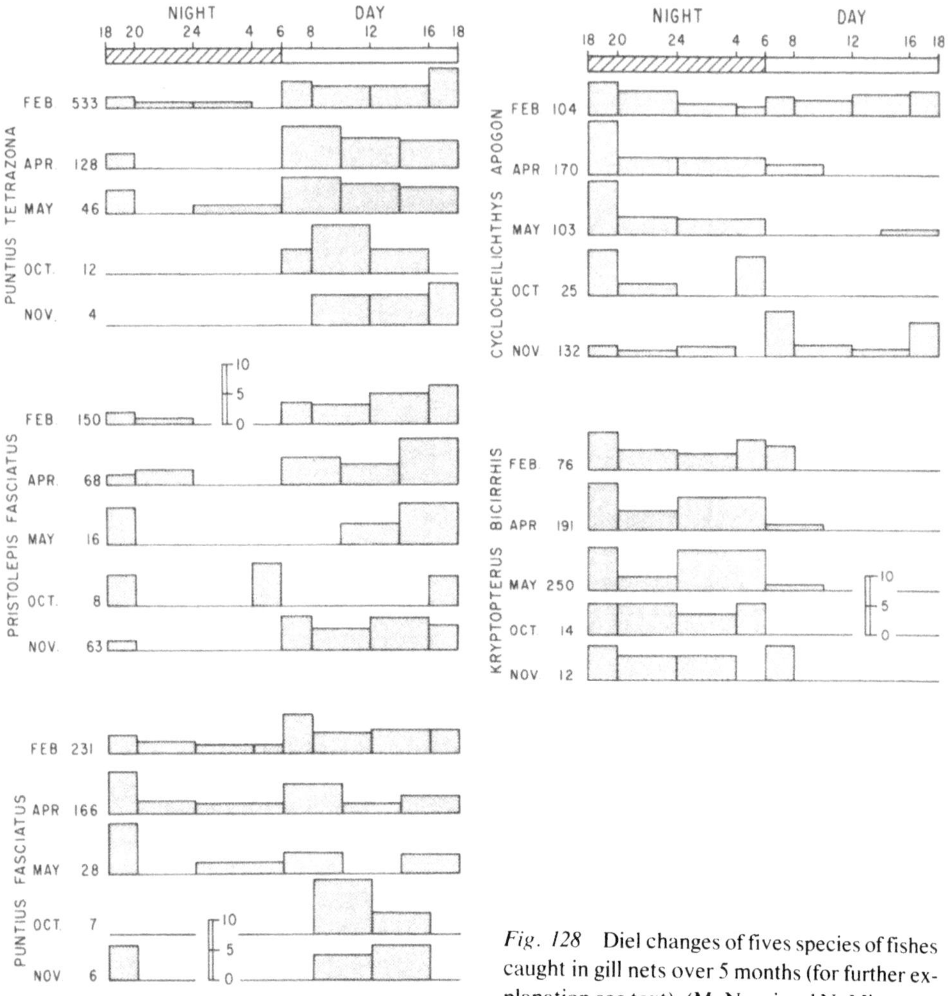

Fig. 128 Diel changes of fives species of fishes caught in gill nets over 5 months (for further explanation see text). (M. Nagai and N. Mizuno)

Out of the six species defined earlier as nocturnal only *Oxygaster hypophthalmus* and *Kryptopterus bicirrhis* were confirmed as nocturnal species (Fig. 134b). The rest of the species do show a slight bias towards nocturnal activity (Table 154). In the two former species two peaks, at dawn and dusk, were observed. However, during the wet season more individuals were obtained just after dawn and this may have resulted in an artefact, i.e. an increase in the ratio of the daytime catches in this period (Fig. 134b).

Thus it is seen that various species of fishes exhibited various trends of

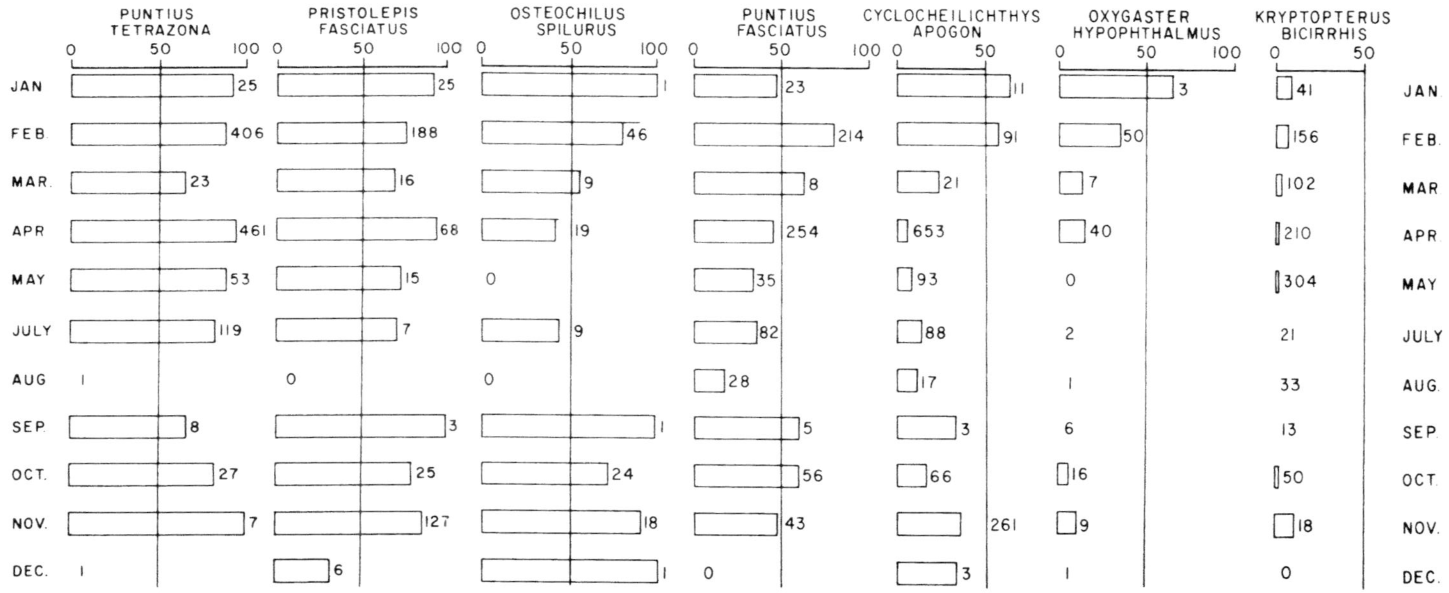

Fig. 129 Monthly changes in the ratio of numbers of individuals of seven species of fishes caught in the daytime using gill nets to the total number caught in a 24-hour period. (N. Mizuno, M. Nagai and T. Mizuno)

diurnal activity. Even within a single species, trends in diurnal activity varied with time. No single factor can be attributed to be responsible for such complex activity of these fishes. However, it is speculated that feeding activity and changes in water level may play a major role in their diurnal activity; the latter factor has been discussed by Lowe-McConnell (1975) for riverine fishes.

Several other species of fishes were also caught but their numbers were too small for detail analysis. However, *Xenentodon canciloides* and *Trichopsis vittatus* may be assumed to be diurnal fishes. In contrast, *Notopterus notopterus, Rasbora cephalotaenia, Kryptopterus apogon, Clarias meladerma*

Table 154 Total numbers of several fishes caught with gill nets through a 24-hour period and intermittently between 1971 and 1973. (N. Mizuno et al.)

Species	Number of individuals		
	Total	Daytime	Nighttime
Notopterus notopterus	86	7	79
Oxygaster oxygastroides	994	105	889
O. hypophthalmus	135	28	107
Rasbora cephalotaenia	33	5	28
Luciosoma setigerum	10	7	3
Cyclocheilichthys apogon	1307	230	1077
Puntius fasciatus	748	408	340
P. tetrazona	1131	1014	117
Osteochilus hasselti	36	24	12
O. spilurus	128	90	38
Labiobarbus faestiva	162	55	107
Kryptopterus bicirrhis	945	26	919
K. apogon	55	0	55
Clarias meladerma	2	0	2
Macrognathus aculeatus	3	0	3
Mystus nemurus	7	0	7
M. cavasius	66	2	64
Xenentodon canciloides	20	16	4
Chanda ranga	21	6	15
Trichopsis vittatus	23	21	2
Trichogaster leeri	12	8	4
Luciocephalus pulcher	21	19	2
Channa spp.*	7	0	7
Pristolepis fasciatus	480	390	90
Nandus nebulosus	61	13	48

* *Channa microcephalus*, *Ch. lucius* and *Ch. striatus*.

and *Macrognathus aculeatus*, two species of the genus *Mystus* and *Channa* spp. are presumed to be nocturnal. *Luciosoma setigerum, Osteochilus hasselti, Labiobarbus faestiva, Chanda ranga, Trichogaster leeri* and *Nandus neblosus* did not show any obvious diurnal pattern (Table 154).

7.4 Food habits
(N. Mizuno, R. P. Lim, Y. Shiraishi, M. Nagai and J. I. Furtado)

The feeding and food habits of fishes in Tasek Bera were observed in the field with concurrent analysis of their gut contents. A qualitative examination of the gut contents of 21 species of fishes was made between February and May 1971 (Table 155). A quantitative analysis of the gut content was carried out on fishes caught over a 24-hour period in February 1971 (Table 156).

Only three species, *Osteochilus spilurus, O. hasselti* and *Labiobarbus faestiva* were phytophagous (Table 155). Their gut contents consisted not only of algae such as *Micrasterias, Cosmarium, Staurastrum,* and *Closterium* but also pieces of aquatic macrophytes such as *Utricularia, Nitella* and *Cryptocoryne*. In fact, *O. hasselti* and *Labiobarubus faestiva* were observed grazing on benthic algae and picking off leaves of submerged macrophyte during the day and night. Thus the daily rhythms of these species through gill net catches discussed earlier cannot be substantiated with field observations. The importance of macrophytes as a direct food source cannot be overemphasised as they play a greater role in providing shelter and protection from predation to the fishes. Their role in providing micorhabitats for several invertebrates and periphytic algae which provide food for carnivorous fishes cannot be ignored.

The remaining 17 species of carnivorous fishes can be divided into two groups. The first group consists of piscivorous species such as *Notopterus Chitala, Silurodes hypopthalmus, Kryptopterus apogon, Mystus nemurus* and two species of the genus *Channa*. Their gut content consists of several small fishes such as *Cyclocheilichthys apogon, Puntius tetrazona, P. fasciatus* and *Rasbora* spp.

Furtado et al. (1974) compared the functional morphology of several fishes in Tasek Bera in relation to their feeding ecology. He found that piscivorous fishes like *Channa micropeltes* and *Wallagonia miostoma* differed from those species feeding on small pieces of food material. They had wide gapes, sharp teeth, few tuberculate or thornlike gill rakers, large muscular stomachs and short intestines. Furthermore, *Kryptopterus apogon, Mystus nemurus* and two species of *Channa* were nocturnal in habit (Table 154). From their daily activities and/or from their morphologies *Scleropages formosus, Silurichthys*

Table 155 Gut content of 20 species of fish caught in Tasek Bera. (M. Nagai)

Fish species	Food items																	
	Diptera	Chironomidae	large larva	small larva	pupa	*Eriocera* sp.	Ephemeroptera	*Cloeon* sp.	Trichoptera	Psychomyiidae	Odonata	Other insects	Attached algae	Zooplankton	*Utricularia*	*Nitella*	Detritus	Fish
Notopterus notopterus			*	*	*			*			*	*						
N. chitala																		*
Oxygaster oxygastroides					*													
O. hypophthalmus					*													
Rasbora dorsiocellatus				*		*		*										
Cyclocheilichthys apogon				*	*													
Puntius tetrazona partipentazona								*		*								
P. fasciatus					*													
Osteochilus hasselti													*		*	*	*	
O. spilurus													*					
Labiobarbus faestiva													*	*	*			
Silurodes hypophthalmus																		*
Kryptopterus bicirrhis					*						*	*						
K. apogon																		*
Mystus nemurus																		*
M. cavasius					*			*				*		*				
Channa striatus																		*
Ch. micropeltes																		*
Oristolepis fasciatus				*				*			*							
Nandus nebulosus								*				*						

hasselti, Ompok bimaculatus, Clarias spp. and *Prophagorus niewhofi* can be inferred to be piscivorous. Thus 17 species out of a total of 95 species (18%) can be considered to be piscivorous. The diversity of small fishes is also important in providing food for these larger fishes. The situation in Tasek Bera is similar to that of the fish fauna of the Amazon River (Lowe-McConnell 1975).

The second group consists of 11 small fish species. They feed mainly on aquatic insects especially chironomid larvae and pupae. *Puntius fasciatus* feeds not only on chironomid larvae but also bryozoans (Tables 155 and 156). *Pristolepis fasciatus* is insectivorous as was reported by Furtado et al. (1974). Furtado et al (1974) reported that *Puntius tetrazona* and *P. fasciatus* were

Table 156 Gut content analysis based on abundance of food organisms in 6 species of fish. Figures in parentheses show the total number of food organisms counted in all the guts examined. The food organisms are arranged in relation to their habitats from surface to bottom (from Shiraishi et al. 1972).

Food items	species / No. of specimens	Diurnal				Nocturnal	
		Puntius fasciatus 181	*P. tetrazona partipentazona* 96	*Cyclocheilichthys apogon* 41	*Pristolepis fasciatus* 22	*Notopterus notopterus* 16	*Kryptopterus bicirrhis* 20
Terrestrial small animals		Acarina (35) (ticks) Opilonida (1) (spiders)	Acarina (5) Terrestrial insects (3)		Opilonida (1)		Ant (1)
Zooplankton		Rhizopodea (130) Totatoria (43) Brachiopoda (283) Copepoda (275)		Brachiopoda (4) Copepoda (1)	Copepoda (5)		Brachiopoda (1) Copepoda (3)
Phytoplankton		Cyanophyceae (980) Bacillariophyceae (101) Chlorophyceae (10)	Chlorophyceae (10)				
Aquatic insects		Chironomidae (553) Ephemeroptera (9) Trichoptera (10)	Chironomidae (139) Ephemeroptera (1) Trichoptera (9)	Chironomidae (13) Ephemeroptera (6)	Chironomidae (168) Ephemeroptera (168) Trichoptera (3)	Chironomidae (124) Ephemeroptera (36) Odonata (3) Hemiptera (2)	Chironomidae (18) Ephemeroptera (12) Hemiptera (2)
Shrimps		(1)			(10)	(38)	(2)
Others		Fish scale Bryozoa (3485) Tubificina (1) Hydracarina (3)	Bryozoa (190)	Residue (much)	Fish scales (10)	Fish scales (3)	Fish scales (1)

Table 157 Gut content analysis of three fish species based on the abundance of individuals or by quantitative visual estimation (Furtado et al. 1974)

Species	*Puntius tetrazona*	*Puntius fasciatus*	*Pristolepis fasciatus*
Sample size	20	20	20
Phytoplankters			
Spondylosium sp.	16	++++	
Protococcous	++++	+	
Oscillatoria	++		
Microspora sp.	++		
Spirogyra sp.	+++	++++	
Micrasterias sp.	++	+	
Pleurotaenium	+++	+	
Closterium	24		
Desmidium	++	++++	
Staurastrum	++		
Penium	9		
Navicula	+	+	
Diatoma sp.	++++	+	
Ulothrix sp.	++		
Fragillaria sp.	10		
Plant tissues			
Tracheid elements	47	+	
Cyperaceae (seeds)			55
Brown decayed matter		+	
Zooplankters			
Cladocerans	313	18	14
Copepods	65	13	
Ostracods		30	2
Protozoans	5		
Shrimps	3		
Insects			
Formicidae (ants)	1		
Ephemeroptera (nymph)			175
Odonata (nymph)			10
Dystiscidae (larva)			12
Chironomidae			82
Aquatic insects?	1		

Visual quantitative estimation:
++++ = abundant; ++ = moderately abundant; + = not frequent.

plankton feeders (Table 157), which contradicts the results obtained in the present study where the fishes were found to be insectivorous (Table 156).

This apparent difference could be due to the season during which the fishes were caught, for they were observed to pick on lumps of attached algae drifting in the water during the wet season. Furtado et al. (1974) may have collected their samples during the wet season. Also fishes are known to change their diet with seasons as discussed by Larkin (1956).

Most of the fish caught in the study area feed on microorganisms. As such the microflora and fauna play an important role as food sources for these fishes; plankton and insects are eaten by small fishes which are in turn eaten by large fishes.

Although Tasek Bera has an abundance of detrital organic material and the luxuriant growth of aquatic macrophytes these resources are poorly utilised by the fishes. Only two species of fishes, *Osteochilus hasselti* and *Labiobarbus faestiva*, feed on a small fraction of aquatic macrophytes. *O. hasselti* and *Cyclocheilichthys apogon* are presumed to feed on detrital material, as these were found in their guts mixed with algae, pieces of fresh macrophytes and insects (Tables 155 and 156). No fish were found to feed wholly on emergent macrophytes. Thus several niches within Tasek Bera have not been exploited by fishes.

Lowe-McConnell (1975), in reviewing the information on the food of fish from Africa, South America and Southeast Asia, found a longitudinal succession of dominant food resources for stream fish communities. Fish in the head waters depend primarily on allochthonous material. In the middle zone they were predatory omnivores feeding on benthic invertebrates especially insects. In the lower reaches of streams they were mainly specialised in feeding on mud containing detritus and microorganisms. The fish community in the study area corresponds to the middle zone of streams as defined by Lowe-McConnell (1975). This is not unusual as Tasek Bera is situated somewhat at the midpoint of the Sungai Bera. Forest areas, however, are inhabited by species like *Hemiramphodon pogonognathus* and *Luciocephalus pulcher* which feed on allochthonous material. This resembles the community of the headwaters of Lowe-McConnell's classification.

7.5 Seasonal growth rates of several fishes
(N. Mizuno, R. P. Lim, M. Nagai and J. I. Furtado)

A review regarding some aspects of growth of tropical freshwater fishes, based mainly on their annual growth rates is given by Lowe-McConnell (1975). However, little information was presented with regard to seasonal changes in growth rates mainly because of the scarcity of such data in tropical

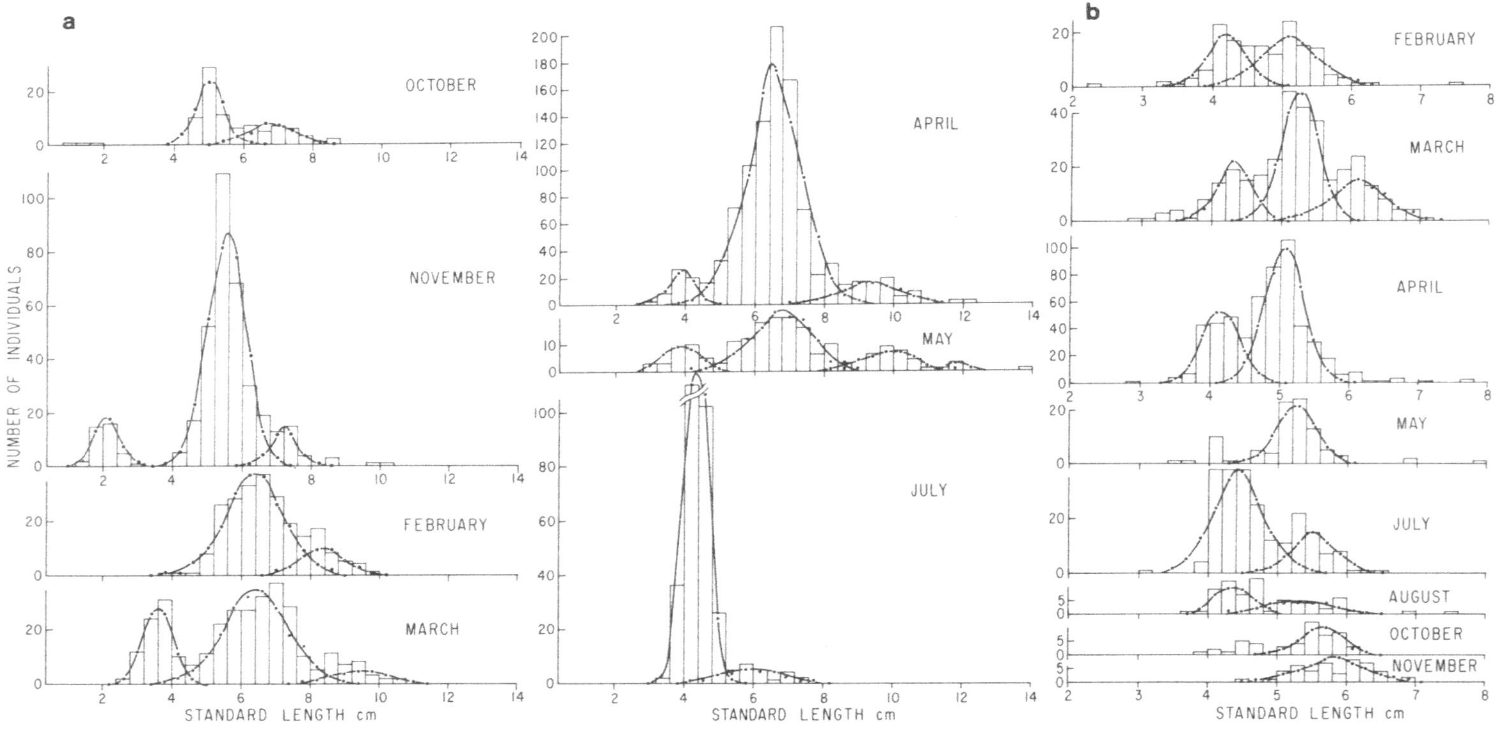

Fig. 130 Length frequency distribution and normal curves fitted with the aid of probit analysis of (a) *Cyclocheilichthys apogon*, and (b) ***Puntius fasciatus***. (M. Nagai and N. Mizuno)

regions. Thus no generalisation on the seasonality in growth rates of fishes can be made. It is pertinent then to study the seasonal growth changes of the more dominant fishes in Tasek Bera. The Petersen method was used in this study.

Four species of fishes were studied, namely *Cyclocheilichthys apogon, Oxygaster oxygastroides, Puntius fasciatus* and *P. tetrazona partipentazona*. All fishes caught, using various fishing gear over three years from 1971 to 1973, were measured for their standard lenghts. The monthly changes in size frequency distribution were analysed using Cassie's probit analysis method. Approximate normal curves were derived from empirical data (Figs. 130a and b). Based on these curves the mean lenghts of fishes for each month were extrapolated and depicted as monthly growth curves (Figs. 131a and b).

From the four growth curves two conclusions can be inferred. The first is that these fishes grow rapidly during the wet season (September-March) and slowly during the dry season (April-August) (Figs. 131 a and b). Similar

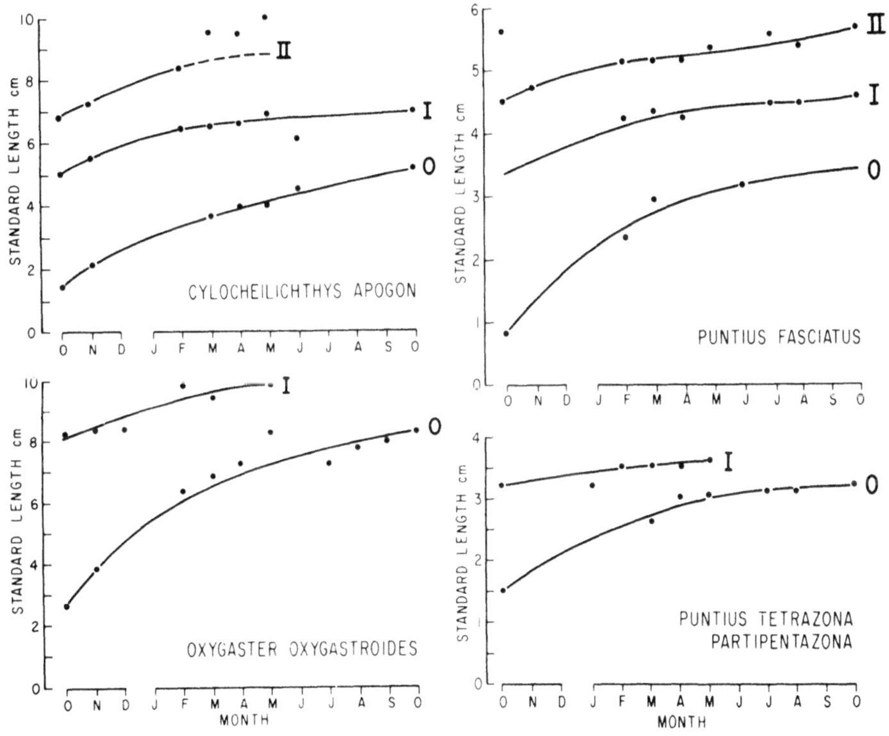

Fig. 131 Growth curves of *Cyclocheilichthys apogon, Oxygaster oxygastroides, Puntius fasciatus* and *P. tetrazona partipentazona*(for further explanation see text). (N. Mizuno and M. Nagai)

results were reported by Chevey and Le Poulain (1940) for fishes in the Mekong Delta, whose growth rates were reduced during the dry season leaving annual rings on their scales (cited by Lowe-McConnell 1975).

The second conclusion is related to the breeding period of fishes in Tasek Bera. The breeding period of these fishes is considered short as the sizes of the different age classes are quite distinct. The breeding period is also restricted to increasing water level or the early wet season. This conclusion was further reinforced from observations of numerous fish fry in the study area between October and November (Fig. 132). This corresponds to the generalisation that ostariophysian fishes of tropical seasonal rivers spawn during the initial period of the wet season (Lowe-McConnell 1975). Thus environmental conditions play an important role in regulating the seasonality of growth rates and restricting spawning of the fishes in Tasek Bera. This feature is similar to those of temperate regions.

Based on the mean lenghts from the growth curves for November the size increments of the four species and their growth ratios for a single year were calculated (Table 158). As a comparison the increments and growth ratios of small cyprinid fishes of temperate North America are given (Carlander 1969).

The ratios of size increments ranged from 0.7 to 1.2 for individuals between

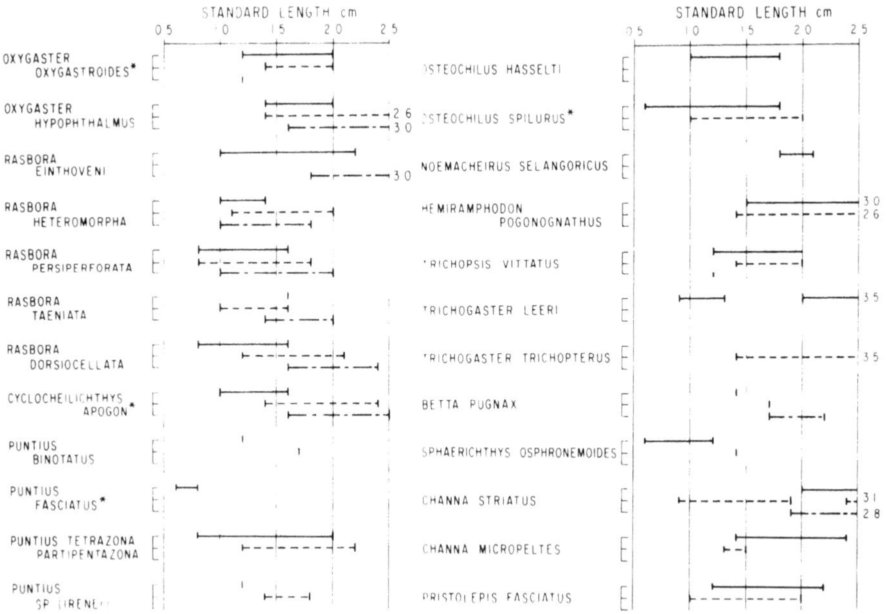

Fig. 132 Range of standard lengths of some fish fry obtained with scoop and cast nets in late October (solid lines), early (broken lines) and late (chain lines), November 1971. (N. Mizuno)

Table 158 Comparison of yearly increment in standard and total length (cm) of several small ostariophysian fishes between Tasek Bera and temperate North America. (N. Mizuno)

Fish species	Age in years			Increment	Ratio
	0	1	2		
Tasek Bera					
Puntius tetrazona	1.8	3.3		1.5	0.83
Oxygaster oxygastroides	3.8	8.3		4.5	1.19
O. hypophthalmus	2.3	4.0		1.7	0.74
Puntius fasciatus		3.1	4.7	1.6	0.52
Cyclocheilichthys apogon		5.5	7.2	1.7	0.31
Kryptopteus bicirrhis		4.8	6.7	1.9	0.40
North America					
Notropis coccogenis	2.8	4.9		2.1	0.75
Rhinichthys atratulus	1.8	3.5		1.7	0.94
Orthodon microlepidotus	5.6	14.5		8.9	1.59
Notropis coccogenis		4.9	6.7	1.8	0.37
N. galacturus		3.9	6.4	2.5	0.64
N. rubricroceus		3.8	5.0	1.2	0.32
Hybopsis biguttata					
female		7.1	9.6	2.5	0.35
male		8.3	11.7	3.4	0.41
H. leptocephala					
female		4.0	6.2	2.2	0.55
male		4.7	7.7	3.0	0.64

* Data of total length selected from Carlander's 'Handbook of freshwater fishery biology, Volume 1' (1969).

age 0 to 1 years and from 0.3 to 0.5 for individuals between age 1 to 2 years. These results are comparable with those of the cyprinids of North America (Table 158). Fryer and Iles (1972) studying 12 species of *Tilapia* in Africa reported a wider range of growth ratios of 0.3 to 1.0 for individuals of the age class 1 to 2 years. However, since these fishes generally attain larger sizes in their longer life cycles, it seems reasonable to infer that they show more rapid growth in their young stages than the small fishes shown in Table 158. Fryer and Iles (1972) also found that growth in *Tilapia* spp. was affected by seasonal fluctuations in environmental conditions. From the little data available it can be generalised that there is little difference in the overall growth rates of fishes from temperate and tropical regions. This is due to the fact that growth of both tropical and temperate fishes is greatly reduced during the dry period and the winter respectively.

Lowe-McConnell (1975) vividly describes the wet seasons of tropical seasonal river as periods when feeding, fattening, growing and spawning is optimum for nearly all species of fishes. In contrast the dry season is a period of little feeding due to a lack of available food (except of large piscivorous fishes), retarded growth and high mortality. High mortality could be density dependent as fish begin to crowd in the open water as other parts of the aquatic ecosystem begin to dry up during the dry season. Such a phenomenon may prevail in Tasek Bera during the dry season.

8. General discussion

J. I. Furtado and S. Mori

The Tasek Bera is a lowland freshwater inundated riparian system at its climax phase due to intensive denudation, extensive stream capture, low stream orders, the formation of benthic peat, the absence of distinct river channels, the occurrence of reverse flows especially during the northeast monsoon, and inundations of the order of 1-5 m reaching the crowns of the swamp-forest trees (Sections 2.1 and 2.2). This riparian swamp system has a history of about 4.500 years B.P. with a gradual colonisation of hydrophytes, and a sudden change from a lowland river depositing silt to a riparian swamp depositing autochthonous peat (Section 2.2). This transition from a river to a riparian swamp appears to be of biogenic origin, to the damming effects of *Pandanus*. The Tasek Bera is a remnant of a once extensively distributed forested riparian system in tropical southeast Asia, known as 'Rassau' rivers because the flowing waters are lined by *Pandanus* (Corner 1940). Such riparian swamps form 10-20% of the lowland landscape and, together with coastal, peat and mangrove swamps, have been transformed and utilized by man for wet-rice cultivation which has prevailed in this region for at least 2.000 years. In view of the scarcity of riparian swamps such as the Tasek Bera, it is necessary to understand their role at the land-water interface especially in regulating water discharge through the rivers, its purification and the recovery of nutrients, and the importance of the inundation or floods for essential ecological processes such as enhancing biological production and stimulating reproduction of riparian fishes (e.g. cyprinids and *Scleropages*).

Only recently (about 600 years B.P.) has the Tasek Bera been colonised by man, presumably by the Semalai and their predecessors (Section 2.2). The impact of these tribal peoples has been minimal both in the catchment area where they practised shifting agriculture, and in the swamp-forest where they cleared patches of swamp-forest to enhance *Pandanus* and *Lepironia* regen-

Furtado, J. I. and Mori, S., eds., Tasek Bera. ISBN 90 6193 100 2.

eration for domestic use, and the abundance of food fishes. The Semalai thus consider the riparian swamp as their 'mother', and have hitherto protected and managed the Tasek Bera basin through inter-tribal hostilities and extensive land use. Since the grouping of the Semalai into permanent settlements around Pos Iskandar during the 1948 Emergency Period, and intensive land development in its vicinity since the Second Malaysia Plan, the future of the Tasek Bera as a monument of natural and national heritage is being threatened.

The foregoing studies embodied in this monograph concern only a small fraction of the Tasek Bera, namely the open water and the *Pandanus-Lepironia* littoral regions. The structure and dynamics of the forested littoral region or the *Eugenia* swamp-forest, which constitutes 67% of the Tasek Bera have not been studied. The *Pandanus-Lepironia* littoral region is a successional phase of the *Eugenia* swamp-forest, due to the transformation by aborigines. Thus, some aspects of the studies on the *Pandanus-Lepironia* littoral may be applicable to the *Eugenia* swamp-forest littoral, especially metabolic processes like production and nutrient uptake. Nevertheless, the complex structure of the *Eugenia* swamp littoral signifies a rich diversity of ecological units and linkages which mediate the metabolic processes of this riparian swamp ecosystem. Of particular importance is the need to determine the role of the swamp-forest in regulating the flow of water and nutrients from the landscape to the river channels, and the role of seasonal inundation or floods in regulating ecosystem metabolism such as production, decomposition, reproduction and nutrient transfers. In addition, a number of ecological problems need further study such as (1) the origin of 'black' water or 'air hitam', (2) the causes for low oxygenation of the waters, (3) the factors contributing to the accumulation of benthic organic detritus, (4) the pathways of nutrient transfers, (5) the nutrients and biotic populations limiting productivity, (6) the pathways of secondary and tertiary (including fish) production, (7) the migration of animals especially riverine fishes within the habitat compartments of the Tasek Bera and between the Tasek Bera and the rivers downstream, (8) the biotic structure of the communities in the different zones or habitats, (9) the mechanisms of population regulation in the different habitats (especially of the dominant species), and (10) the ecological factors contributing to an intense speciation especially of desmids, rotifers, aquatic insects and freshwater fishes.

8.1 Food chain
(T. Mizuno and J. I. Furtado)

The Tasek Bera is characterised by two distinct food chains, the producer and the decomposer food chains (Fig. 133). In the producer food chain, the macrophytes of the *Eugenia* swamp-forest, the emergents *Lepironia* and *Pandanus* species in the disturbed littoral, the submerged macrophytes such as *Utricularia*, *Cryptocoryne*, *Blyxa* and *Nitella*, and algae. Although the algae are rich in diversity especially of desmids and as attached algae or periphyton, the forest and emergent macrophytes dominate biological production. Except for the contribution of the minor macrophytes (Chapter 5), there is no good assessment on the primary production of the swamp-forest. In the decomposer food chain, the organic detritus contributed by the *Eugenia* swamp-forest and the other macrophytes is the basis for decomposition (Section 4.2.1). The litter production by this riparian swamp-forest is comparable to that of lowland tropical forests being about 10 tonnes $\cdot$ $ha^{-1} \cdot y^{-1}$ (Furtado et al. 1980). The importance of fungi and invertebrates in facilitating organic decomposition in comparison to bacteria, is due to the limitation of nitrogen and phosphorus for heterotrophic bacterial activity (Section 4.1). Organic decomposition is slow being further limited by available oxygen resulting in detrital accumulation (Section 4.2), and being of the order of 1.4 years (Furtado et al. 1980).

Secondary producers occur at three levels in the Tasek Bera. Above water, defoliating insects such as the grasshopper *Oxya* on *Lepironia*, and herbivorous vertebrates, are presumably important. Except for the limited studies on *Oxya* (Section 6.1.6), there is no information on the structure and production of this consumer compartment. In the water, secondary production occurs in the grazing and detritus food chains. The limnetic zooplankton appear unimportant (Section 6.2.1); however the attached animals (both herbivorous and detritivorous), detritivorous shrimps and benthic invertebrates, and herbivorous and detritivorous fishes, appear abundant and important in these pathways (Chapters 6 and 7), with the detritivores dominating. Except for the limited studies on attached invertebrates, shrimps and chironomids, there is no information on the structure and production of these aquatic consumer compartments.

The tertiary producers comprise carnivorous invertebrates, fishes and other vertebrates, both above and in water. There is no information on the structure and production of these compartments, and on the transfer of materials and energy they effect between the water and the atmosphere of land. Furthermore, the impact of intense predation pressure on the rich

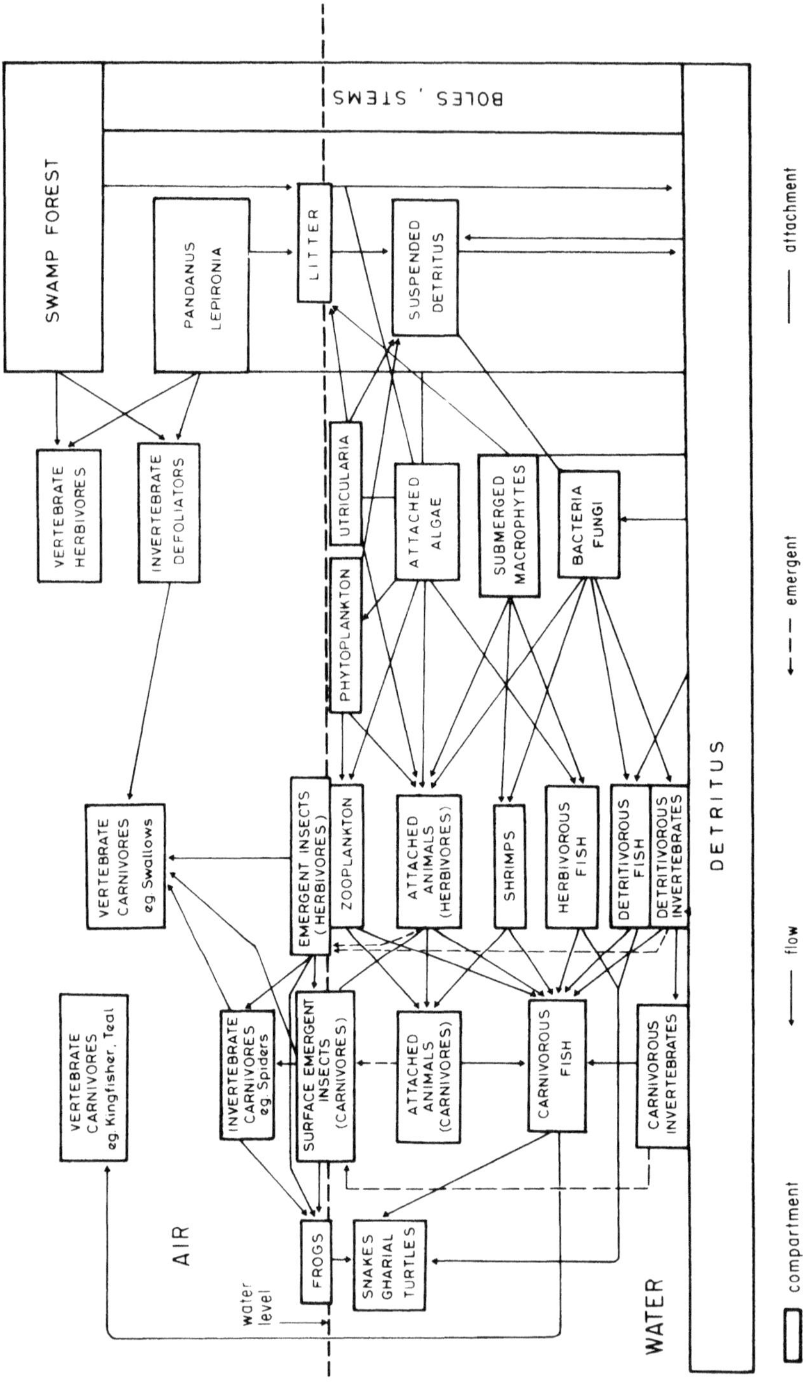

Fig. 133 Diagrammatic presentation of the components of the swamp ecosystem at Tasek Bera. (T. Mizuno and J. I. Furtado)

speciation of freshwater fishes and aquatic insects is not known.

The litter returns substantial amounts of nutrients to the swamp: 85.04 $kg \cdot ha^{-1} \cdot y^{-1}$ N, 1.87 $kg \cdot ha^{-1} \cdot y^{-1}$ P, 30.55 $kg \cdot ha^{-1} \cdot y^{-1}$ K, 12.47 $kg \cdot ha^{-1} \cdot y^{-1}$ Na, 129.23 $kg \cdot ha^{-1} \cdot y^{-1}$ Ca and 37.16 $kg \cdot ha^{-1} \cdot y^{-1}$ Mg (Furtado et al. 1980). The importance of these to trophic production and dynamics, particularly the limiting effects of phosphorus, is not known.

The Tasek Bera is thus a heterotrophic system with and extremely complex food chain that is poorly understood in terms of energy and nutrient flows, symbiotic linkages and speciation.

8.2 Biogeochemical cycling

The high litter production of about 10 $tonnes \cdot ha^{-1} \cdot y^{-1}$ (Furtado et al. 1980) contributes an abundance of nutrients to the swamp: 85.04 $kg \cdot ha^{-1} \cdot y^{-1}$ N, 1.87 $kg \cdot ha^{-1} \cdot y^{-1}$ P, 30.55 $kg \cdot ha^{-1} \cdot y^{-1}$ K, 12.47 $kg \cdot ha^{-1} \cdot y^{-1}$ Na, 129.23 $kg \cdot ha^{-1} \cdot y^{-1}$ Ca and 37.16 $kg \cdot ha^{-1} \cdot y^{-1}$ Mg. Nutrient release is rapid for the first 6 to 8 weeks (Section 4.2.4). And yet, the waters are poor in nutrients (Chapter 3). The nutrients are thus rapidly taken up and stored by the macrophytes. The nature of the availability and uptake of nutrients is not fully known. The importance of fungi over bacteria and invertebrates in decomposition needs further investigation. The differential uptake and storage of nutrients by the macrophytes, as between the woody and non-woody and between the dicotyledons and monocotyledons (Section 4.2.4), needs further investigation. Nervertheless, the macrophytes in the forested and non-forested littoral regions appear important in purifying the waters and being able to cope with organic loading. The capacity of the swamp-forest needs to be examined further as it has implications in pollution control.

The impact of seasonal inundations and dry periods in mediating the processes of decomposition, nutrient transfer from land to water, nutrient loss and nutrient uptake is not fully known. The differential availability of nutrients especially phosphorus would impinge on production and the dynamics of the food chain.

8.3 Productivity
(I. Ikusima, J. I. Furtado, S. Mori, T. Mizuno and Y. Tezuka)

The productivity of the *Pandanus-Lepironia* littoral and the limnetic zones were studied whereas the production of the swamp-forest is not known. The littoral zone is endowed with a rich algal flora comprising Cyanophyta (20 species), Pyrrophyta (5), Bacillariophyta (61), Chrysophyta (7), Euglenophyta (24), Chlorophyta (265), Charophyta (2), and Rhodophyta (3) (Chapter 5). Among these the Chlorophyta, Bacillariophyta and Cyanophyta are dominant, with the Desmidiales constituting 70% of the taxa. This floristic composition is similar to the Indo-Malayan-North Australian Phytogeographical region (Krieger 1938) with an affinity to Sumatra, and to that of dystrophic lakes in the temperate regions; and is probably due to the intense competition for light, especially by attached forms. In contrast to this diversity, the population size or biomass (standing stock) and productivity are low. The gross phytoplankton productivity of 10 mg $C \cdot m^{-3} \cdot day^{-1}$ in the stream channel and 70 mg $C \cdot m^{-3} \cdot day^{-1}$ in the littoral region (Sections 5.5 and 5.6) is comparable to that of tropical lakes (Talling 1965, Ganf 1970, Lewis 1974, Beadle 1974). The saturated light intensity is 15 klx in the photosynthetic curve at 31-32°C, and the assimilation number of 1.3-1.4 under saturated light intensity support this. The decrease in photosynthetic rate first occurs at 20 klx by about 60% of the maximum value. The high total CO_2 and low pH (5.5) suggests an abundance of free CO_2 but a paucity of HCO_3^- for plant fixation. Attached algae at the surface in the open waters, mainly diatoms such as *Frustulia* and *Eunotia* and filamentous green algae such as *Hyalotheca*, *Desmidium*, *Mougeotia* and *Oedogonium*, attain a maximum biomass of 2 g dry $wt \cdot m^{-2}$. At the bottom surface, these algae, mainly *Bulbochaeta*, *Mougeotia*, *Spirogyra* and *Desmidium*, and fungi, attain a biomass of 8-10 g dry $wt \cdot m^{-2}$. The phytoplankton in the open water zone has a lower standing crop (0.7-1.9 mg chl.-$a \cdot m^{-3}$) than in the littoral region (1.3-3.5 mg chl.-$a \cdot m^{-3}$), due to exfoliation of attached algae in the latter. The low productivity and biomass of phytoplankton thus appears to be due to nutrient deficiency.

Attached algae in the littoral region consist mainly of *Desmidium* and *Diatoma* on *Utricularia* shoots, and *Bulbochaeta*, *Spirogyra* and *Mougeotia* and fungi on *Lepironia* reeds. These algae attain a biomass of 37 g dry $wt \cdot m^{-2}$ at the surface of *Lepironia* reeds, and 100 g dry $wt \cdot m^{-2}$ at the bottom. They have a photosynthetic production of 38 mg $C \cdot h^{-1} \cdot m^{-2}$ (reed surface area) (or 100 mg $O_2 \cdot h^{-1} \cdot m^{-2}$) under direct light conditions on a fine day (mean horizontal radiation 0.86 $cal \cdot cm^{-2} \cdot min^{-1}$), and 5.7 mg $C \cdot h^{-1} \cdot m^{-2}$ (or 15 mg $O_2 \cdot h^{-1} \cdot m^{-2}$) under artificially low light intensity (RLI 0.9%). These attached algae are better developed in the littoral region because of substratum availability. They have a short life cycle, and are in low abundance during the dry and the wet seasons. They exfoliate cells in thin sheets from the substrata, due to the action of water currents; and, together with macrophytes, contribute about 400 g dry wt of $litter \cdot m^{-2}$ to the open water especially at high water levels. These epiphytes take about three months to recover after a major flood, as in 1971. The production/respiration (P/R) ratio for attached algae is large in the surface waters and decreases to 0.09 at 2 m depth, indicating that net production is still positive at this depth. Net production is small in the water column because of decomposition at the bottom.

Batrachospermum has a net production (1.68 g $O_2 \cdot m^{-2} \cdot day^{-1}$), 1.3-1.8 times higher than in the temperate region. *Utricularia flexuosa* has a biomass of 370 g dry wt $\cdot m^{-2}$ and a standing crop of 1.9 g chl.-a $\cdot m^{-2}$. Its young apical shoots have a gross photosynthetic rate of 50 mg $O_2 \cdot g^{-1} \cdot h^{-1}$ at light saturation of 20 klx, and a dark respiration rate of 2.2 mg $O_2 \cdot g^{-1} \cdot h^{-1}$. It is thus twice as productive as temperate submerged macropohytes (Ikusima 1973). Its estimated maximum daily net production (in pure stands) of 12 g dry matter $\cdot m^{-2} \cdot day^{-1}$ is less than that of *Eichhornia crassipes* (Westlake 1963). It has a solar energy conversion efficiency (to net production) of about 2.7% which is probably maximum for tropical aquatic macrophytes.

Pandanus helicopus occurs in stands of up to 0.7 ha, and at a density of 4 trunks $\cdot m^{-2}$. Its trunk elongates about 40 cm $\cdot y^{-1}$ with a defoliation of about 38 old leaves $\cdot y^{-1}$. It thus produces litter at the rate of 2.3 kg $\cdot m^{-2} \cdot y^{-1}$, and calcium at 20 g $\cdot m^{-2} \cdot y^{-1}$; these values of dry matter production and calcium supply correspond to those of some poor tropical forests (Rodin and Bazilevich 1967).

Lepironia articulata occurs in extensive pure stands at a density of 140 culms or reeds $\cdot m^{-2}$ and 30 clones $\cdot m^{-2}$. It has an above-ground biomass of 700 g $\cdot m^{-2}$ (190-1,427 g $\cdot m^{-2}$). Below ground, well-developed rhizomes, roots and stems account for about 20% of the total biomass, and have a length of 7 m $\cdot m^{-2}$. The plant propagates vegetatively producing a new culm every 51-55 days. Each culm matures in about seven months, and has a mean longevity of more than 227 days. *L. articulata* has a gross production of 4.89 g dry matter $\cdot m^{-2} \cdot day^{-1}$, which is lower than other macrophytes. It has a low (0.5%) solar energy conversion efficiency (to gross primary production). It produces abundant litter to the benthos (2.24 g $\cdot m^{-2} \cdot day^{-1}$ or 8.2 tonnes $\cdot ha^{-1} \cdot y^{-1}$) (Ikusima 1978).

The *Lepironia-Pandanus* littoral region is a regenerating phase of the *Eugenia* swamp-forest, as indicated by the abundance of tree stumps in the bottom. The Semalai people periodically burn this littoral region presumably to enhance its production. The *Lepironia* stand recovers in more than a year. *Lepironia* is important for dissolved oxygen exchange as substantiated by a higher daily exchange of dissolved oxygen in the water with *Lepironia* (0.45 to −1.00 g $O_2 \cdot m^{-2} \cdot day^{-1}$) than without *Lepironia* (0.39 to −1.02 g $O_2 \cdot m^{-2} \cdot day^{-1}$). The supply of oxygen from air to water (0.38 g $O_2 \cdot m^{-2} \cdot day^{-1}$) is important in oxygenating the water column particularly in the dark period.

The overall effect of this modified littoral region is in increasing the concentration of dissolved oxygen downstream, as from 2.5 mg $O_2 \cdot l^{-1}$ at 25°C at Pos Iskandar to 4.1 mg $O_2 \cdot l^{-1}$ at 29°C at Tanjong Kuin. The aquatic macrophytes, especially the *Utricularia* and microphytes, are thus intensively utilised by a rich variety of animals.

In the *Eugenia* swamp-forest, production is essentially by the dominant emergent macrophytes. Submerged macrophytes such as *Cryptocoryne griffithii* have a high biomass (470 $g \cdot m^{-2}$) and leaf area index (4.2 $m^2 \cdot m^{-1}$) but a low gross photosynthetic production (7.4 mg $O_2 \cdot g^{-1} \cdot h^{-1}$) because of their shade adaptation.

8.4 Conservation and management (S. Mori and J. I. Furtado)

The unique structure and dynamics of the Tasek Bera raise several conservation and management issues for this relic inundated freshwater riparian swamp ecosystem in Peninsular Malaysia, particularly because of the premium value on these lowlands for rural settlements and agro-industrial development:

1) The community metabolism of the Tasek Bera is poorly understood in several respects. The 'black water' colouration represents the balance between production and decomposition, indicating its dystrophic nature. Littoral production which dominates the swamp appears maximal during the monsoon seasons, especially the northeast monsoon, suggesting the importance of nutrient transfer from the catchment area to the swamp, and from the decomposer to the producer compartments, through flushing for stimulating productivity. Secondary (and possibly tertiary) production is also optimal during this period as seen in the breeding of several fish species such as *Scleropages formosus*. A clearer understanding of the dynamics of this swamp system would enhance the transformation and optimisation of similar ecosystems (e.g. riparian ricefields) for future human development.
2) The ecological factors governing speciation in the Tasek Bera and the tropics in general is poorly understood. Tasek Bera is unusually rich in certain plant (e.g. 250 spp. algae especially desmids) and animal species (e.g. 150 spp. fishes especially cyprinoids, siluroids and anabantoids, 33 spp. odonates especially coenagrionids and libellulids, 24 spp. aquatic heteropterans), which constitute an important gene pool for potential exploitation and management, or for potential domestication and aqua-

culture. The processes governing biotic evolution, especially predation in such tropical ecosystems, could provide clues essential for enhancing human cultural evolution.

3) The Tasek Bera and its catchment area are rich in wildlife, both plants and animals. There are for example about 200 species of birds and many large mammals. It is necessary to protect the Tasek Bera area because of the dynamic interactions between land and water systems. While this protection may enhance research, it may facilitate a low level of tourism. It is proposed that an area of at least 200 km^2 be protected for biological purposes, to include the swamp and the adjacent terrestrial systems. Since the swamp is intimately dependent on the catchment area, a zone at least 1,000 m wide should be protected around the swamp.

4) The Tasek Bera can be used economically for the Semalai aborigines who practice subsistence agriculture. They could be introduced to wet rice cultivation at the downstream region of the swamp where *Lepironia* stands abound. Some form of drainage and irrigation system would be necessary to control the water level for rice cultivation. This location of such an agricultural development would minimise the impact of agrochemicals used on the swamp system. In addition a variety of wildlife, both plants (e.g. rattans, tembusu, minyak damar) and animals (e.g. jungle fowl, wild pig, sambar deer), can be harvested on a controlled basis.

5) The Tasek Bera and its catchment area is best conserved as an aboriginal area, national monument and nature reserve, with the controlled exploitation and development of natural resources. The swamp ecosystem and its catchment area are vulnerable at two points: (a) at its northern boundary where it is in contact with the Sungai Bera, where it experiences reverse flow during flood conditions especially the northeast monsoon; and (b) at its southern boundary where the Sungai Palong meets the Tasek Dampar, where it receives water from the Sungai Muar system during floods. Hence, a buffer zone is needed at these two points to protect the Tasek Bera from potential pollution emanating from the Pahang-Bera and the Muar-Palong river systems.

Appendix I. List of contents of Data Book 1 (T. Mizuno 1972) and Data Book 2 (J. I. Furtado 1974).

Data Book 1 (T.Mizuno 1972)

Data Book 2 (J. I. Furtado 1974)

Page

Appendix II. Pollen and spores determined from the Tasek Bera sediments

This list attempts to show the probable source taxa of each of the pollen and spore types determined from the Tasek Bera sediments. It was not possible to study the pollen of all the species in the Malaysian Flora, and this list must therefore be regarded as rather preliminary.

In numerous instances, pollen types observed occur in certain taxa which, on ecological grounds, were unlikely to have contributed pollen to the site. These instances are mentioned in parentheses. It is assumed that pollen originated from taxa occurring at this time in the flora of southern west Malaysia, except in a few cases where the presence of certain pollen types can be explained only in terms of long-distance dispersal from elsewhere or in local extinction.

Notes on the ecology of source taxa are also presented, together with a brief description of like forms. These notes are of necessity brief and incomplete, and are included only to aid those lacking detailed knowledge of the ecology of the Malaysian flora. These notes have been taken freely from the widespread literature on the ecology of west Malaysian vegtation, and also from personal observations. Space does not permit acknowledgement of observations in this list. Foremost sources of information on the ecology of the relevant taxa are the works of Corner (1940), Holttum (1968), Ridley (1907, 1922-27), Wyatt-Smith (1963, 1964), the available volumes of the series 'A tree flora of Malaya' (eds. T. C. Whitmore and Kochummen) and Flora Malesiana (ed. van Steenis). Notes on the pollination mechanisms of source taxa are given when this is considered important (e.g. for anemophilous taxa which produce proportionately large quantities of pollen).

In the attribution of pollen types to ecological groups, attribution has been made only to those habitats from which it is thought most of the pollen observed has been derived. The source taxa of pollen types may by no means be rigourously restricted to the habitats suggested.

Abbreviations used for ecological groups:
F = Forest; NF = Non-Forest; R = Riparian; S = Swamp.

Pollen type	Source taxa	Ecol. group	Source ecology
Angiosperms			
Acalypha	*Acalypha* spp.	NF	Small herbs or shrubs; abundant on waste ground.
Aglaia comp	*Aglaia* spp. *Amoora* spp.	F+R	Common trees throughout lowland forests, being especially characteristic along rivers (*Aglaia* spp.) and in swamp-forest (*Amoora* spp.).
Alnus	*Alnus* spp.	?F	Warm temperate to temperate trees. It is probable that pollen of this type is windblown from Indo-China where *Alnus* spp. are indigenous. Another possibility is that it is reworked from tertiary deposits. Laboratory contamination is considered unlikely.
Antidesma	*Antidesma* spp.	F	Widespread forest trees.
Ardisia crispa comp	*Ardisia crispa* and some other *Ardisia* spp.	F	Common forest trees.
Arenga	*A. pinnata* or *A. westerhouti*	F+NF	*A. pinnata* is cultivated widely. *A. westerhouti* is widespread in lowland areas.
Arisaema comp	Probably *Arisaema* spp. Other *Araceae* may have pollen of this type	?S	*Arisaema* spp. are often common along river banks.
Austrobuxus nitidus	*Austrobuxus nitidus*	R+S	Common trees of swamp-forest and riparian communities.
Baeckia comp	*Baeckia* spp.	F+NF	Small trees/shrubs. Common on mountain ridges and on poor soils in lowland areas. Common around Tasek Bera.
Barringtonia	*Barringtonia* spp.	F+NF+R+S	Common trees, particularly of coastal and riparian vegetation and secondary forest.
Boea comp	*Boea* spp. and possibly some other genera of the Gesneriaceae	F	Common forest herbs.

Pollen type	Source taxa	Ecol. group	Source ecology
Burseraceae	The genera *Canarium*, *Dacryodes* and *Santiria*	F	Burseraceae forms the second most important family of forest trees in west Malaysia. *Dacryodes* spp. and *Santiria* spp. occur frequently in swamp-forests.
Calophyllum	*Calophyllum* spp.	F+NF+R+S	*calophyllum* spp. are common in most west Malaysian vegetation types, but are particularly characteristic of swamp-forests.
Campnosperma	*Campnosperma* spp.	F+NF+S	Frequent in primary, regrowth and swamp-forests, but particularly characteristic of the latter.
Canthium dicoccum comp	*C. confertum*, *C. dicoccum* and *C. glabrum*	F+R	*Canthium* spp. are small trees, occurring throughout west Malaysia. *C. dicoccum* is often found gregariously along tidal rivers. *C. confertum* frequents streamsides.
Canthium sp.	A *Canthium* sp. other than the three species mentioned above	F	Most *Canthium* spp. other than those mentioned above occur mostly in forest or along forest margins.
Caryota	*Caryota* spp.	F+NF+S	The probable source of this pollen is *C. mitis* which occurs most commonly in forest, but also frequents some regrowth communities and occurs in swamp-forest at Tasek Bera.
Claoxylon spp.	*Claoxylon* spp.	F	Many species are common in lowland forest.
Compositae	Possibly *Erigeron*	*NF*	*Non-forest herbs.*
Croton comp	*Croton* spp., *Crodiaeum* spp. (possibly other genera of the Crotonoideae)	F+NF	Largely small trees or shrubs of primary and regrowth communities.
Cyperaceae	*Lepironia articulata* and other Cyperaceae	S	*Lepironia articulata* occupies extensive areas at Tasek Bera. Other Cyperaceae spp. are mainly swamp herbs but sometimes occur in regrowth communities.
Daemonorops comp	Many *Calamus* spp. and *Daemonorops* spp.	F	*Calamus* spp. and *Daemonorops* spp. are common climbers throughout lowland and montane forests.
Dillenia	*Dillenia* spp.	F+R+S	Tree and shrubs, common on river banks and in forsts, swamps and regrowth communities.

Pollen type	Source taxa	Ecol. group	Source ecology
Dipterocarpus	*Dipterocarpus* spp.	F	Large trees of low altitude primary forests. Some species are characteristic of riversides, e.g. *D. oblongifolius*.
Dissochaeta	*Dissochaeta* spp.	F+NF	*Dissochaeta* spp. are predominantly epiphytic shrubs, commonly occurring on forest edges.
Dombeya	?*Dombeya* spp.	?F	Small trees or shrubs, occurring rarely in west Malaysia.
Drypetes comp	Probably *Drypetes* sp.	F	Common trees of lowland forests.
Duabanga	*Duabanga grandifolia*	F	Common trees of lowland and montane forests in northern and central west Malaysia. Particularly abundant in foot hills, and at higher elevations, along streams.
Durio zibethinus comp	*D. zibethinus*, *D. carinatus* and possibly some other *Durio* spp.	F+NF	*D. zibethinus* is widely cultivated for its fruit. The species is not, however, indigenous, being native to Borneo. *D. carinatus* frequents swamp-forests; many *Durio* spp. occur scattered throughout lowland forests. Pollinated by bats.
Elaeocarpus comp	Predominantly *Elaeocarpus* spp.	F+NF+S	*Elaeocarpus* spp. are common in swamp vegetation at Tasek Bera. They are also characteristic of forests and regrowth vegetation.
Elatostema comp	*Elatostema* spp.	F+NF	Common herbs of forest floor and seal vegetation.
Elytranthe	*Elytranthe* spp.	F+NF	Parasitic shrubs on forest and non-forest trees.
Engelhardia	*E. serrata* (other *Engelhardia* spp.)	F+R	*E. serrata* occurs commonly along rivers and sometimes, in swamp-forests. Other *Engelhardia* spp. are largely confined to montane areas. Probably largely wind pollinated.
EugeniA comp	*Eugenia* spp. and some other Myrtaceous genera, e.g. *Tristania*	F+S	*Eugenia* spp. are widespread in most vegetation types in west Malaysia. At Tasek Bera, they are particularly important in swamp-forest. *Tristania* spp. are also characteristic of dry-land and swamp-forests.
Fagraea sim	*Fagraea* spp.	F+S	Common trees of dry-land and swamp-forests.
Flagellaria indica	*F. indica*	S	Climbing herb, especially characteristic of open swamp-forest, or herbaceous swamp communities.

Pollen type	Source taxa	Ecol. group	Source ecology
Garcinia griffithii comp	*G. griffithii* and possibly some other species	F	Common forest trees, particularly abundant in lowland forest.
Garcinia merguensis comp	*G. eugeniifolia*, *G. merguensis* and possibly *G. rostrata* (pollen of the latter species was not examined)	F	Trees of lowland and forests.
Gardenia	*Gardenia* spp.	F+R	Common streamside trees, present also in very damp forest.
Glochidion	*Glochidion* spp.	NF+S	Common trees of swamps and regrowth vegetation.
Gramineae	Most Gramineae spp.	NF	Mainly non-forest herbs, becoming dominant in degraded regrowth communities, e.g. lalang. Some species occur in swamps. It is thought, however, that most pollen observed here originated from dry-land communities. Largely wind pollinated. Many species are autogamous.
Gynotroches	Mainly *G. axillaris*, possibly some confusion with other similar pollen types, eg. *Uncaria* comp. *Saurauia* comp	F+NF+S	Common tree of regrowth and swamp-forests.
Haloragis sim	Probably not *Haloragis*. The real source of this five-porate pollen type is not yet known	?S	*Haloragis* spp. are swamp herbs.
Hopea comp (4 colpate)	Some *Hopea* spp. and *Shorea* spp.	F	Large trees of low altitude primary forests.
Ilex	*Ilex* spp.	F+NF+S	Trees, widespread throughout west Malaysia, particularly in mountains, in swamp vegetation and in regrowth communities.
Jussieua	*Jussieua* spp.	S	Common swamp herbs.

Pollen type	Source taxa	Ecol. group	Source ecology
Koompassia	*K. malaccensis* (*K. excelsa*)	F	*K. malaccensis* is extremely widespread in lowland forests, being west Malaysia's third commonest tree. *K. excelsa* frequests streamsides but is restricted to north and central Malaya.
Korthalsia sim	*Korthalsia* spp.	F	Common rattans of forests.
Linociera comp	*Linociera* spp. and possibly other Oleaceous genera	F+S	*Linociera* spp. occur commonly in lowland forests. *L. paludosa* is common at Tasek Bera in swamp vegetation.
Lithocarpus comp	*Lithocarpus* spp. and *Castanopsis* spp.	F	Common trees of lowland and montane forests.
Livistona comp	*Livistona* spp.	F+S	*L. kingiana* frequents lowland swamp-forests and valley bottoms. Several other species occur, often locally, in montane forests. Some species are frequently planted.
Lophopetalum	*Lophopetalum* spp. (⅓ Indet No. 88)	S	Common trees of swamp-forests.
Macaranga comp	*Macaranga* spp. and several *Mallotus* spp.	F+NF+R+S	Common trees of non-forest areas, often pioneering cleared areas. Some species occur also in swamps.
Mallotus comp	*M. macrostachyus* and possibly some other *Mallotus* spp.	F+NF	*Mallotus* spp. are characteristic of both forest and non-forest communities. *M. macrostachyus* is common in regrowth forest at Tasek Bera.
Medinilla comp	*Medinilla* spp. and possibly in related genera	F	Common epiphytes of forests, forest margins and swamp vegetation.
Meliaceae/ Sapotaceae undiff.	Possibly predominantly from *Palaquium* spp. but also other genera of both Meliaceae and Sapotaceae	F	Common trees of lowland forests.
Melastoma malabathricum comp	*M. malabathricum* and some other *Melastoma spp. (Terminalia, Dactylocladus)*	NF+S	Common shrubs of open spaces. Present in herbaceous swamp vegetation at *Tasek Bera*.

Pollen type	Source taxa	Ecol. group	Source ecology
Mimusops comp	*Mimusops* spp. and other Sapotaceae	F	Forest trees.
Morinda comp	*Morinda* spp.	F+NF	Common trees of forests and regrowth communities.
Myrica	*M. esculenta* (*M. javanica*)	F+NF	*M. esculenta*, a shrub or small tree, occurs widely in open areas and degraded forest in southern west Malaysia. *M. javanica* is restricted to montane areas. Pollinated by wind.
Myrsinaceae	Predominantly *Ardisia* spp. and *Maesa* spp.	F	Common widespread trees and shrubs.
Nauclea comp	*Anthocephalus* spp., *Nauclea* and *Neonauclea* spp.	F+NF	*Nauclea* spp. occur commonly in low-*land and montane forest. Anthocephalus* spp. and *Neonauclea* spp. occur in forests but are most characteristic of regrowth vegetation.
Nepenthes	*Nepenthes* spp.	F+NF+S	Common climbing herbs, particularly characteristic in open areas, on swamps and in forests.
Oncosperma	*Oncosperma horridum* (*O. tigillarium*)	F	Large palms, *O. horridum* is common in lowland and hill forests. *O. tigillarium* occurs only within proximity to the coast.
Oroxylum indicum comp	*O. indicum*	NF+R	Frequents seral communities and riparian fringes. Pollinated by bats.
Palmae undiff.	Palmae spp.	?F	Probably forest palms.
Pandanus	*Pandanus* spp. principally *P. helicopus*	S	*P. helicopus* is dominant in swamp vegetation at Tasek Bea, also being common along tidal rivers. Other *Pandanus* spp. occur in varying degrees of abundance in many vegetation types. Pollinated by wind.
Parkia	*Parkia* spp.	F+NF	Common lowland forest trees, often encouraged by man for its fruits. Bat pollinated.
Payena sim	*Payean* spp. and possibly related genera	F	Common trees throughout lowland forests.

Pollen type	Source taxa	Ecol. group	Source ecology
Pentace	Some *Pentace* spp. e.g. *P. curtisii*, *P. polyalthia*	F	Widespread lowland forest trees.
Pilea comp	*Pilea* spp.	F+NF	Common herbs of forest floor and also in seral vegetation.
Pinanga	*Pinanga* spp.	F	Common understorey palms.
Pithecellobium comp	*Pithecellobium* spp., *Albizzia* spp.	F	Common lowland forest trees.
Phyllanthus nutans comp	*P. nutans*, possibly other *Phyllanthus* spp.	NF	Small shrubs and herbs of open ground.
Planchonella comp	*Planchonella* spp. and possibly related genera	F	Widespread forest trees.
Pometia	*Pometia* spp.	F+R	Common trees of lowland forests and streamsides.
Poranthera comp	*Poranthea* spp. and possibly related genera	NF	Shrubs of regrowth vegetation.
Proteaceae/ Sapindaceae types I and II		?F	Probably forest trees.
Psychotria comp	Psychotria spp.	F+NF	Herbs and shrubs of forests and regrowth vegetation.
Pterospermum	*Pterospermum* spp.	F+R	Common forest trees. *P. javanica* frequents seral communities along rivers, especially on freshly accumulating mud-banks.
Quercus comp	*Quercus* spp., possibly also some *Myrsine* spp.	F	Widespread forest trees, especially characteristic of montane forests. Pollinated by wind.
Saurauia comp	Predominantly *Saurauia* spp.	F+NF+R	Common understorey trees of forests, river banks and regrowth communities.
Shorea comp	*Hopea* spp., *Shorea* spp. and *Vatica* spp.	F	Large, common trees often forming dominant component of low altitude, primary forests.
Schoutenia	*S. accrescens*, *S. glomerata* other *Schoutenia* spp.)	F+R	Both *S. accrescens* and *S. glomerata* frequent lowland forest. *S. glomerata* occurs commonly along Saraca streams.
Symplocos	*Symplocos* spp.	F+NF	Common trees of both forests and regrowth vegetation.
Trema	*Trema* spp.	F+NF	Common non-forest trees. Sometimes present in primary forest.

Pollen type	Source taxa	Ecol. group	Source ecology
Uncaria comp	Possibly *Uncaria* spp.	F+NF+S	Common climbing herbs of open areas and swamp-forests.
Utricularia flexuosa	*Utricularia flexuosa*	S	One of dominant aquatic species at Tasek Bera.
Utricularia sp.	Other *Utricularia* spp. esp. *U. aurea*	S	*U. aurea* and other *Utricularia* spp. are widespread at Tasek Bera.
Urticaceae/ Moraceae undiff.	Many genera of Moraceae and Urticaceae, eg. *Artocarpus*, *Pipturus*, *Poikilospermum*	F+NF+R+S	Representatives of Moraceae and Urticaceae occur in practically all life forms and in almost every ecological niche.
Walsura comp	*Walsura* spp. and related genera	F	Widespread forest trees.
Gymnosperms			
Dacrydium	*Dacrydium* spp.	F	Common on high mountains. Wind pollinated.
Pinus	?*Pinus merkusii*, ?*P. insularis*	NF	Not native to west Malaysia. Pollen possibly windblown from Sumatra or the Philippines, or from plantations. Some specimens possibly reworked from tertiary deposits. Wind pollinated.
Podocarpus neriifolius comp	*P. polystachyus* (*P. neriifolius*)	F	*P. polystachyus* occurs infrequently in lowland forest, and commonly in swamp-forest and coastal associations. *P. neriifolius* is common in mountains. Wind pollinated.
Pteridophyta			
Athyrium comp	*Athyrium* spp.	F	Characteristic terrestrial ferns of forests.
Cyathea comp	*Cyathea* spp.	F+NF	*Cyathea* spp. are particularly characteristic of montane vegetation but frequent most regrowth communities.
Cyclosorus sect. *Sphaerostephanos*	*C. callosa*, *C. megaphyllus* and related spp.	F+NF	Common terrestrial ferns of forest communities and regrowth vegetation.
Davallia comp	*Davallia* spp.	F+NF	Epiphytes of forests and regrowth vegetation.
Dicranopteris comp	*Dicranopteris linearis* and possibly other *Dicranopteris* spp.	NF+S	Abundant thicket forming ferns. Common in regrowth vegetation. Also present in swamp vegetation at Tasek Bera.

Pollen type	Source taxa	Ecol. group	Source ecology
Filices undiff. (predominantly spores which have lost their exosporium)	Many genera of Dennstaedtiaceae, Polypodiaceae, Thelypteridaceae	F+NF+S	The source taxa of these spores probably occur in every ecological niche.
Filices areolate	Predominantly *Nephrolepis* spp. Also some species of Polypodiaceae	F+NF+S	*Nephrolepis* spp. are characteristic of forests and regrowth vegetation. Also common on swamps.
Gleichenia comp	*Gleichenia* spp., possibly some *Dicranopteris* spp.	NF	Abundant thicket forming ferns, characteristic of regrowth vegetation.
Hymenophyllum comp	*Hymenophyllum* spp. and *Trichomanes* spp.	F	Forest epiphytes.
Lindsaya comp	Predominantly *Lindsaya* spp.	F+NF	Ferns of forests and regrowth communities.
Lycopodium cernuum	*L. cernuum*	NF+S	Common in regrowth vegetation and on swamps.
Lycopodium clavatum group	*L. clavatum* and related spp.	F+NF	Herbs of primary and regrowth vegetation.
Lycopodium phlegmaria group	*L. phlegmaria*	F+S	Common epiphytes of dry-land and swamp-forest trees.
Lycopodium sp.	*Lycopodium* spp.	?F+N+F+S	?Herbs of primary, regrowth or swamp vegetation.
Lygodium reticulatum comp	*L. reticulatum*	NF	Climbing herbs of secondary vegetation.
Lygodium scandens	*L. scandens*	NF+S	Climbing herbs of secondary vegetation.
Platycerium sim	Predominantly *Platycerium* spp.	F+NF	Epiphytes on primary and regrowth forest trees.
Pteris (four morphological types)	*P. longipinnula*, *P. tripartita*, *P. vittata* and possibly other species.	F+NF	Common ferns particularly in disturbed areas. Some species occur in primary forests.
Selaginella megastachys group (sp. III)	*Selaginella* spp.	F+NF	Herbs of forest and non-forest.
Selaginella planata comp	Predominantly *S. planata*	F+NF	Herbs of forest and non-forest.
Stenochlaena areolaris	*S. areolaris*		Epiphyte on *Pandanus* spp. Presently restricted to the Philippines and New Guinea. A single spore observed, due to large size unlikely to be windblown. In view of the preference of this species for *Pandanus, it may previously have been indigenous to the area.*
Stenochlaena palustris	*S. palustris*		Climbing fern, common in disturbed forest or open areas.
Taenitis blechnoides comp	*T. blechnoides*		Widespread in forests.

References

Alexander, C. P., 1931. The crane-flies (Tipulidae, Diptera), Deutsche Limnologische Sunda Expedition. Arch. Hydrobiol. Suppl., 9: 135-191.

Alfred, E. R., 1963. Notes on a collection of fresh-water fishes from Penang. Bull. Nat. Mus. Singapore, 32: 143-154.

Afred, E. R., 1964a. Notes on a re-examination of some Bleeker type specimens of Indo-Malayan fresh-water fishes. Part 2. Abramidinae and Rasborinae. Bull. Nat. Mus. Singapore, 32: 128-134.

Alfred, E. R., 1964b. Some colourful fishes of the genus *Puntius*, Hamilton. Bull. Nat. Mus. Singapore, 32: 135-142.

Alfred, E. R., 1964c. Some comments on the type specimens of Malayan fishes described by George Dunker. Bull. Nat. Mus. Singapore, 32: 165-166.

Alfred, E. R., 1966a. The fresh-water fishes of Singapore. Zoologische Verhandelingen, 78: 1-68.

Alfred, E. R., 1966b. The fresh-water food fishes of Malaya. 2. *Xenentodon cancila*. Fed. Mus. J. Kuala Lumpur, 11: 77-79.

Alfred, E. R., 1969a. Conserving Malayan fresh-water fishes. Malay. Nat. J., 22(2): 69-74.

Alfred, E. R., 1969b. The Malayan Cyprinoid fishes of the family Homalopteridae. Zoologische Mededelingen, 43(18): 213-237.

Alliwill, P. M., 1968. The loss of elements from decomposing litter. Ecology, 49: 142-145.

Anderson, H. H., 1889. Notes on Indian Rotifers. J. Asiat. Soc. Beng., 58: 345.

Anderson, J. A. R. and Muller, J., 1975. Palynological study of a Holocene peat and a Miocene coal deposit from N.W. Borneo. Rev. Palaeobotan. Palynol., 19: 290-351.

Ang, K. J., 1971. Seasonality and dynamics of littoral zooplankton of some small lakes and the production of one of them: *Simocephalus serrulatus* (Koch). Ph.D. thesis, University of Waterloo, Waterloo, Canada. 259 pp.

Aruga, Y. and Monsi, M., 1963. Chlorophyll amount as an indicator of organic matter productivity in bio-communities. Plant & Cell Physiol., 4: 29-39.

Arumugam, P. T., 1971. Limnological studies of a reservoir, Subang Lake, Klang, with special reference to algal blooms and pollution. B.Sc. Ecology honours thesis, University of Malaya, Kuala Lumpur, Malaysia. 71 pp.

Balss, H., 1937. Potamoniden (Decapoda Brachyura) der Philippinen und des Malaisischen Archipels. Int. Rev. Hydrobiol., 34: 143-187.

Barton, 1965. Drainage patterns in the Malay Peninsula. (Unpublishing report).

Beadle, L. C., 1974. The inland water of tropical Africa, an introduction to tropical limnology. Longman, New York. 365 pp.

Beattie, J., Bricker, C. and Garvin, D., 1961. Photolytic determination of trace amounts of organic material in water. Anal. Chem., 33: 1890-1892.

Beauchamp, R. S., 1953. Sulphates of African inland water. Nature, 171: 769-771.

Beeton, A. M., 1957. Relationship between Secchi disk readings and light penetration in Lake Huron. Trans. Amer. Fish. Soc., 87: 73-79.

Behre, K., 1956. Die Süsswasseralgen der Wallacea-Expedition. Arch. Hydrobiol. Suppl., 23: 1-104.

Benninghoff, W. S. and Kapp, R. O., 1962. Suggested notation to indicate identification status of fossil pollen. Pollen Spores, 4: 332.

Bishop, J. E., 1971. Limnology of a small Malayan river, Sungei Gombak. Ph.D. thesis, University of Malaya, Kuala Lumpur, Malaysia, 504 pp.

Bishop, J. E., 1973. Limnology of a small Malayan river, Sungai Gombak. Dr. W. Junk Pub., The Hague.

Blache, J., 1950. Considération sur le Plancton de surface des Eaux Douces du Cambodge. Service de la pêche et de la chasse en Indochine, Phnom Penh, 18-36.

Blache, J., 1951. Aperçu sur le plancton des eaux douces du Cambodge. Cubium publié par l'Association des Amis du Laboratoire des pêches Coloniales, 6: 62-94.

Black, A. P., 1963. Characteristics of colored surface water. Am. Water Works Assoc. Jour., 55: 753-769.

Black, A. P., Singley, J. E., Whittle, G. P. and Maulding, J. S., 1963. Stoichiometry of the coagulation of color causing organic compounds with ferric sulfate. Am. Water Works Assoc. Jour., 55: 1347-1366.

Brandt, R. A. M., 1968. Description of new non-marine molluscs from Asia. Arch. fur. Mollusken Kunde, 98(5-6): 213-284.

Brehm, V., 1933. Die Cladoceren der deutschen Limnologischen Sunda-Expedition. Arch. Hydrobiol., Suppl., 11: 631-771.

Brehm, V., 1938. Die Cladoceren der Wallacea-Expedition. Internat. Revs. ges. Hydrobiol. Hydrogr., 38: 99-124.

Brehm, V., 1953a. Bemerkungen zu den tiergeographischen Verhältnissen der indischen Süsswasser-Caladnoiden. Oesterr. Zool. Zeit., Bd. 4, Heft 4/5.

Brehm, V., 1953b. Indische Diaptomiden, Pseudodiaptomiden und Cladoceren. Oesterr. Zool. Zeit., Bd. 4, Heft 3: 241-345.

Brittan, M. R., 1954. The Cyprinid fish genus *Rasbora* in Malaya. Bull. Raffles Mus. Singapore, 25: 129-156.

Brooks, J. L. and Deevey, E. S., Jr. 1963. New England. In Frey, D. J. (ed.), Limnology of North America. University of Wisconsin Press, Madison, pp. 117-172.

Bullock, J. A. and Furtado, J. I., 1968. Population assessment and movement in *Ptilomera dramas* Breddin (Hemiptera-Gerridae). Arch. Hydrobiol., 64(2): 121-130.

Carey, I., 1976. Orang Asli, the aboriginal tribes of Peninsula Malaysia. Oxford University Press, Kuala Lumpur. 376 pp.

Carlander, K. D., 1969. Handbook of freshwater fishery biology, Vol. 1. Iowa State Univ. Press, Ames, Iowa.

Carter, G. S. and Beadle, L. C., 1930. The fauna of the swamps of the Paraguayan Chaco in relation to its environment. I. Physico-chemical nature of the environment. Linn. (Zool.), 37: 205-258.

Carter, G. S., 1955. The papyrus swamps of Uganda. W. Heffer and Sons, Cambridge. 25 pp.

Cheng, L. and Fernando, C. H., 1969. A taxonomic study of the Malayan Gerridae (Hemiptera: Heteroptera) with notes on their biology and distribution. Oriental Insects, 3(2): 97-160.

Chevey, P. and Poulain, F. Le, 1940. La pêche dans les eaux douces du Cambodge. Mém. Inst. Océnogr. Indoch., 5: 1-193.
Christman, R. F. and Gassemi, M., 1966. Chemical nature of organic color in water. Am. Water Works Assoc. Jour., 58: 723-741.
Christman, R. F. and Minear, R. A., 1971. Organics in lakes. In Faust, S. J. and Hunter, J. V. (eds.), Organic compounds in aquatic environments. Marcel Dekker, New York, pp. 119-143.
Chung Sooi Keong, 1973. Geological map of West Malaysia. Scale 1: 500,000, 7th ed.
Clarke, G. L., 1941. Observation on the transparency in the south western section of the North Atlantic Ocean. J. Mar. Res., 4: 221-230.
Coe, R. L., Freeman, P. and Mattingly, P. F., 1950. Handbooks for the identification of British insects, Diptera. 2. Nematocera: families Tripulidae to Chironomidae. Royal Entomological Soc. London. IX. 2., London.
Corbet, P. S., 1962. A biology of dragonflies. Witherby, London. 247 pp.
Corner, E. J. H., 1940. Wayside trees of Malaya. Govt. Printer, Singapore. 2 vols.
Dale, W. L., 1959. The rainfall of Malaya, Part I. J. Trop. Geogr., 13: 23-37.
Dale, W. L., 1960. The rainfall of Malaya, Part II. J. Trop. Geogr., 14: 11-28.
Dale, W. L., 1963. Surface temperatures in Malaya. J. Trop. Geogr., 17: 57-71.
Dickinson, C. H. and Pugh, G. J. F. (eds.), 1974. Biology of plant litter decomposition. Vol. 1 and 2. Academic Press, London.
Dumanski, J. and Ooi, C. H., 1966. Reconnaissance soil survey of the Temerloh-Gemas region, Malaya. Malay Soil Survey Rep. No. 5. 67 pp.
Dunn, I. G., 1967. Diurnal fluctuations of physico-chemical conditions in a shallow tropical pond. Limnol. Oceanogr., 12: 151-153.
Erdtman, G., 1943. An introduction to pollen analysis. Chronica Botanica, Waltham, Mass., 293 pp.
Faegri, K. and Iversen, J., 1964. Text-book of modern pollen analysis. Munksgaard, Copenhagen. 2nd rev. ed. 237 pp.
Fernando, C. H. and Lim, A. W., 1964. A guide to the freshwater crabs (Potamonidae) of Malaya. Department of Zoology, University of Singapore, No. 6.
Finnegan, S., 1933. A new species of mite parasite on the spider *Liphistius malaynus*, from Malaya. Proc. Zool. Soc., Lond., 2: 413-417.
Flenley, J. R., 1967. The present and former vegetation of the Wabag Region of New Guinea. Unpublished Ph.D. thesis. Australian National University.
Fotiyev, A. V., 1968. Nature of water humus. Geochemistry, 179: 183-186.
Fowler, H. W., 1938. A list of the fishes known from Malaya. Fish. Bull. Singapore, 1: 2-268.
Fritsch, F. E., 1931. Some aspects of the ecology of freshwater algae. J. Ecol., 19: 233-272.
Fryer, G. and Iles, T. D., 1972. The Cichlid fishes of the Great Lakes of Africa, their biology and evolution. Oliver & Boyd, Edinburgh. 1-641.
Furtado, J. I., 1966. Studies on Malayan Odonata. Ph.D. thesis, University of Malaya, Kuala Lumpur, Malaysia. 280 pp.
Furtado, J. I., 1969. Ecology of Malayan odonates: Biotope and association of species. Ver. Internat. Verein. Limnol., 17: 865-887.
Furtado, J. I., 1974. Checklist of fishes collected from Tasek Bera. Malaysian-Japanese IBP (PF) Research Data on Tasek Bera No. 2., pp. 265-266. (restricted circulation).
Furuya, Y., 1971a. On some larvae and pupa of Trichoptera in Thailand. Rep. Nara Hydrobiology, 4: 12-17. (in Japanese).
Furuya, Y., 1971b. On *Nymphula* sp., Lepidoptera, from Malaysia. Rep. Nara Hydrobiology, 4: 17-18. (in Japanese).

Ganapati, S. V., 1955. Diurnal variations in dissolved gases, hydrogen ion concentration and some of the important dissolved substances of biological significance in three temporary rock pools in stream beds at Mettur Dam. Hydrobiologica, 7: 283-303.
Gates, D. M., 1965. Energy exchange in the biosphere. Harper & Row, New York, and John Weatherhill, Inc., Tokyo. 151 pp.
George, M. G., 1961. Diurnal variations in two shallow pounds in Delhi, India. Hydrobiologia, 18: 265-273.
Giltay, L. S., 1935. Liste des Arachnides d'Extrême-Orient et des Indes orientales recueillis, en 1932, par S.A. R. le prince Leopold de Belgique. Bull. Mus. Roy. Hist. Nat. Berg., 11(20): 1-15.
Gobbett, D. J. and Hutchison, C. S., 1973. Geology of the Malay Peninsula. Wiley-Interscience. 438 pp.
Gobbett, D. J. and Tjia, H. D., 1973. Tectonic history. In: Gobbett, D. J. and Hutchinson, C. S. (eds.), Geology of the Malay Peninsula. Wiley-Interscience, pp. 305-334.
Golterman, H. L. and Clymo, R. S. (eds.), 1969. Methods for chemical analysis of freshwater (IBP Handbook No. 8), Blackwell Scientific Publ., Oxford. 213 pp.
Habe, T., 1964. Freshwater Molluscan fauna of Thailand, Nat. Life. Southeast Asia III, Further taxonomic notes on the Malayan species. Bull. Raffles Mus., 26: 67-119.
Hamilton, A. C., 1972. The interpretation of pollen diagrams from highland Uganda. Palaeo-ecology of Africa, 7: 45-149.
Handa, N., 1966. Distribution of dissolved carbohydrate in the Indian Ocean. Jour. Oceanogr. Soc. Japan, 22: 16-22.
Hauer, J., 1938. Die Rotatorien von Sumatra, Java und Bali nach den Ergebnissen der deutschen Limnologische Sunda-Expedition. Arch. Hydrobiol. Suppl. XV: 296-384, 507-602.
Hayashi, H., Okino, T. and Aoyama, K., 1972. The balance of organic matter in a water column in a lake. I. Examinations into the property of the concept of water column, using an artificial water column separated by vinyl sheet. Japan J. Limnol., 33: 51-59.
Herre, A. W. C. T., and Myers, G. S., 1938. A contribution to the ichthyology of the Malay Peninsula. Part 2. Fresh-water fishes. Bull. Raffles Mus. Singapore, 13: 53-74.
Herre, A. W. C. T., 1940a. New species of fishes for the Malay Peninsula and Borneo. Bull. Raffles Mus. Singapore, 16: 5-26.
Herre, A. W. C. T., 1940b. Addition to the fish fauna of Malaya and notes on rare or little known Malayan and Bornean fishes. Bull. Raffles Mus. Singapore, 16: 27-56.
Ho, S. C., 1970. A limnological and productivity study of the Ampang impounding reservoir, Kuala Kumpur. B.Sc. Ecology honours thesis, University of Malaya, Kuala Lumpur, Malaysia. 59 pp.
Hodder, B. W., 1968. Man in Malaya. Univ. of London Press. 144 pp.
Hogetsu, K., Ichimura, S., Hori, S., Oshima, Y. Kasanaga, H., Ono, H. and Takada, K., 1954. Phytoecological studies of Ozegahara moor. (in Japanese). In Ogura, Y. (ed.), 'Report of general surveys on the Ozegahara Moor'.
Holttum, R. E., 1968. A revised flora of Malaya. Vol. 2, Ferns of Malaya. Govt. Printer, Singapore, 2nd ed. 563 pp.
Hora, S. L., 1941. Notes on malayan fishes in the collection of the Raffles Museum, Singapore. Part 2 and 3. Loaches of the families Cabitidae and Homalopteridae. Bull. Raffles Mus. Singapore, 17: 44-64.
Hora, S. L. and Gupta, J. L., 1941. Notes on Malayan fishes in the collection of the Raffles Museum, Singapore. Part 1. Catfishes of the families Siluridae, Bagridae, Amblycepidae,

Akysidae, Sisoridae, Chacidae, Schilbeidae and Clariidae. Bull. Raffles Mus. Singapore, 17: 12-43.
Hutchinson, G. E. and Loffler, H., 1956. The thermal classification of lakes. Proc. Nat. Acad. Sci. Wash., 42: 84-86.
Hutchinson, G. E., 1957. A treatise on limnology. Vol. 1. Geography, physics and chemistry. John Wiley, New York. Part. 1: 672 pp., Part 2: 1015 pp.
Hutchinson, G. E., 1967. A treatise on limnology. Vol. II. John Wiley, New York. 1115 pp.
Hynes, H. B. N., 1961. The invertebrate fauna of a Welsh mountain stream. Arch. Hydrobiol., 57: 344-388.
Hynes, H. B. N., 1970. The ecology of running waters. Liverpool University Press. 555 pp.
Hynes, H. B. N. and Kaushik, N. K., 1969. The relationship between nutrients and protein production in submerged autumnal leaves. Verh. Internat. Verein. Limnol., 17: 95-103.
Ichimura, S., 1956. On the ecological meaning of transparency for the production of matter in phytoplankton community of lake. Bot. Mag. Tokyo, 69: 219-226.
Ikusima, I., 1965. Ecological studies on the productivity of aquatic plant communities I. Measurement of photosynthetic activity. Bot. Mag. Tokyo, 78: 202-211.
Ikusima, I., 1967. Ecological studies on the productivity of aquatic plant communities III. Effect of depth on daily photosynthesis in submerged macrophytes. Bot. Mag. Tokyo, 80: 57-67.
Ikusima, I. 1970. Ecological studies on the productivity of aquatic plant communities IV. Light condition and community photosynthetic production. Bot. Mag. Tokyo, 83: 330-341.
Ikusima, I., 1972. Productivity of aquatic plant communities, Vol. I. Aquatic macrophytes. Kyoritsu Shuppan, Tokyo, (in Japanese). 98 pp.
Ikusima, I., 1973. Primary production of Lake Bera. Proceedings of the East Asian Regional Seminar for IBP, Kyoto, Japan. pp. 73-81.
Ikusima, I., 1978. Primary production and population ecology of the aquatic sedge *Lepironia articulata* in a tropical swamp. Tasek Bera, Malaysia. Aquatic Botany, 4: 269-280.
Imamura, T., 1964. Water-mites (Acari, Hydrachnellae) from (Thailand. Nat. Life Southeast Asia III: 67-79.
Inger, R. F., and Chin, P. K., 1962. The fresh-water fishes of North Borneo. Fieldiana (Zoology), 45: 1-268.
Iversen, T. M., 1973. Decomposition of autumn-shed beech leaves in a springbrook and its significance for the fauna. Hydrobiol., 72: 305-312.
Jensen, V., 1974. Decomposition of Angiosperm tree leaf litter. In Dickinson, C. H. and Pugh, G. J. (eds.), Biology of plant litter decomposition. Vol. I. Academic Press, London, pp. 69-104.
John, D. M., 1973. Accumulation and decay of litter and net production of forest in tropical West Africa. Oikos, 24: 430-435.
Johansen, O. A., 1937a. Aquatic Diptera, Part III. Chironomidae: Subfamilies Tanypodinae, Diamesinae, and Orthadadiinae. Mem. Cornell Univ. Agric. Exp. Sta., 205: 1-81.
Johansen, O. A., 1937b. Aquatic Diptera Part IV. Chironomidae: Subfamily, Chironomiae. Mem. Cornell Univ. Agric. Exp. Sta., 210: 1-56.
Johnson, D. S., 1957. A survey of Malayan freshwater life. Malayan Nature J., 12: 57-65.
Johnson, D. S., 1958. Subspecific and infra-specific variation in some freshwater prawns of the Indo-pacific region. Proc. Cent. Bicent. Congr. Biol., Singapore, 259-267.
Johnson, D. S., 1961. Notes on the freshwater Crustacea of Malaya. I. The Atyidae. Bull. Raffles Mus. Singapore, 26: 120-153.
Johnson, D. S., 1963. Distributional and other notes on some freshwater prawns (Atyidae and Palaemonidae) mainly from the Indo-west Pacific region. Bull. Nat. Mus., 32: 5-30.

Johnson, D. S., 1964. Key to Malayan Cladocera. Department of Zool., Univ. of Singapore. No. 6.
Johnson, D. S., 1966. Some factors influencing the distribution of freshwater prawns in Malaya. Proc. Symp. Crustacea. Part I. pp. 418-433.
Johnson, D. S., 1967a. On the chemistry of freshwaters in southern Malaya and Singapore. Arch. Hydrobiol., 63(4): 477-496.
Johnson, D. S., 1967b. Distributional patterns of Malayan freshwater fish. Ecology, 48(5): 722-730.
Johnson, D. S., 1968. Malayan blackwaters. Proc. Symp. Recent Adv. Trop. Ecology, 303-310.
Johnson, D. S., 1969. Non-Penaeid prawns of inland waters, including brackish waters, in western Malaysia and Singapore. Nat. Res. Mal. Sing. Proc. 2nd Symp. on Sci. Tech. Res. 1967. Strems. II. Univ. of Malaya.
Jorgenson, S., 1967. A method of absolute pollen counting. New Phytol., 66: 489-493.
Jothy, A. A., 1968. Preliminary observations of disused tin-mining pools in Malaysia and their potentials for fish production. Indo-Pacific Fish. Council, IPFC/Tech. Paper 69/11.
Jozuka, K., Morita, O., Ando, H., Yamabe, T. and Mashiko, K., 1963. Studies on the calcium uptake by marine teleosts of Labridae. Ann Rept. Noto. Mar. Lab. Fac. Sci., Univ. of Kanazawa, 2: 11-16.
Jutling, B. van, 1953. On a collection of non-marine Mollusca from Malaysia in Raffles Museum, Singapore, with an appendix on cave shells.
Karunakaran, S., 1969. Studies on the bionomics and taxonomy of Singapore Chironomidae. M.Sc. thesis, University of Singapore, 404 pp.
Kaushik, N. K. and Hynes, H. B. N., 1968. Experimental study on the role of autumn-shed leaves in aquatic environments. J. Ecol., 56: 229-243.
Kaushik, N. K. and Hynes, H. B. N., 1971. The fate of the dead leaves that fall into streams. Arch. Hydrobiol., 68: 465-515.
Kawai, T., 1961. A Plecoptera nympha from Thailand. Nat. Life Southeast Asia, 1: 199-201.
Kawai, T., 1968. Stoneflies (Plecoptera) from Thailand and India with descriptions of one new genus and two new species. Oriental Insects, 2(2): 107-139.
Kemmerer, A. J. and Neuhold, J. M., 1969. A method for gross primary productivity measurements. Limnol. Oceanogr. 14: 607-610.
Khan, A. A., Siddiqui, A. Q. and Nazir, M., 1970. Diurnal variations in a shallow tropical freshwater fish pond on Shahjahanpur, U.P. (India). Hydrobiologia, 35: 297-304.
Kiefer, Fr., 1933. Die freilebenden Copepoden der Binnengewässer von Insulinde. Arch. Hydrobiol. Suppl., 519-621.
Kira, T., Ogawa, H., Yoda, K. and Ogino, K., 1967. Comparative ecological studies in three main types of forest vegetation in Thailand. IV. Dry matter production, with special reference to the Khao Chong rain forest. Nature & Life in S.E. Asia, 5: 149-174.
Kira, T. and Shidei, T., 1967. Primary production and turnover of organic matter in different forest ecosystems of the Western Pacific. Jap. J. Ecol., 17: 70-87.
Kirita, H., 1971. Re-examination of the absorption method of measuring soil respiration under field conditions. IV. An improved absorption method using a disc of plastic sponge as absorbent holder. Jap. J. Ecol., 21: 119-127. (in Japanese with English summary).
Klinge, H. and Ohle, W., 1964. Chemical properties of rivers in the Amazonian areas in relation to soil conditions. Verh. Internat. Verein. Limnol., 15: 1067-1076.
Klinge, H. and Rodrigues, A. W., 1968a. Litter production in an area of Amazonian Terra Firm forest. Part I. Amazoniana, 1: 287-302.

Klinge, H. and Rodrigues, A. W., 1968b. Litter production in an area of Amazonian Terra Firm forest. Part II. Amazoniana, 1: 303-310.

Kobayashi, J., 1958. Chemical studies of the rivers in Southeast Asian countries – on the water quality of Thailand. Nogaku-kenkyu, 46: 63-112.

Kobayashi, J., 1959. Chemical investigation on river waters of southeastern Asiatic countries. Report I – The quality of waters of Thailand. Ber. Ohara Instituts Landwirtsch. Biologie, 11: 167-233.

Kobayashi, J., 1960. A chemical study of the average quality and characteristics of river waters of Japan. Biologie, 11: 313-358.

Kononova, M. M., 1966. Soil organic matter. Pergamon Press, New York, pp. 351-352.

Koyama, T., 1953. Measurement and analysis of gases in sediments. J. Earth Sci. Nagoya Univ., 1: 107-118.

Koyama, T., 1954. Distribution of carbon and nitrogen in lake muds. J. Earth Sci. Nagoya Univ., 2: 5-14.

Koyama, T., 1964. Biogeochemical studies on lake sediments and paddy soils and the production of atmospheric methane and hydrogen. Recent researches in the fields of hydrosphere, atmosphere and nuclear geochemistry. SUGAWARA Festival volume. Maruzen Co. Ltd., Tokyo, pp. 143-177.

Koyama, T. and Tomino, T., 1967. Decomposition process of organic carbon and nitrogen in lake water. Geochem. J., 1: 109-124.

Koyama, T. and Tomino, T., 1968. Mineralization process of organic matter in lake water. Bull. Misaki Mar. Biol. Inst. Kyoto Univ., 12: 111-123.

Krieger, W., 1932. Die Desmidiacean der deutschen Limnologischen Sunda-Expedition. Arch. Hydrobiol. Suppl., 11: 129-230.

Kumada, K., 1974. Chemistry of soil organic matter. Paper presented at the Seminar on the Utilization of Cereal crop residues conducted by FFTC during July 7-13, Tokyo.

Kumada, K., Sato, O., Ohsumi, Y. and Ohta, S., 1967. Humus composition of mountain soils in Central Japan with special reference to the distribution of P type humic acid. Soil Sci. Plant Natr., 13: 151-158.

Lanchester, W. F., 1901. On the Crustacea collected during the Skeat Expedition to the Malay Peninsula. Proc. Zool. Soc. Lond., 11: 534-574.

Larkin, P. A., 1956. Interspecific competition and population control in freshwater fish. J. Fish. Res. Bd. Canada, 13(3): 327-342.

Law, W. M., 1961. The Batu Arang Coalfield, Selangor. Thesis, Dept. of Geology, Univ. of Malaya, Kuala Lumpur, Malaysia.

Leamy, M. L. and Panton, W. P., 1966. Soil survey manual for Malayan conditions. Div. Agric. Bull. No. 119, Min. Agric. Co-op., Malaysia. 226 pp.

Lenz, F., 1941. Jugendstadien der sectio chiromominae (Tendi pedini) conectentes (Subf. Chironominae = Tendipedinae). Zusammenfassung und Revision. Arch. Hydrobiol. 38: 1-69.

Lewis, W. M., Jr., 1974. Primary production in the plankton community of a tropical lake. Ecol. Monogr., 44: 377-409.

Likens, G. E., Bormann, F. H., Johnson, N. M. and Pierce, R. S., 1967. The calcium, magnesium, potassium and sodium budgets for a small forested ecosystem. Ecology, 48: 772-785.

Lim, R. P., 1976. Community description, population dynamics and production of Cladocera with special reference to the littoral region of Pinekurat Lake, Ontario. Ph.D. thesis, University of Waterloo, Waterloo, Canada. 350 pp.

Lim, R. P. and Furtado, J. I., 1974. Habitat types and environmental factors at Tasek Bera. Malaysian-Japanese IBP (PF) Research Data on Tasek Bera No. 2, p. 3. (Restricted circulation).

Livingstone, D. A., 1955. A lightweight piston sampler for lake deposits. Ecology, 36: 137-139.

Livingstone, D. A., 1963. Chemical composition of rivers and lakes. Ch G. Data of Geochemistry Fleischer, M., U.S. Govt. Printing Office, Washington, 64 pp.

Lonnerblad, G., 1930. Über die Sauerstoffabsorption des Bodensubstrats in einigen Seentypen. Bot. Notiser, 1930: 53-60.

Lowe-McConnell, R. H., 1969. Speciation in tropical freshwater fishes. Biol., J. Linn. Soc., 1: 51-75.

Lowe-McConnell, R. H., 1975. Fish communities in tropical freshwaters. Longman, London, 1-337.

Lui, L. C., 1969. A key to the Cladocera in the fish ponds of the tropical Fish Culture Research Institute. T.F.C.R.I. Report. 1969. 45-51.

Mackereth, F. J. H., 1963. Some methods of water analysis for limnologists. Freshwat. Biol. Assoc. Sci. Publ., No. 12. 70 pp.

Manskaya, S. N. and Drozdova, T. V., 1968. Geochemistry of organic substances. Pergamon Press, New York, 155-163.

Marlier, G., 1967. Ecological studies on some lakes in the Amazon valley. Amazoniana, 1: 91-115.

Martin, J. P., 1950. Use of acid rose bengal and streptomycin in the plate method for estimating soil fungi. Soil Sci., 69. 125-???

Martin, P. S., 1963. The last 10,000 years: A fossil pollen record of the American Southwest. Univ. of Arizona Press. Tuscon, Arizona, 87 pp.

Mashiko, K. and Jozuka, K. 1962. Studies on the uptake by teleost fishes, I. Ca uptake by the crucian carp. Sci. Rept. Kanazawa Univ., 8: 107-126.

Mathews, C. P. and Kowaleczewski, A., 1969. The disappearance of leaf litter and its contribution to production in the River Thames. J. Ecol., 57: 543-552.

McLachlan, A. J., 1969. The effect of aquatic macrophytes on the variety and abundance of benthic fauna in a newly created lake in the tropics (Lake Kariba). Arch. Hydrobiol., 66: 212-231.

Menon, A. G. K., 1954. Notes on Malayan fishes in the collection of the Raffles Museum, Singapore. Bull. Raffles Mus. Singapore, 25: 5-26.

Merton, L. F. H., 1962. A visit to Tasek Bera. Malay. Nat. J., 16: 103-110.

Michell, D. S. (ed.), 1974. Aquatic vegetation and its use and control. UNESCO. 135 pp.

Midwood, R. B. and Felbeck, Jr., G. T., 1968. Analysis of yellow organic matter from fresh water. Am. Water Works Assoc. Jour., 60: 357-366.

Mizuno, T. and Mori, S., 1970. Preliminary hydrobiological survey of some Southeast Asian inland waters. Biol. J. Linn. Soc., 2: 77-117.

Monsi, M. and Saeki, T., 1953. Über den Lichtfaktor in den Pflanzengesellschaften und seine Bedeutung für die Stoffproduktion. Jap. J. Bot., 14: 22-52.

Morisita, M., 1959. Measuring of the dispersion of individuals and analysis of the distributional pattern. Mem. Fac. Sci. Kyushu Univ., Ser. E., 2: 215-235.

Morley, R. J., 1976. Vegetation change in West Malaysia during the Late Quaternary period: A palynological study of selected lowland and lower montane sites. Unpublished Ph.D. thesis, Univ. of Hull. 507 pp.

Nose, T., 1962. On the emerging behaviour of *Chironomus plumosus* larvae. Tansui-kempo, 11: 11-18. (in Japanese with English synopsis).

Nye, P. H., 1961. Organic matter and nutrient cycles under moist tropical forest. Pl. Soil, 13: 333-346.

Oana, S., 1957. Bestimmung des Argon im besonderen Hinblick auf gelöste Gase in natürlichen Wassern. J. Earth Sci. Nagoya Univ., 5: 103-124.
Ohle, W., 1934. Chemische und physikalische Untersuchungen norddeutscher Seen. Arch. Hydrobiol., 26: 386-464.
Ohle, W., 1955. Die Ursachen der restanten Seeneutrophierung. Verh. Internat. Verein. Limnol., 12: 373-382.
Okuma, C., 1968. Preliminary survey on the spider-fauna of paddy field in Thailand. Mushi, 42(8): 89-118.
Okuma, C. and Wongsiri, T., 1973. Second report on the spider-fauna of paddy field in Thailand, 47(1): 1-17.
Olsen, S., 1966. Recent trends in the determination of orthophosphates in water. In Chemical environment in the aquatic habitat. Proc. I.B.P. Symp. Amsterdam and Nieuwersluis, 10-16 Oct.
Olson, J. S., 1963. Energy storage and the balance of producers and decomposers in ecological systems, 44: 322-331.
Pearsall, W. H., 1922. A suggestion as to the factors influencing the distribution of free floating vegetation. J. Ecol., 9: 241-253.
Peck, R. M., 1974. A comparison of four absolute pollen preparation techniques. New Phytol., 37: 567-687.
Pennak, R. W., 1953. Freshwater invertebrates of the United States. Ronald Press Co., New York, 769 pp.
Petr, T., 1968. Population changes in aquatic invertebrates living in two water plants in a tropical man-made lake. Hydrobiologia, 32: 449-486.
Poole, H. H. and Atkins, W. R. G., 1929. On the penetration of light into seawater. J. Mar. Biol. Ass. U.K., 14: 177-198.
Poore, M. E. D., 1968. Studies in Malaysian rain forest. I. The forests on Triassic sediments in Jengka forest reserve. J. Ecol., 56: 143-196.
Prakash, A. and Rashid, M. A., 1968. Influence of humic substances on the growth of marine phytoplankton: Dinoflagellates. Limnol. Oceanogr., 13: 598-607.
Prakash, A. and Rashid, M. A., 1973. Influence of humic substances on the growth of marine phytoplankton: Diatoms. Limnol. Oceanogr., 18: 516-524.
Prowse, G. A., 1957. An introduction to the desmids of Malaya. Malayan Nat. J., 11: 42-58.
Prowse, G. A., 1958. The Eugleninae of Malaya. Gard. Bull. Singapore, 16: 136-204.
Prowse, G. A., 1960. New and unusual Flagelata in Malaya. Proc. Cent. Bicent. Congr. Biol. Singapore, 292-298.
Prowse, G. A., 1962. Further Malayan freshwater Flagelata. Gard. Bull. Singapore, 19: 105-145.
Prowse, G. A., 1969. Some new desmid taxa from Malaya and Singapore. Gard. Bull. Singapore, 24: 337-353.
Rao, T. S. S. and Rao, V. C., 1962. Studies on diurnal variations in the hydrobiological conditions in the Waltan Coast. J. Mar. Biol. Ass. India., 4: 23-43.
Ratnasabapathy, M., 1974. The species composition and phytogeographical significance of the algal flora of Tasek Bera. Paper presented at the IBP synthesis Meeting, 12-18 August 1974. Kuala Lumpur.
Ratnasabapathy, M. and Kumano, S., 1974. Desmids from Tasek Bera, West Malaysia. Bull. Jap. Soc. Phycol., 22: 22-28
Reitsma, Tj., 1969. Size modification of recent pollen grains under different treatments. Rev. Palaeobotan. Palynol., 9: 175-203.

Rhode, W., 1949. The ionic composition of lake waters. Verh. Internat. Verein. Limnol., 10: 377-386.
Richardson, 1947. An outline of the geomorphical evolution of British Malaya. Geol. Mag., 84(3): 129-144.
Richards, P. W., 1957. The tropical rain forest. Cambridge Univ. Press, London. 450 pp.
Rodin, L. E. and Bazilevich, N. I., 1967. Production and mineral cycling in terrestrial vegetation. Oliver & Boyd, Edinburgh, pp.
Roux, J., 1930. Notes sur quelques crustacés décapodes dulca quicoles de l'Archipel. Indo. Australian. Rev. Suisse Zool., 37: 353-362.
Roux, J., 1934. New freshwater decapod Crustaceans from the Malay Peninsula. Bull. Raffles Mus. Singapore, 9: 28-33.
Roux, J., 1936. Second note upon freshwater crustaceans from the Malay Peninsula. Bull. Raffles Mus. Singapore, 12: 29-43.
Ruttner, F., 1931. Hydrographische und Hydrochemische Beobachtungen auf Java, Sumatra and Bali. Arch. Hydrobiol. Suppl., 8: 197-454.
Sands, W. N., 1930. Crabs in paddy fields. Malay Agric. J., 18: 568.
Schindler, D. W., 1971. Light, temperature, and oxygen regimes of selected lakes in the Experimental Lakes Area, northwestern Ontario. J. Fish. Res. Bd. Canada, 28: 157-169.
Schmidt, G. W., 1969. Vertical distribution of bacteria and algae in a tropical lake. Int. Revue Ges Hydrobiol., 57: 791-797.
Scott, A. M. and Prescott, G. W., 1961. Indonesian desmids. Hydrobiologia, 17: 1-132.
Scrivenor, J. B., 1931. The geology of Malaya. MacMillan, London. 217 pp.
Shapiro, J., 1957. Chemical and biological studies on the yellow organic acids of lake water. Limnol. Oceanogr., 2: 161-169.
Shapiro, J., 1964. Effect of yellow organic acids on iron and other metals in water. Am. Water Works Assoc. Jour., 56: 1062-1082.
Shapiro, J., 1966. The relation of humic colour to iron in natural waters. Verh. Internat. Verein. Limnol., 16: 477-484.
Shinozaki, K., 1953. On the generalization of the logistic curve. I. General discussions. J. Osaka City Med. Center, 2: 143-137. (in Japanese, with English summary).
Shiraishi, Y., Mizuno, N., Nagai, M., Yoshimi, K., and Nishiyama, K., 1972. Studies on the diel activity and feeding habit of fishes at Lake Bera, Malaysia. Jap. J. Ichtyol., 19(4): 295-306. (in Japanese with English summary).
Shirota, A., 1966. The plankton of South Vietnam – freshwater and marine plankton. O.T.C.A., 1-462.
Shirota, A., 1968. Studies of Chironomids. Tokyo. (in Japanese).
Shnitzer, M., 1971. Metal organic matter interactions in soils and waters. In Faust, S. J. and Hunter, J. V. (eds.), Organic compounds in aquatic environments. Marcel Dekker, New York, pp. 297-315.
Shnitzer, M. and Khan, S. U., 1972. Humic substances in the environment. Marcel Dekker, New York, pp. 71-77.
Shu, Y. K., 1969. Some N.W. trending faults in the Kuala Lumpur and other areas. Newsl. Geol. Soc. Malaysia, 17: 1-5.
Siolo, H., 1964. General features of the limnology of Amazonia. Verh. Internat. Verein. Limnol., XV., 1053-1058.
Sioli, H., 1967. Studies in Amazonian waters. Atas do Simposio Sobre a Biota. Amazoniana, 3: (Limnolgia), 9-50.

Sioli, 1968a. Principal biotopes of primary production in the waters of Amazonia. In Misra, R. and Gopal, B. (eds.), Proc. Symp. Recent Adv. Trop. Ecol. 591-600.
Sioli, H., 1968b. Hydrochemistry and geology in the Brazilian Amazon region. Amazonia, 1: 267-277.
Sioli, H. Shwabe, G. H. and Klinge, H., 1969. Limnological outlooks on landscape-ecology in Latin America. Trop. Ecol., 72-82.
Smedley, N., 1931. A freshwater crab *Paratelphusa sexpunctatam* in the Malayan Peninsula. Bull. Raffles Mus. Singapore, 5: 68-70.
Smith, H. M., 1945. The freshwater fishes of Siam or Thailand. Bull. U.S. Nat. Mus., 188: 1-662.
Soepadmo, E., 1974. Check list of plants in the swamp forest and lowland dipterocarp forest at Tasek Bera. Malaysian-Japanese IBP (PF) Research Data on Tasek Bera No. 2, pp. 130-131. (restricted circulation).
Sreenivasan, A., 1964. The limnology, primary production and fish production in a tropical pond. Limnol. Oceanogr., 9: 391-396.
Sreenivasan, A., 1966. Limnology of tropical impoundments, I. Hydrological features and fish production in Stanley Reservoir, Mettur Dam. Int. Revue Ges. Hydrobiol., 51: 295-306.
Stauffer, P., 1968. The Kuala Lumpur fault zone: A proposed major strike slip fault across Malaya. Newsl. Geol. Soc. Malaysia, 15: 2-4.
Stone, B. C., 1971. Additions to the Malayan Flora. Malay. Nat. J., 24: 91-94.
Strickland, J. D. H. and Parsons, T. R., 1968. Practical handbook of sea-water analysis. Fish. Res. Bd. Canada, Ottawa. Bull., 1967.
Strickland, J. D. H., Holm-Hansen, O., Eppley, R. W. and Linn, R. J., 1969. The use of deep tank in plankton ecology. I. Studies of the growth and composition of phytoplankton crops at low nutrient levels. Limnol. Oceanogr., 14: 23-34.
Sufi, S. M. K., 1956. Revision of the Oriental fishes of the family Mastcembelidae. Bull. Raffles Mus. Singapore, 27: 93-146.
Sugawara, K., 1939. Chemical studies in lake metabolism. Bull. Chem. Soc. Japan, 14: 375-451.
Swingle, H. S., 1963. Methods of analysis for water, organic matter and pond bottom soils. Auburn University, Auburn, Alabama.
Talling, J. F., 1957. Diurnal changes of stratification and photosynthesis in some tropical African waters. Proc. Roy. Soc. B., 147: 57-83.
Talling, J. F., 1966. The annual cycle of stratification and phytoplankton growth in Lake Victoria (East Africa). Int. Revue Ges. Hydrobiol., 51: 545-621.
Talling, J. F., 1969. The incidence of vertical mixing and some biological and chemical consequences in tropical African Lakes. Verh. Internat. Verein. Limnol., 17: 998-1012.
Talling, J. F. and Rzoska, J., 1967. The development of plankton in relation to hydrobiological regime in the blue Nile, J. Ecol., 55: 637-662.
Talling, J. F. and Talling, I. B., 1965. The chemical composition of Africa lake waters. Int. Revue Ges. Hydrobiol., 50: 421-463.
Tezuka, Y. and Kawai, A., 1975. Bacterial production. In Mori, S. (ed.), Productivity of Communities in Japanese inland waters. (in press). Univ. of Tokyo Press.
Thienemann, A., 1944. Bestimmungstabellen für die bis jetzt bekannten Larven und Puppen der Orthocladinen (Diptera, Chiromidae). Arch. Hydrobiol., 39: 551-664.
Thompson, K., 1973. Production of *Cyperus papyrus*. Read at IBP Synthesis Meeting. Univ. College of Wales, Aberystwyth, U.K.
Tokunaga, M. and Stone, A., 1964. Insects of Micronesia, Diptera, Chironomidae, Simuliidae. Bernice P. Bishop Mus., Insects of Micronesia, 12: 485-627.

Tweedie, M. W. F., 1938. Note on *Paratelphusa maculata* De Man. Bull. Raffles Mus. Singapore, 14: 203-205.

Tweedie, M. W. F., 1952. Notes on Malayan fresh water fishes. 3. The Anabantid fishes. 4. Some new and interesting records. 5. Malay names. Bull. Raffles Mus. Singapore, 24: 63-95.

Tweedie, M. W. F., 1956. Notes on Malayan fresh water fishes. 6. A method of collecting small fishes, especially loaches. 7. Cobitidae and Cyprinidae in the Raffles Museum collection. Bull. Raffles Mus. Singapore, 27: 56-64.

Ueno, M., 1939. Plankton studies in Formosan inland waters. Philippine Jour. Sci., 69(1): 35-67.

Ueno, M., 1961. Mayflies of Thailand. Nat. Life Southeast Asia, 1: 207-208.

Ueno, M., 1964. Hydrobiological studies in Southeast Asia (in Japanese). Study of Southeast Asia, 2(2): 52-66.

Ueno, M., 1966. Freshwater zooplankton of Southeast Asia. Study of Southeast Asia, 3(5): 94-109.

Ueno, M., 1969. On the Malayan Bathynellid once again. Bull. Nat. Sci. Mus. Tokyo, 12(1): 143-150.

Vaas, K. F. and Sachlan, M., 1953. Limnological studies on diurnal fluctuations in shallow ponds in Indonesia. Verh. Internat. Verein. Limnol., 12: 309-319.

Vejabhongse, N. P., 1937. A note on the habits of a mayfly and the damage caused by its nymphs. Jour. Siam Soc. Nat. Hist., Suppl., 11: 53-56.

Verma, M. N., 1967. Diurnal variation in a fish pond in Seoni, India. Hydrobiologia, 30: 129-137.

Vollenweider, R. A., 1960. Beiträge zur Kenntnis optischer Eigenschaften der Gewässer und Primarproduktion. Mem. Inst. Ital. Idrobiol., 12: 201-244.

Vollenweider, R. A., 1969. (ed.) A manual on methods for measuring primary productivity in aquatic environments. Blackwells, Oxford. 213 pp.

Weber, M. and Beaufort, L. F. de, 1916. The fishes of the Indo-Australian Archipelago, 3. Ostariophysi. Leiden, The Neth., 1-417.

Wenz, W. and Hass, F., 1968. Archiv für Molluskenkunde. Frankfurt am Main.

West, R. G., 1968. Pleistocene geology and biology. Longmans, London. 377 pp.

Westlake, D. F., 1963. Comparisons of plant productivity. Biol. Rev., 38: 385-425.

Westlake, D. F., 1975. Primary production of freshwater macrophytes. Vol. 3, Photosynthesis and productivity in different environments. Cambridge University Press, London, pp. 189-206.

Williams, W. D., 1964. Some chemical features of Tasmanian inland waters. Aust. J. Mar. Freshw. Res., 15: 107-122.

Willoughby, L. G., 1974. Decomposition of litter in fresh water. In Dickinson, C. H. and Pugh, G. J. F. (eds.), Biology of plant litter decomposition Vol. II, Academic Press, London. pp. 659-691.

Wilson, G. T. and Mackleod, N., 1974. A critical appraisal of empirical equations and models for the prediction of the coefficient of reaeration of deoxygenated water. Water Research, 8: 341-366.

Wolf, B., 1937. Animalium cavernarum catalogus. 's-Gravenhage, 2, catalogue des animaux des grottes, 1: 616.

Wright, H. E. and Patten, H. L., 1963. The pollen sum. Spores, 5: 445-450.

Wyatt-Smith, J., 1964. A preliminary vegetation map of Malaya with descriptions of the vegetation types. J. Trop. Geogr., 18: 200-213.

Wycherley, P. R., 1967. Rainfall probability tables for Malaysia. Rubber Res. Inst. Plant. Man., 12.

Yoda, K., 1967. Comparative ecological studies on three main types of forest vegetation in Thailand. III. Community respiration. Nature and Life in S.E. Asia, 5: 83-148.
Yoda, K., Kira, T., Ogawa, H. and Hozumi, K., 1963. Self-thinning in overcrowded pure stands under cultivated and natural conditions. Jour. Biol. Osaka City Univ., 14: 107-129.
Yoda, K. and Kira, T., 1969. Comparative ecological studies on three main types of forest vegetation in Thailand. V. Accumulation and turnover of soil organic matter, with notes on the altitudinal soil sequence on Khao (Mt.) Luang, peninsular Thailand. Nat. Life in S.E. Asia, 6: 83-110.
Yoshihara, T., 1951. On the fitting, the summation, and application of the logistic curve. J. Tokyo Univ. Fish, 38: 181-195.
Yoshimura, S., 1938. Dissolved oxygen of the lake waters of Japan. Rep. Tokyo Bunrika Daig. Sect. C., 8: 63-277.
Zafar, A. R., 1959. Taxonomy of lakes. Hydrobiologia, 13: 281-299.

Subject index

GPSR Compliance

The European Union's (EU) General Product Safety Regulation (GPSR) is a set of rules that requires consumer products to be safe and our obligations to ensure this.

If you have any concerns about our products, you can contact us on ProductSafety@springernature.com

In case Publisher is established outside the EU, the EU authorized representative is:

Springer Nature Customer Service Center GmbH
Europaplatz 3
69115 Heidelberg, Germany

Batch number: 10824490

Printed by Printforce, the Netherlands